CW01279060

Sources and Studies in the History of Mathematics and Physical Sciences

Sources and Studies in the History of Mathematics and Physical Sciences was inaugurated as two series in 1975 with the publication in Studies of Otto Neugebauer's seminal three-volume History of Ancient Mathematical Astronomy, which remains the central history of the subject. This publication was followed the next year in Sources by Gerald Toomer's transcription, translation (from the Arabic), and commentary of Diocles on Burning Mirrors. The two series were eventually amalgamated under a single editorial board led originally by Martin Klein (d. 2009) and Gerald Toomer, respectively two of the foremost historians of modern and ancient physical science. The goal of the joint series, as of its two predecessors, is to publish probing histories and thorough editions of technical developments in mathematics and physics, broadly construed. Its scope covers all relevant work from pre-classical antiquity through the last century, ranging from Babylonian mathematics to the scientific correspondence of H. A. Lorentz. Books in this series will interest scholars in the history of mathematics and physics, mathematicians, physicists, engineers, and anyone who seeks to understand the historical underpinnings of the modern physical sciences.

More information about this series at http://www.springer.com/series/4142

Dirk Grupe

Stephen of Pisa and Antioch: Liber Mamonis

An Introduction to Ptolemaic Cosmology and Astronomy from the Early Crusader States

Dirk Grupe
München
Bayern, Germany

ISSN 2196-8810 ISSN 2196-8829 (electronic)
Sources and Studies in the History of Mathematics and Physical Sciences
ISBN 978-3-030-19233-4 ISBN 978-3-030-19234-1 (eBook)
https://doi.org/10.1007/978-3-030-19234-1

This Springer imprint is published by the registered company Springer Nature Switzerland AG.
The registered company address is: Gewerbestrasse 11, 6330 Cham, Switzerland

Preface

This book is based on the first of the two parts of my doctoral dissertation, which was submitted in 2013 under the title *The Latin Reception of Arabic Astronomy and Cosmology in Mid-Twelfth-Century Antioch: The Liber Mamonis and the Dresden Almagest*. The subject of the study was a pair of medieval Latin texts on Ptolemaic astronomy which had a short time earlier been identified as the products of a coordinated attempt to introduce Arabic knowledge from various disciplines into Latin Europe. Produced in the Crusader principality of Antioch during the second quarter of the twelfth century, the texts comprise the earliest-known Latin translation of Ptolemy's *Almagest* and a related account on Ptolemaic astronomy and cosmology, entitled *Liber Mamonis*, which was composed by a famous translator of Arabic medicine, Stephen the Philosopher, also known as Stephen of Pisa or Stephen of Antioch. With his *Liber Mamonis*, Stephen of Pisa (and Antioch) produced the first translation, partly revised and richly commented, of the Arabic cosmography *On the Configuration of the World* by Ibn al-Haytham (eleventh century).

The present edition aims to make Stephen's work on the celestial sciences accessible to a wider readership. The edition is based on the only known manuscript of the *Liber Mamonis*, MS Cambrai, Médiathèque d'Agglomération, A 930. When preparing the edition, I was able to make valuable use of an unpublished edition of the Cambrai manuscript by Prof. Richard Lemay (†2004). In the same unpublished work, Prof. Lemay also showed for the first time the dependency of the *Liber Mamonis* on Ibn al-Haytham's cosmography. Many of the emendations and comments suggested by Lemay have been included in the present book (indicated in the edition by an attached (L)). Some changes to Lemay's transcription have been made, mostly in the technical parts of the work. In particular, Stephen's thoughts about the inferior planets, which largely escaped Lemay's attention, required new treatment.

For the study of the *Liber Mamonis*, it is fortunate that the Arabic text of Stephen's main source, Ibn al-Haytham's treatise *On the Configuration of the World*, has been available for some years in a critical edition together with an English translation by Prof. Tzvi Langermann. A much awaited reprint of Langermann's book, with an updated preface, was published in 2016. To facilitate the comparison of the *Liber Mamonis* with Stephen's Arabic source, where possible, the numbering of passages which Langermann applied to the Arabic text has been inserted in this edition.

Although Ibn al-Haytham's arguments can be studied in English from Langermann's translation, a translation of the *Liber Mamonis* is a necessary addition. One reason is that Stephen does not follow the Arabic wording closely and on many occasions gives the text a different meaning or a different character. Another reason is that only about half of the *Liber Mamonis* is taken from Ibn al-Haytham; the other half consists of independent comments which Stephen added to the text. It is often in these additions to the work that the individuality of Stephen's thinking becomes apparent, and I hope it will be of interest both to historians specialised in medieval astronomy and cosmology and to a wider readership as well.

The second part of my PhD dissertation, on the Dresden *Almagest*, is currently in preparation to be published as another volume. It contains the discovery of a previously unknown Arabic translation of Ptolemy's *Almagest*, which was produced by the ninth-century mathematician Thābit ibn Qurra. Based on a Latin version of Ptolemy's *Almagest* in MS Dresden, SLUB, Db. 87, I argue in my dissertation that abnormities in this text were not made at random but result from a coherent revision of Ptolemy's arguments and that these changes were already made in the Arabic source tradition of the Latin Dresden text. I argue further that this revised Arabic *Almagest* was not an isolated work which incidentally came into the hands of the Latin translator, but it had an influence also on others, especially Arabic, astronomical works. At a third stage, I prove that Thābit ibn Qurra was the creator of this revised *Almagest*; based on this identification, I speculate in my thesis that Thābit ibn Qurra's *Almagest* possibly also survived in Arabic, in a fragment in MS Jaipur, Maharaja Sawai Man Singh II Museum Library, 20. Before publishing my edition of the Dresden *Almagest*, I wished to consult the Arabic copy in MS Jaipur 20 and, if possible, use it to improve the edition of the Latin text. While this has become possible recently, the first part of my thesis, on the *Liber Mamonis*, has now become the first part to appear in print, and I hope that the second part will follow shortly. For the meantime, I thank the Maharaja Sawai Man Singh II Museum Trust in Jaipur for a most pleasant research stay at the Palace Library, which was offered to Dr. Mª José Parra and me in autumn 2018 and during which we could extensively study the Jaipur *Almagest*.

At this point, it is my pleasure to express my gratitude to the people who have supported my studies and contributed to the realisation of the present book. Important help came from Dr. Josefina Rodríguez Arribas, Dr. Barbara Obrist, Prof. Taro Mimura, Prof. José Luis Mancha, Dr. S. Mohammad Mozaffari, Prof. Andreas Kühne, Prof. Menso Folkerts, and others, who I thank deeply for their kind collaboration. Prof. Charles Burnett drew my attention to what would become the subject of my doctoral research. The Deutscher Akademischer Austauschdienst, the Arts and Humanities Research Council, and the Dr. Günther Findel-Stiftung generously supported my activities as a PhD student. The inclusion of the manuscript reproductions in this book was made possible by the kind permission of the Médiathèque d'Agglomération de Cambrai and the Biblioteca Palatina in Parma, now part of the Complesso monumentale della Pilotta, upon concession of the Ministero per i Beni e le attività culturali. Despite having a busy schedule as a patent attorney, Dr. Miles Haines kindly accepted to proofread my English translation of the Latin text. My warmest gratitude goes to Dr. Mª José Parra for her unceasing support over the past months when the collected material on the *Liber Mamonis* was being brought into a publishable form.

München, Germany Dirk Grupe

Contents

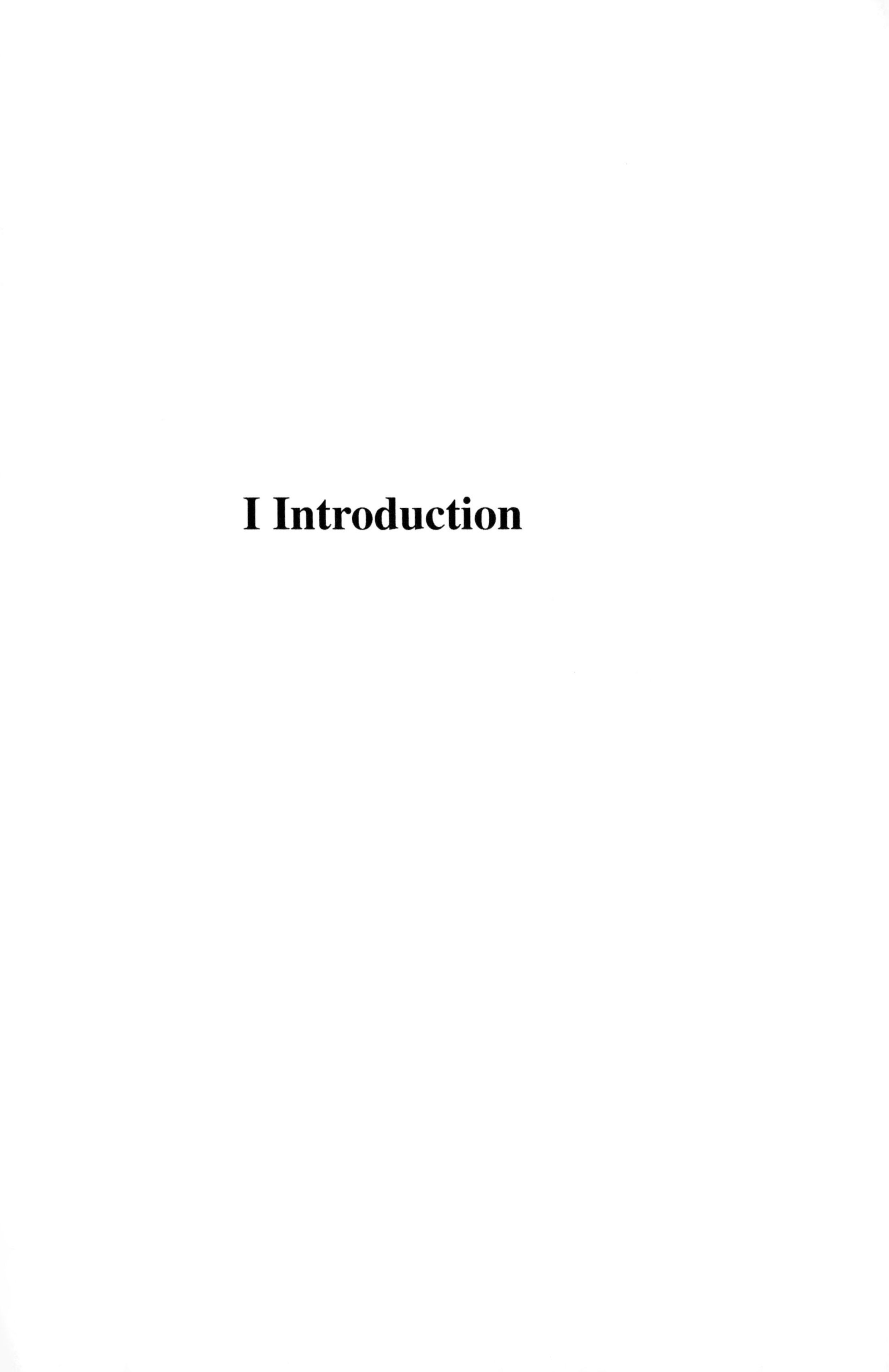

I Introduction

Chapter 1

Studies in Arabic Astronomy in the Early Crusader States

1.1 The Crusader States and the Translation Activities of the Twelfth Century

The arrival of Ptolemaic astronomy in Latin Europe from Greek and Arabic sources in the twelfth century stands for one of the epoch-making changes in the history of European science. The rediscovery of Ptolemy's ideas by medieval Latin scholars was not an incidental or an isolated event; rather it took place in a period of extensive intellectual transfer, which had begun in the late eleventh century and had involved many parts of the Mediterranean world, across continents, cultures, and religions. Especially on the part of western Europeans, one incentive to seek foreign knowledge at that time was the growing awareness that substantial parts of the classical Greek learning of Antiquity had become lost in the Latin tradition. A well-known example is the works of Aristotle, of which only fragments had survived in western libraries. And while the traditional canon of the seven liberal arts still defined the framework of learning in medieval Europe, the lack of authoritative standard works was keenly felt—especially in the mathematical sciences, among them astronomy, since for centuries the works of Euclid and Ptolemy had also been inaccessible to Latin readers. These texts continued to be transmitted in the Greek-speaking parts of the eastern Mediterranean, and they were even more eagerly studied in Arab countries, where

D. Grupe, *Stephen of Pisa and Antioch: Liber Mamonis*, Sources and Studies in the History of Mathematics and Physical Sciences,
https://doi.org/10.1007/978-3-030-19234-1_1

Greek science and philosophy had been introduced in the ninth century and where they had laid the foundations for many independent intellectual developments.

The twelfth and thirteenth centuries were times of vigorous translation activity, as Latin scholars left their familiar environments in the search for Greek or Arabic copies of the desired texts and, by translating them into Latin, gradually filled the gaps in the libraries back home. Inevitably, on their travels these men also encountered modern scientific advances which the foreign custodians of the classical texts had made in the meantime. As a result, by the beginning of the fourteenth century not only had most of the available scientific knowledge of Greek Antiquity been recovered for the Latin world, but the works of later foreign scientists and commentators now formed the basis for a new Latin engagement with the sciences.[1]

This interest in restoring the knowledge of classical Antiquity and the simultaneous revival of traditions in other fields, such as the liturgy or the return to a jurisdiction based on the Justinian *Corpus iuris civilis* in many parts of Europe, established the idea of a European "renaissance" during the twelfth century.[2] It was also a period of departure, motivated by many modern developments, and which itself gave rise to others. As for the mentioned transfer of scientific knowledge, for example, current practical needs led to an increased interest in Greek medicine and mathematical sciences, which were the main focus of attention for the first translators[3]; and together with the classical knowledge recent contributions to these disciplines were also enthusiastically adopted. The increased transfer of knowledge also coincided with and reflected recent trends, such as the greater mobility of people and institutions, and in turn promoted the establishment of new forms of institutionalised learning in Europe.

While it was primarily Greek knowledge that was sought, the classical texts were received mostly through Arabic rather than Greek sources. This can be partly explained by the Arabs' more active engagement with the sciences, but it also appears to have been a question of access to a particular source that determined a

[1] A comprehensive study of the Latin acquisition of Greek and Arabic learning in the twelfth and thirteenth centuries remains a desideratum. Until now, the most complete accounts of translated works and their translators are to be found in Steinschneider, *Die europäischen Übersetzungen aus dem Arabischen*, and Haskins, *Studies*. Useful overviews are given in Haskins, *Renaissance*, ch. 9, and d'Alverny, 'Translations and translators'. The important role of Hebrew translations for the transmission of Arabic science to Europe is documented in Steinschneider, *Die hebraeischen Übersetzungen des Mittelalters*. The internal European motivation for the translations, in contrast to a stimulus from outside triggered by the intensified contacts with Arabs and Byzantines, is discussed, e.g., in Burnett, 'Coherence', and Kluxen, 'Der Begriff der Wissenschaft'.

[2] Most important is Haskins, *Renaissance*.

[3] Prominent examples of early Arabic-Latin translations on medicine and mathematical sciences are the works by Constantine the African (late 11th c.; medicine), Adelard of Bath and Hermann of Carinthia (both 1h. 12th c.; mathematics, astronomy and astrology).

translator's choice. As it turned out, the predestined places for intellectual transfer were multilingual environments, where Latin students of the new sciences had the chance to come into direct contact with masters of the required languages, and eventually with masters of the sciences themselves. A Greek element had remained strong in southern Italy, whereas with regard to Arabic learning, similar sites included former Arab territories that had recently fallen under Latin rule: examples at the beginning of the twelfth century were parts of the Iberian Peninsula, Sicily and Crusader Syria.

Concerning the different regions of Arabic-Latin contacts, the predominance of translations from Spain and Sicily is in stark contrast to the vanishingly small amount of texts that were obtained or translated in the Crusader East. The multiethnic and multilingual societies of the Crusader states, at the seam of the Byzantine and Arabic cultures, might have been expected to have provided a fertile ground for cultural transfer; but the sparse evidence of any such exchange between local Arabs and Frankish Crusaders corresponds with a general feeling that the nature of the Crusader states gave little opportunity for academic activity or fruitful intercultural contact. On the side of the Europeans, an ideology of holy warfare seemed to exclude *a priori* any willingness to adopt the local culture. In addition, westerners who were attracted to the Crusader states have been deemed adventurers who were spiritually rather than intellectually driven, while for those Europeans with an interest in learning, the libraries in Spain were much easier to reach and much safer than the Levant.[4]

Similar observations have been made with regard to the Arabic element in the different regions. In Norman Sicily, where Arab intellectuals continued to hold important positions as doctors or in the public administration, Arabic- and Latin-speaking professionals cooperated on a daily basis. In Spain, the fall of Toledo even placed a centre of Arabic learning in Christian hands, where local Arabic-speaking Jews and Mozarabs often acted as intermediaries transferring Islamic science to the Latin conquerors. In the Crusader states, by contrast, large parts of the Muslim elite had either been killed in the early stages of the conflict or preferred emigration to living under Frankish rule; furthermore, no equally important centre of Arabic scholarship was captured. As regards the availability of Arabic learning and an environment for its communication, the conditions in the

[4]E.g., Haskins, *Studies*, p. 130: "Plainly the Crusaders were men of action rather than men of learning, and there was little occasion for western scholars to seek by long journeys to Syria that which they could find nearer home in Spain." Similarly, Kedar, 'The subjected Muslims of the Frankish Levant', p. 173f.: "clerics interested in theological and philosophical speculation, the application of dialectic to Roman and canon law, the study of scientific tracts translated from the Arabic or the Greek—in short, clerics attracted by the new intellectual trends of the age, stayed behind in Europe. [...] In other words, the clerics who settled in the Frankish Levant [...] were not interested in, or capable of, intellectual give-and-take with Oriental Christian or Muslim scholars."

Crusader states thus appeared far less conducive than those in the West. In fact, modern historiography has often judged the contacts in the East to have been negligible for the transfer of Arabic sciences to Europe[5]; on the other hand, in central Spain the city of Toledo became the dominant centre for Arabic-Latin translations from the middle of the twelfth century onwards. This later concentration around a single, highly productive translation school in Toledo defined the standards for Latin translations of Arabic scientific and philosophical texts.[6] It was also in Toledo, in the late twelfth century, that Gerard of Cremona produced the most influential Latin translation of Ptolemy's astronomical compendium, *Syntaxis mathematica*—better known by its Arabic title, *Almagest*—which for the following centuries became the leading reference for astronomical studies in Europe.

In recent years, though, the view just sketched of the Crusader states as an academic and intellectual desert has come under question from various angles. An increased interest in the social history of the Crusader states as multiethnic societies has revealed a more complex situation with regard to Arabic-Latin interactions in the East, and a continuing Byzantine influence.[7] Economic life also produced regular contacts between Christians and Muslims at various social levels, which probably led to a better knowledge of the Arabic language among the Franks than has generally been assumed.[8] In addition, the realities in the Levant often brought Christian and Muslim leaders into political or even military cooperation.[9] While the existence of intellectual transfer in the Crusader states has rarely been totally dismissed, the contacts in the East have recently been attributed more importance, especially with regard to the starting period of the translation movement before the development of the centres in the West.[10] Significant exchanges are known to

---

[5]E.g., Haskins, *Renaissance*, p. 15: "In any case the Crusades fail us as a cause of the Latin Renaissance, for it began well before the First Crusade, and the two movements scarcely touch." Similarly, Irwin, *Islam and the Crusades 1096–1699*, p. 235: "no Arab books were translated into Latin or French in the Latin East [...]. They had no interest in each other's scholarship or art. The important cultural interchanges had taken place earlier and elsewhere. Arabic learning was mostly transmitted to Christendom via Spain, Sicily, and Byzantium." On the supposed absence of astronomical studies in the East see Mercier, 'East and West contrasted in scientific astronomy', p. 340: "with the possible exception of Abraham Ibn Ezra [there is] essentially no support from the point of view of mathematical astronomy for the notion that the Crusades provided an avenue for the transmission to the Latin West."

[6]Burnett, 'Coherence'.

[7]See Prawer, 'Social classes in the Crusader states: the "minorities" '; *idem*, 'Social classes in the Latin Kingdom: the Franks'; Kedar, 'The subjected Muslims of the Frankish Levant'.

[8]Attiya, 'Knowledge of Arabic'. See also Williams, 'Philip of Tripoli's translation of the pseudo-Aristotelian Secretum Secretorum', pp. 85–89. A more critical view is taken by Hillenbrand, *The Crusades—Islamic Perspectives*, p. 10.

[9]Köhler, *Allianzen und Verträge*.

[10]Kedar, 'The Subjected Muslims of the Frankish Levant', p. 174; Hiestand, 'Un centre intellectuel en Syrie du Nord?'.

have taken place in the cosmopolitan city of Antioch, which had been a part of the Byzantine empire until its capture by the Seljuks in 1085, and only 13 years later, in 1098, had become the capital of one of the first Crusader states. Adelard of Bath was an early visitor. Shortly afterwards, in 1126/7, an important Latin translation of Arabic medicine was produced there by Stephen of Pisa, who thanks to this work also became known as Stephen of Antioch or Stephen the Philosopher. In the early thirteenth century it is further documented that Philip of Tripoli travelled to Antioch to make a translation of the pseudo-Aristotelian *Secretum secretorum* from Arabic into Latin.[11]

These few ascertained cases may be seen as exceptions to the rule.[12] However, the fact that over an extended period western translators decided to travel to Antioch in order to access particular Arabic sources, and even produce translations there, reveals that the city provided well for academic activities at the highest level. It shows that, even after a century of Frankish rule, Arabic texts which were apparently unavailable in the West could be found in libraries in the Crusader states; and, like the Mozarabs and the Jews in Spain, local Oriental Christians could act as teachers of Arabic and as intermediaries of Arabic knowledge for the Latins.[13] Furthermore, especially in the time after the First Crusade, the trading city of Pisa provided important connections for the exchange of goods and ideas with the East[14]: there was a Pisan base in Byzantium, which provided direct access to Greek sources, and also a Pisan quarter in Antioch. In addition to this channel between Pisa and Antioch, we now know not just that the aforementioned Pisan-born Stephen the Philosopher (of Pisa and Antioch) produced his medical translation (entitled *Regalis dispositio*) of the Arabic *Kitāb al-malakī* during his stay in Syria but that he later turned his attention to astronomy. In these activities, Stephen apparently did not work alone in Antioch but had contact with other translators. So the *Regalis dispositio* of 1127 does not stand isolated as a "fruit of the First Crusade",[15] as was long believed; rather, it is a part of the first attempt at a systematic Latin acquisition of Arabic sciences over a wider range.[16]

In addition to the medical work *Regalis dispositio*, Stephen of Pisa and Antioch was associated with three texts on Ptolemaic astronomy. A treatise on Arabic

[11]Haskins, *Renaissance*, pp. 130–140.

[12]Haskins, *Renaissance*, p. 282f.; Kedar, 'The subjected Muslims of the Frankish Levant', p. 174.

[13]See Williams, 'Philip of Tripoli's translation of the pseudo-Aristotelian Secretum Secretorum', p. 93f.: "In the Crusader Levant [...] we see several of the preconditions being met for scholarly exchange and translation: centers of learning for both Western and indigenous scholars, local intermediaries to serve as language teachers, multiple sources for Arabic books. Whatever reasons may be adduced to explain how few translations may have been done here and how little intellectual interaction took place, perhaps more was accomplished here than we have supposed until now."

[14]Haskins, *Renaissance*, p. 64.

[15]Rose, *Verzeichniss*, p. 1059.

[16]Burnett, '‘Abd al-Masīḥ of Winchester'; *idem*, 'Antioch as a link'; *idem*, 'Transmission'.

cosmology, entitled *Liber Mamonis* and extant in a single copy (MS Cambrai, Médiathèque d'Agglomération, A 930), as well as a collection of astronomical tables and rules with the title *Regule canonis* (now lost) can be attributed to Stephen himself. Closely linked to these works is an Arabic-Latin translation of Ptolemy's *Almagest* by a certain ʿAbd al-Masīḥ Wittoniensis, which is preserved in a single fragmentary copy (MS Dresden, SLUB, Db. 87, fols 1r-71r).[17]

Judging from the poor state of preservation of the astronomical works from Antioch, the texts do not appear to have found a wide circulation in Europe. Aside from incidental circumstances that often determined the success of a translation, a wider spread of the Antiochian texts was possibly prevented by the strong impact that translations from Toledo would soon have on the European reception of Arabic science.[18] Nonetheless, the texts add significantly to the importance of the Crusader states as places of early systematic transfers from Arabic. They also bear witness to an intellectual life there that was more active than previously believed.

1.2 The Astronomical Works from Antioch

With ʿAbd al-Masīḥ Wittoniensis' translation of the *Almagest*, the Antiochian texts comprise the earliest-known attempt to produce a Latin version of Ptolemy's 'great compendium' of astronomy. In addition, Stephen of Pisa and Antioch's *Liber Mamonis* provides us with one of the first accounts of Ptolemaic astronomy and cosmology by a Latin scholar, several decades before studies of a similar level of scholarship can be found in the West. Due to the pioneering character of their work, the Antiochian translators needed to develop several of the prerequisites for dealing with mathematical astronomy in Latin language and notation by themselves. Their texts thus represent a precedent to the corresponding achievements of translators working in the West. Furthermore, unlike the later productions from the Iberian Peninsula, the translations from Antioch derived from eastern Arabic traditions. This implied a different situation with regard to the available sources and it also had an influence on the way in which Stephen and his co-workers received this material. Indeed, from the eleventh century onwards, when the contradictions and insufficiencies in the Ptolemaic theory had been intensively discussed in the Arabic literature, astronomical studies in the western and the eastern parts of the Arab world developed in different ways: while in the Maghrib and al-Andalus the deficits identified in the fundamental theories held back the interest in mathematical astronomy, the East saw a flourishing of new astronomical ideas

[17] Burnett, 'Transmission'.

[18] Cf. Burnett, 'Transmission', p. 42. For another largely unknown Latin translation of Ptolemy's *Almagest* before Gerard of Cremona's, see Haskins, *Studies*, pp. 159–64, and Haskins, Lockwood, 'The Sicilian translators of the twelfth century'.

which departed from Ptolemaic doctrines.[19] The Latin translators in Spain thus drew on less progressive astronomical traditions, and had little contact with living teachers.[20] The translations from Antioch therefore provide a valuable example of the effect that a different, eastern environment might have on the reception of Arabic astronomy and cosmology by Latin scholars. They also provide insights into the spread of astronomical literature and its use in a particular setting in the Islamicate East, in a period from which little evidence on astronomical studies has survived in the Arabic sources.[21]

1.2.1 *Regule canonis* and *Liber Mamonis*

Stephen of Pisa and Antioch concentrated his studies in Arabic science first on medicine and later on astronomy and natural philosophy. In doing so, he did groundbreaking work in different fields of Arabic learning which, though widely separated, attracted the attention of many of the early Latin translators. Yet Stephen's interest was not limited to the importation of factual knowledge from selected disciplines; rather, in his works he propagates, and takes the initiative for, a large-scale acquisition of foreign learning with the aim of fundamentally reforming scholarship in Europe (see below, Sects. 4.1, 4.2 and 4.5). On the one hand, the *Liber Mamonis* shows Stephen's attempt to introduce Ptolemaic astronomy in an accurate and understandable form to an untrained Latin readership (see below, Sects. 3.3 and 4.1). At the same time, however, he is already trying to initiate an independent Latin engagement with astronomy, supplied with a set of new tools and encouraged by his own example as a critical investigator (see below, Sects. 3.2, 4.3 and 4.4). Thus, Stephen's achievements in the celestial sciences were not just of a quality equal to his more celebrated work on medicine, but they also show a considerable degree of far-sightedness and intellectual independence.

In his early medical work, *Regalis dispositio*, Stephen made a case for very strict standards of translation. In sharp contrast, his later cosmological treatise, *Liber Mamonis*, appears as a free compilation based on several mostly unmentioned sources (see below, Sects. 2.1, 2.3, 3.1 and 5.1). Stephen's main authority for the *Liber Mamonis* is the Arabic cosmography *On the Configuration of the*

[19] See, e.g., Morelon, 'General survey of Arabic astronomy', p. 17f., and Mercier, 'East and West contrasted in scientific astronomy'. These publications also provide concise accounts of the most important authors and works of Arabic astronomy and their transmission to Europe. A survey of the sciences in Islamic Iberia can be found in Samsó, Forcada, *Las Ciencias de los Antiguos en al-Andalus*.

[20] Mercier, 'East and West contrasted in scientific astronomy', p. 325.

[21] The dissertation on which this book is based concentrated on these aspects. The following is a brief summary.

World by Ibn al-Haytham; however, the numerous insertions he makes reveal the simultaneous influence of other Arabic astronomers as well as non-astronomical sources (see below, Sects. 3.1, 4.5 and 5.1). Stephen enriched the compilation further with comments of his own and he reworked the text systematically into a guide to Ptolemaic astronomy—especially the summarised astronomy which he had published before, in a lost collection of astronomical tables and instructions entitled *Regule canonis* (see below, Sects. 3.1, 4.1 and 4.3). Stephen's liberal use of sources in the *Liber Mamonis* is paralleled by an equally free language in which he wrote the text (see below, Sect. 3.2). As far as the information in the *Liber Mamonis* allows us to estimate, the astronomical tables which Stephen had used for the *Regule canonis* most probably derived from the early Arabic Mumtaḥan tradition (see below, Sect. 5.1).

Stephen conceived the *Liber Mamonis* as a complementary work to his astronomical tables, *Regule canonis*. While the *Regule canonis* supplied the reader with tools to easily predict the positions of celestial bodies, the *Liber Mamonis* provided an illustrative, reader-friendly explanation of the procedures involved, based on a depiction of the physical universe. Although the works appeared very different in character, Stephen uses various ways to indicate that they effectively deal with the same subject; for instance, there are repeated references to the *Regule canonis* in corresponding contexts in the *Liber Mamonis* and also adaptations of Ibn al-Haytham's cosmography to the concepts of mathematical astronomy (see below, Sects. 3.2, 4.3 and 4.4).

To provide a foundation for his teaching, Stephen refers to the Greek tradition by presenting Ptolemy as the most respectable, though not always correct, authority in the celestial sciences. Nevertheless, a direct use of Ptolemy cannot be identified either for the *Regule canonis* or the *Liber Mamonis* (see below, Sect. 5.1). In particular ʿAbd al-Masīḥ Wittoniensis' translation of the *Almagest* does not seem to have been available to Stephen when he wrote his astronomical works.

1.2.2 The Latin Translation of Ptolemy's *Almagest*

The extant fragment of ʿAbd al-Masīḥ Wittoniensis' Arabic-Latin translation of the *Almagest*, in MS Dresden, SLUB, Db. 87, fols 1r-71r, copied around 1300, preserves only the first four of the original thirteen books of Ptolemy's work.[22] Moreover, none of the tables, which constitute an important part of the *Almagest*, are present in the Dresden copy. The translator's name appears in colophons at the end of each book, in the genitive form 'wittoniensis (or 'wuttomensis' or any combination of these readings; the colophon of Book II has 'wintomiensis') ebdelmessie'. The name, Ebdelmessie, is a Latinisation of ʿAbd al-Masīḥ, meaning 'Servant of

[22]Schnorr von Carolsfeld, *Katalog*, p. 305; Björnbo, 'Die mittelalterlichen lateinischen Übersetzungen', p. 392; Heiberg, 'Noch einmal die mittelalterliche Ptolemaios-Übersetzung', p. 215f.; Haskins, *Studies*, p. 109.

the Messiah', and indicates an Arab Christian origin, whereas 'wittoniensis' is commonly interpreted as relating to the city of Winchester.[23]

ʿAbd al-Masīḥ translated the *Almagest* from an Arabic version produced in the late ninth century by the famous polymath and translator Thābit ibn Qurra (836–901).[24] Thābit's version of Ptolemy became popular in the Islamic East, but it does not seem to have had much influence in the Maghrib or al-Andalus, where instead other Arabic translations of the *Almagest* were in use.[25] Thābit's version of the *Almagest* included several systematic changes relative to the Greek original. These modifications are largely preserved in ʿAbd al-Masīḥ's Latin translation, which rendered Thābit's text accurately and with much understanding of the content.

A close relation between ʿAbd al-Masīḥ's Latin translation of the *Almagest* and Stephen of Pisa and Antioch's works is clear, due to the similarities in the numerals and the identical use of distinctive technical terms.[26] In fact, ʿAbd al-Masīḥ's astronomical terminology corresponds almost entirely to Stephen's (see below, Glossary), and although he translated from Arabic, as Stephen did, he avoids any transliterations of Arabic terms and uses several Greek expressions. In the colophons of his translation, ʿAbd al-Masīḥ also refers to his profession as a translator of scientific texts by the Greek term 'philophonia' ('love of languages'), in imitation of, and also in contrast to, 'philosophia'. Furthermore, a system of alphanumerical numerals that is known from Stephen's medical and astronomical works—essentially a Latin abjad (see Table 2.1 on p. 19)—is also used by ʿAbd al-Masīḥ, though in a modified form: ʿAbd al-Masīḥ replaced Stephen's

[23]It is therefore also read as 'wintoniensis'; Kunitzsch, *Der Almagest*, p. 9, n. 30. Kunitzsch, like the previous investigators of the Dresden *Almagest*, believed that this text was translated from a Greek source. An Arabic origin was first suggested by Richard Lorch, *Sector-Figure*, p. 356f., and reported by Burnett, 'ʿAbd al-Masīḥ of Winchester', esp. p. 161, who also speculates about the possible connections between this presumably eastern translator with Winchester.

[24]The existence of an Arabic translation of Ptolemy's *Almagest* by Thābit ibn Qurra (alone) was long denied by modern researchers, until it could be proved thanks to ʿAbd al-Masīḥ's Latin translation of it; Grupe, 'The Thābit-Version', where I also show that Thābit's translation of Ptolemy was intensively used by Naṣīr al-Dīn al-Ṭūsī (13th c.) and the Persian mathematician al-Nasawī (first half 11th c.). After my identification of Thābit's version of the *Almagest*, several other Arabic manuscripts could be safely associated with it. They include a large fragment of the Arabic text from Book I until the beginning of Book VI, in MS Jaipur, Maharaja Sawai Man Singh II Museum Library, 20, and an abridgement apparently of the entire text, in MS Tehran, Majlis-Senate, 1231; presented to the public in November 2015, in my conference paper 'Thābit ibn Qurra's version of the Almagest and its reception in Arabic astronomical commentaries'. On the same occasion I have also demonstrated the use of Thābit's translation of Ptolemy's *Almagest* by Avicenna, al-Abharī and other Persian astronomers. These findings were possible thanks to the collaboration of Dr. Mª José Parra and the material provided to me by Prof. Richard Lorch and Prof. Paul Kunitzsch, to whom I express my sincerest gratitude.

[25]For a discussion of other Arabic versions of the *Almagest* see Kunitzsch, *Der Almagest*, esp. pp. 17–34.

[26]Burnett, 'Transmission', p. 29f.

letters for the tens with symbol attachments to the letters representing the units.[27] This shorthand for the tens caused much confusion among later copyists of ‹Abd al-Masīḥ's text and was probably the cause for the total omission of the number tables in the Dresden copy. However, since ‹Abd al-Masīḥ's modified notation of the numbers is clearly developed from Stephen's, it supports the assumption that his translation of the *Almagest* is later, not earlier, than Stephen's astronomical works.[28]

1.2.3 General Observations

Stephen did not just conceive of the astronomical *Regule canonis* and the cosmological *Liber Mamonis* as two corresponding works; they also formed what can be seen as a self-contained introduction to the celestial sciences. It is thus unclear whether Stephen had originally intended that a translation of Ptolemy's *Almagest* should at some moment supplement these writings. Nonetheless, with or without ‹Abd al-Masīḥ's translation of the *Almagest*, Stephen of Pisa and Antioch provided Latin Europe with the first comprehensive yet concise set of texts on the methods and the doctrines of Ptolemaic astronomy and cosmology and on their practical application.

Writing for a European public that was still unfamiliar with Ptolemaic astronomy, Stephen himself must have been largely untrained in this discipline when he moved from Italy to Syria. Stephen arrived in the East as a doctor and a translator of medicine, with the intention of improving Constantine the African's recent translation of the medical work *Kitāb al-malakī* (see below, Sect. 2.2). During this stay in Syria Stephen most likely acquired a respectable degree of understanding of astronomy as well. One can conclude from this that Antioch was an ideal setting for a student keen to acquire a profound knowledge of astronomy and natural philosophy and eventually to make independent scientific speculations at an advanced level (see below, Sect. 4.4).

The kind of learning that flourished in twelfth-century Syria is reflected by the range of sources that influenced the Antiochian translations. Significantly, for his translation of the *Almagest* ‹Abd al-Masīḥ used the Arabic version by Thābit ibn Qurra alone instead of one of the more literal Arabic translations of the Greek, such as the ones by al-Ḥajjāj or Isḥāq ibn Ḥunayn with corrections by Thābit ibn Qurra. ‹Abd al-Masīḥ seems to have considered Thābit's text superior to the other versions; or, and this is equally plausible, Thābit's version may have been the only one available to him in Antioch.

[27] Grupe, *The Latin Reception*, pp. 83–88.

[28] The heavily corrupted state of the numerals in the Dresden manuscript misled Burnett, who assumed that these numerals represent a more primitive stage and concluded that ‹Abd al-Masīḥ's translation of the *Almagest* preceded Stephen's writings; Burnett, 'Transmission', p. 29.

On the other hand, Stephen's creative use of Ibn al-Haytham's *On the Configuration*, especially for teaching mathematical astronomy, and his criticism of several parts of the Ptolemaic tradition were most probably inspired by corresponding trends in his immediate environment. While a variety of Arabic texts was obviously available in Antioch, Stephen's advanced understanding and his independent approach to the astronomical literature make it practically certain that his studies benefited from contact with members of a learned community. The evidence suggests that ʿAbd al-Masīḥ Wittoniensis may have been one such local expert.

Chapter 2

The *Liber Mamonis*, Its Author and His Main Source

2.1 The Cosmological Tract in MS Cambrai 930

In an unpublished study from the late 1990s, Richard Lemay presented his discovery that the cosmological text in the Latin manuscript no. 930 of the Médiathèque (formerly Bibliothèque municipale or B. communale) of Cambrai contains a commented Latin translation of Ibn al-Haytham's Arabic cosmography, *On the Configuration of the World* (*Maqāla fī hayʾat al-ʿālam*).[1] Although the Cambrai manuscript had been known to modern researchers since the early twentieth century, the literary dependency of the text remained a mystery. One reason was clearly the fact that the Latin author avoids any explicit reference to his Arabic source. The rubricated heading on the front page of the manuscript reads 'Incipit liber Mamonis in astronomia a Stephano philosopho translatus.'[2] While indicating that the work is mainly a translation by a certain 'Stephen the Philosopher', the chosen title, *Liber Mamonis*, gives no clue as to the origin of the text. Only on two occasions does Stephen the translator indicate (in very general terms) that

[1]In the same unpublished study Lemay also included an edition of the Cambrai text. Lemay's discovery of the Arabic origin of the work was made known by Burnett, 'Stephen, the Disciple of Philosophy', and Gautier Dalché, *La Géographie*, p. 99.

[2]See the reproduction on p. 420.

D. Grupe, *Stephen of Pisa and Antioch: Liber Mamonis*, Sources and Studies in the History of Mathematics and Physical Sciences,
https://doi.org/10.1007/978-3-030-19234-1_2

he took most of his information from 'a certain Arab' or that he 'translated from Arabic.'[3]

Moreover, Stephen the author does not simply translate the Arabic text but mixes it with an extensive commentary of his own. At the same time, indications that would enable the reader to distinguish between the translated parts and Stephen's own comments are few and far between. Stephen introduces a new division of the text, into four books, which has no correspondence in the Arabic, while remnants of the original division of the Arabic text into a sequence of fifteen chapters survives in the Latin only in the form of a number of highlighted initials.

2.1.1 The Manuscript

Preceded by a fly-leaf, the text in MS Cambrai 930 extends over forty-eight folios gathered together into six quaternios. A numbering of the first five quires, probably by the scribe himself, is found in the bottom margins on the last page of each one. As the modern foliation of the manuscript includes the prefixed fly-leaf, the numbering of the quaternios is found on fols 9v, 17v, etc. The manuscript ends prematurely after the sixth quire with an uncertain amount of text missing. Judging from the corresponding text in Ibn al-Haytham's *On the Configuration* and also from the absence of a numbering of the last preserved quire, the missing parts of the Cambrai manuscript probably did not comprise more than a singleton or a bifolio, which had been added after the last quaternio and was later lost. On the verso page of the prefixed fly-leaf the description 'Quidam tractatus de astronomia', followed by the number 'xxii', has been written. Modern library stamps of two different types belonging to the Cambrai library are imprinted on fols 1v, 2r, 13r, 33r, and 49v.

The pages measure 14.8 cm × 21.2 cm with a text block of 10.8 cm × 16.7 cm. A text layout of thirty-four lines per page is used with a line spacing of ca 5 mm and a letter height of 2–3 mm. A later trimming of the parchment can be seen from the numbering of the quires, as parts of the numbers were cut off. Despite the economical use of space, the text has a very balanced appearance thanks to its careful arrangement and a neatly written minuscule (cf. the plates on p. 420 and 421). The black script has been given consistent rubrication in the book headings, sentence initials and numerals. Also inserted into the text are fifteen diagrams, drawn in red ink and often with black labels (cf. plate 2 on p. 421). The figures vary in size, complexity, and accuracy. Their degree of completion, their technical quality and the conclusiveness of their placing relative to the corresponding parts of the text generally deteriorate towards the end of the manuscript.

[3]See p. 324:3f.: 'Arabem quendam plurimum secuti sumus, in hoc quoque permultum sequemur,' and p. 340:4: 'de Arabico transtulimus.'

The manuscript has been estimated to date from the late twelfth century,[4] but it may well be slightly older. Besides the incomplete state of some diagrams, typical copying mistakes reveal that the manuscript is not the composer's autograph. The Cambrai manuscript is at present the only known exemplar of the *Liber Mamonis*. However, the entry 'liber mamnonis', or 'liber marmionis', in a late twelfth century catalogue of the Abbey of Whitby indicates that at one time the work may also have reached northern England.[5]

2.1.2 The *Liber Mamonis* in Modern Studies

A consistent picture of the origin of the *Liber Mamonis* and the circumstances of its production has emerged only in recent years. Stephen of Pisa and Antioch's authorship of the text was questioned as late as 2000 (by Lemay), and the first detailed comparison of a passage in Stephen's text with the Arabic cosmography *On the Configuration of the World* was made in 2009 (by Gautier Dalché).[6] While the number of publications dealing with the *Liber Mamonis* is comparatively small, some of them contain assumptions which have later proved false or misleading. So, in the light of our present knowledge, it will be useful to summarise here the previous research on the *Liber Mamonis*.

The first scholar to draw attention to MS Cambrai 930 was Charles Haskins, in 1924, in his famous *Studies in the History of Mediaeval Science*.[7] For a long time, Haskins' discussion remained the most comprehensive study on the subject. He believed that the *Liber Mamonis* was mainly an independent work by its Latin author. Providing transcriptions of selected passages, Haskins drew attention to the author's repeated criticism of Macrobius and his indebtedness to an unnamed Arab. Haskins also noted the appearance of Greek terms in the text and the absence of Arabic expressions, and mentioned the use of alphanumerical numerals partly in the manner of a Latin abjad; Haskins admitted that he did not fully understand other numerals in the manuscript, which represent larger numbers. He concluded that the *Liber Mamonis* was a work from the earliest period of a reception of Arabic astronomy in Europe, written by an author who intended to replace the traditional, Platonist theories of Macrobius with the advanced learning of the Arabs in Western Europe. Haskins considered Stephen of Pisa and Antioch to be possibly the author of the text, but he remained sceptical about this on the grounds of certain unspecified internal evidence.[8]

---

[4] Haskins, *Studies*, p. 99.

[5] See Burnett, 'Antioch as a link', p. 10, note 34, referring to Haskins.

[6] Lemay already roughly indicated translated portions in his unpublished edition of the *Liber Mamonis*.

[7] Haskins, *Studies*, pp. 98–103.

[8] Haskins, *Studies*, pp. 103 and 135.

In 1950, Richard Hunt recognised parallels between Haskins' report of alphanumerical numerals in the *Liber Mamonis* and Stephen of Pisa and Antioch's possible authorship, and a description of a copy of the *Rhetorica ad Herennium* kept in Milan.[9] The Milan *Rhetorica* contained the same abjad numerals that Haskins had identified in the Cambrai manuscript. A subscription in the Milan text also reports that it was written 'in the year 1121 after the passion of the Lord for Stephen the treasurer of (*or*: at) Antioch.'[10] These parallels with Haskins' report induced Hunt to identify the 'philosopher Stephen' of the *Liber Mamonis* with the addressee Stephen in the Milan *Rhetorica*. Moreover, the same unusual dating as in the Milan *Rhetorica*, 'a passione domini', is also known from Stephen of Pisa and Antioch's medical translation, *Regalis dispositio*, for the year 1127, only six years later than the date in the Milan text.[11] This identical system of dating and the reference to Antioch as the place of work of the Stephen of the Milan *Rhetorica* and also in Stephen of Pisa and Antioch's *Regalis dispositio*, in combination with the use of abjad numerals in the Milan *Rhetorica* and the *Liber Mamonis*, finally led Hunt to associate all three texts, i.e., the *Liber Mamonis*, the Milan copy of the *Rhetorica ad Herennium* and the *Regalis dispositio*, with the same person: the Pisan-born Stephen the Philosopher, who is known to have worked as a translator in Antioch.

Haskins' analysis of the numerals in the Cambrai manuscript was continued by Richard Lemay, who discussed the text in two articles published in 1987 and 2000.[12] For numbers up to 360, the use of a Latin abjad, described correctly by Haskins, could be confirmed (cf. Table 2.1).[13] Lemay further noticed that the higher numbers are written in a place-value notation using a partly Latinised form of Hindu-Arabic numerals of the Eastern type.[14] From the five numbers in the

[9]Hunt, 'Stephen of Antioch', referring to a description of MS Milan, Ambrosiana, Cod. E. 7 sup., by Sabbadini, 'Spogli Ambrosiani Latini', pp. 272–76.

[10]'Scribsit hunc rethoricorum librum //// scriba Stephano thesaurario antiochie anno a passione domini millesimo centesimo vicesimo primo'; Sabbadini, 'Spogli Ambrosiani Latini', p. 272.

[11]Cf. Haskins, *Studies*, p. 133f. The dating 'from the passion of the Lord' has been interpreted differently in the modern literature, as it may mean either that the years were counted from the year 33 CE or merely that the calendar year started at Easter. Hunt, 'Stephen of Antioch', p. 173, leaves the question undecided; Sabbadini, 'Spogli Ambrosiani Latini', p. 272, follows the first interpretation; Rose, *Verzeichniss*, p. 1059, and Haskins, *Studies*, p. 133f., in their discussions of the *Regalis dispositio*, and also Talbot, 'Stephen of Antioch', p. 38, argue in favour of the second interpretation, for which Burnett, 'Antioch as a link', pp. 7 and 67–69, gives convincing support.

[12]Lemay, 'De la scolastique à l'histoire'; *idem*, 'Nouveautés fugaces'.

[13]Examples of the alphanumerical numerals can be found in the reproduction below, on p. 421, where in the fourth line a rubricated 'e' stands for the number '5'; the captions in the diagram contain further examples, e.g., '*ka* gradibus *i* sexagenariis *ke* secundis' = 11;9,15°. Some of the letters of the alphanumerical system were commonly used in medieval writing as abbreviations or as labels in mathematical diagrams. Lemay therefore suggests a connection between the lack of success of the alphanumerical notation and its susceptibility to various misinterpretations; Lemay, 'Nouveautés fugaces', p. 381f.

[14]Lemay, 'De la scolastique à l'histoire', p. 471; *idem*, 'Nouveautés fugaces', p. 379.

a	-	1	k	-	10	t	-	100
b	-	2	l	-	20	u	-	200
c	-	3	m	-	30	x	-	300
d	-	4	n	-	40			
e	-	5	o	-	50			
f	-	6	p	-	60			
g	-	7	q	-	70			
h	-	8	r	-	80			
i	-	9	s	-	90			

Reading example: *ukf* = 216

Table 2.1: The alphanumerical numerals in the *Liber Mamonis*

i/t - 1	p - 2	Ψ - 3	'quia' - 4	g - 5	'et' - 6	u - 7	a/d - 8	q - 9

Reading example: *tupa* = 1728

Table 2.2: Latinised forms of Hindu-Arabic numerals of the Eastern type in the *Liber Mamonis*

Liber Mamonis that are written in Hindu-Arabic numerals,[15] Lemay was able to reconstruct a scribal assimilation of the original Arabic forms to their Latinised appearance in the manuscript.[16] The correspondences shown in Table 2.2 could finally be identified.

In his articles, Lemay also suggested that the *Liber Mamonis* by 'Stephen the Philosopher' in the Cambrai manuscript may in fact be a copy of Hermann of Carinthia's undiscovered work *Astronomia*.[17] This rather unconvincing hypothesis was not followed by Patrick Gautier Dalché, who in the meantime studied

[15]Cf. below, p. 252:1f.

[16]Lemay's observations were later supplemented by Burnett, who noticed that the eastern Hindu-Arabic numerals of '4' and '6' also became Latinised in the *Liber Mamonis* into the Tyronian notes of 'quia' and 'et'; Burnett, 'Transmission', p. 29. Haskins had transcribed such a '6' wrongly as 'et', although he considered the possibility that it was a numeral; *Studies*, p. 102, note 129.

[17]In his earlier article, Lemay discusses the Cambrai manuscript as part of a more sweeping argumentation, in which he tries to attribute the so-called 'Sicilian' translation of the *Almagest* to Hermann of Carinthia. In the course of the argument, Lemay assumes that the *Liber Mamonis* is Hermann of Carinthia's *Astronomia*, which in that case would have been misattributed to Stephen shortly after its completion. Apart from the similar period of production, however, and Hermann of Carinthia's vague references to an astronomical work of his in the *De essentiis*, Lemay fails to provide convincing support for this hypothesis; see Lemay, 'De la scolastique à l'histoire', p. 465. Also in his later article Lemay comes to speak about Hermann's alleged authorship of the Cambrai text. But, apart from the assertive heading, "Un *abjad* latin avorté dans l'*Astronomia* de Hermann de Carinthie," and similar imputations in the discussion, Lemay produces no further evidence of this identity; 'Nouveautés fugaces', p. 377f.

the *Liber Mamonis* with an interest in the transmission of Ptolemy's *Geography*. Stephen discusses the *Geography* only briefly in the *Liber Mamonis* but on that occasion reveals detailed knowledge about the subject. Since Stephen could have received this and other information in Antioch through Arabic traditions, Gautier Dalché speculates that he chose the title of his work, *Liber Mamonis*, as a reference to the caliph al-Ma'mūn (813–833), who in the early ninth century promoted cartographical and astronomical studies in Baghdad and also initiated the production of improved Arabic versions of the *Almagest*.[18] In a later supplement, now based on a comparison with Ibn al-Haytham's treatise, Gautier Dalché was able to confirm that Stephen's knowledge of competing geographical doctrines goes beyond the information provided by Ibn al-Haytham. This suggested that scientific knowledge in Antioch also extended to geography.[19]

Possible doubts about Stephen's authorship of the *Liber Mamonis* were finally dispelled by Charles Burnett, who studied the text at a time when its dependency on Ibn al-Haytham's cosmography was still not publicly known. Further to Hunt's arguments, Burnett found other indications that the 'Stephanus philosophus' of the *Liber Mamonis* is indeed the famous author of the *Regalis dispositio*. There are distinctive parallels in form, content and style between the prefaces of Stephen's medical translation, *Regalis dispositio*, and those in the *Liber Mamonis*.[20] In addition, the Latin abjad used in the Cambrai manuscript is also found in one of the oldest known manuscripts of Stephen's *Regalis dispositio*. This confirms Hunt's assumption of a connection between these numerals and the person of Stephen of Pisa and Antioch. Based on the same numerals and an identical use of a rare astronomical terminology, Burnett further showed a close relation between Stephen's *Liber Mamonis* and ʿAbd al-Masīḥ Wittoniensis' Arabic-Latin translation of Ptolemy's *Almagest*.[21] Several statements in the *Liber Mamonis* show that Stephen of Pisa and Antioch indeed had some knowledge of the *Almagest*, which would support this proposed relation between the two texts. Several astronomical parameters and arguments in the *Liber Mamonis* also show that Stephen was influenced by early Arabic astronomers. Following Gautier Dalché, therefore, Burnett

[18] Gautier Dalché, 'Le souvenir de la Géographie de Ptolémée dans le monde latin médiéval', p. 90f.

[19] Gautier Dalché, *La Géographie de Ptolémée en Occident (IVe-XVIe siècle)*, pp. 98–101.

[20] A detailed list of correspondences in the prefaces of both works is given in Burnett, 'Antioch as a link', p. 11, note 38. In the same article a transcription and an English translation of all of Stephen's prefaces is included, pp. 20–60. The similarities make one wonder about the kind of internal evidence that Haskins had in mind when he questioned Stephen's authorship of the *Liber Mamonis*; cf. d'Alverny, 'Translations and translators', p. 439. Haskins' doubts were repeated by Sarton, *Introduction to the History of Science*, p. 236f., who in this regard seems to depend fully on Haskins.

[21] Similarities between the numerals in both texts were noticed by Haskins, who made no attempt at an explanation; *Studies*, p. 109, note 155.

assumed a connection between the Latin title, *Liber Mamonis*, and the name of the early ninth century caliph al-Ma'mūn.[22]

In addition to the texts that were found to use the Latin abjad, further similarities exist in the appearance of the Hindu-Arabic numerals of the Eastern type in another small group of Latin texts. Some of these writings are related to a contemporary of Stephen, the Jewish scholar Abraham ibn Ezra (ca. 1090 till after 1160). One of ibn Ezra's works, the *Pisan Tables*, contains a collection of astronomical values that were calculated specifically for the coordinates of Pisa, the city of Stephen's birth. Moreover, a manuscript today in London (BL, Harley 5402) contains notes dating from around 1160 with instructions for the use of astronomical tables that had been calculated for Lucca, in close proximity to Pisa. These notes once again contain the eastern Hindu-Arabic numerals, whereas a part of the main text of the manuscript, which thus seems to have been written before 1160, contains an astronomical table written in the Latin abjad as used by Stephen of Pisa and Antioch. Thus, the two rare systems of numerals found in the *Liber Mamonis* appear in a second, contemporary manuscript with astronomical content and with a close geographical connection to Pisa. Burnett inferred from this that an important group of Arabic-Latin translators was working in Syria, some decades before the centres of systematic Arabic-Latin translation in the Iberian Peninsula began their production. While the works of these men reached Europe mainly through Pisa, the translator Stephen of Pisa and Antioch emerges as the central figure in this group, especially with regard to its connection with the West.[23]

2.2 'Stephen the Philosopher' and His Work

Before the discovery that Stephen of Pisa and Antioch might be associated with a larger number of texts, he had been known only as the author of a large medical compendium, *Regalis dispositio*.[24] This 'Royal Arrangement' contains a complete Latin translation of ʿAlī ibn al-ʿAbbās al-Majūsī's (Latinised: Haly Abbas, 10th c.)

[22]Burnett, 'Transmission', pp. 30–34.

[23]Burnett, 'Antioch as a link', pp. 15–19; *idem*, 'Transmission', pp. 36–42.

[24]A first systematic discussion of the *Regalis dispositio* and of its author Stephen was given by Valentin Rose in 1905, in a description of the Berlin manuscript of the *Regalis dispositio*; Rose, *Verzeichniss*, pp. 1059–65. Rose's account was followed in 1924 by Haskins' more general treatment of the subject; Haskins, *Studies*, pp. 131–5. Shortly before, an attempt had been made to attribute another medical text, *De modo medendi*, to Stephen, but the evidence for the argument was weak and it had no lasting influence; Ganszyniec, 'Stephanus de modo medendi'. Sarton, *Introduction to the History of Science*, p. 236, included Ganszyniec's views in his own chapter on Stephen, but Talbot, 'Stephen of Antioch', note 13, later rejected the idea as unconvincing. A summary of the research on Stephen of Pisa and Antioch's life and work, including transcriptions of the most relevant testimonies and new findings, is given in Burnett, 'Antioch as a link', and *idem*, 'Stephen, the Disciple of Philosophy'.

Arabic work on medicine, *Kitāb al-malakī*, and was Stephen's first and most influential work. Stephen supplemented his translation of the Arabic text with a medical glossary, often referred to as his *Breviarium* or *Synonyma*, and with prefaces and explicits to individual parts of the work.[25] Stephen's *Regalis dispositio* circulated widely, as is documented by the considerable number of extant manuscripts at various places in Europe; its lasting popularity is indicated by two printed editions made almost four hundred years after the work's completion–one at Venice in 1492, and the other at Lyon in 1523.[26]

In the *Regalis dispositio*, Stephen introduces himself to the reader as 'Stephanus philosophie discipulus' and he mentions Antioch as the place, and 1127 as the year, in which he completed the translation. This makes Stephen the first western scholar known to have made a complete translation of an Arabic scientific text in the East, little less than 30 years after the founding of the first Crusader state.[27] In the prefaces to the *Regalis dispositio*, Stephen also explains that the primary incentive for his translation were the shortcomings of Constantine the African's earlier translation of the same Arabic text, known as *Liber pantegni*. Stephen's *Regalis dispositio* was intended as the completion of (but also a replacement for) the *Liber pantegni*, which had been produced some 40 years earlier and which contains a very free and incomplete translation, mainly of the first, theoretical part of the *Kitāb al-malakī*.[28] Stephen is harsh in his criticisms of Constantine, whom he accuses of having distorted the meaning of the Arabic original and also of claiming for himself the intellectual merits of others. In strong contrast to Constantine's interpretative paraphrase, Stephen's translation is conspicuously literal.[29]

---

[25]Some of Stephen's explicits as well as excerpts from the prefaces of the *Regalis dispositio* are edited in Rose, *Verzeichniss*, pp. 1060–63, Haskins, *Studies*, p. 133, and Burnett, 'Antioch as a link', pp. 67–69. The latter also provides an edition of Stephen's prefaces, on pp. 22–40.

[26]For a catalogue of the known manuscripts of the *Regalis dispositio*, see Haskins, *Studies*, p. 131, notes 8 and 9, supplemented and partially revised by Burnett, 'Antioch as a link', pp. 20–22 and 30. Many of the manuscripts contain only the second half of Stephen's work, on practical medicine, and in some cases his writings have been mixed with texts on related subjects; see, e.g., Rose, *Verzeichniss*, pp. 1059–65.

[27]Cf. Haskins, *Studies*, p. 131. Burnett, 'Antioch as a link', pp. 2–4, discusses the possibility that, before Stephen, Adelard of Bath obtained material for some of his translations during a documented stay in Syria.

[28]Haskins, *Studies*, p. 132; Burnett, 'Antioch as a link', p. 7f.

[29]The literalness of Stephen's version has occasionally been criticised by modern scholars for not providing a better understanding of the original. Steinschneider, 'Donnolo', pp. 333–35, points at Constantine's and Stephen's opposing translation standards and appreciates Constantine's very 'modern' linguistic approach. Similarly, Talbot, 'Stephen of Antioch', on p. 38, concedes that "Stephen's slavishly literal and verbose text closely follows its source" but finds that "Constantine's free paraphrase is easier to understand, gives better sense, and does great justice to the original."

Stephen's concern with linguistic accuracy and its importance to the subject becomes apparent in his *Breviarium*, or *Synonyma*, which he added to his translation as a medical glossary in three languages, Arabic, Latin and Greek, on the basis of Dioscorides' *De materia medica*.[30] In that medical dictionary Stephen is careful not to include any translation that he is uncertain of, and he advises his readers who wish for a better understanding to consult scholars at Salerno or Sicily, many of whom were acquainted with Greek or Arabic. From this latter statement, and from corresponding remarks by Platearius[31] and in the treatise *De aegritudinum curatione* in the so-called Codex Salernitanus, it has been inferred that Stephen was closely connected with the medical schools in southern Italy, where he may have gained his training in medicine.[32] The *Synonyma* also show that Stephen had at least a rudimentary knowledge of Greek.

The biographical information about Stephen in the *Regalis dispositio* is confirmed in a gloss by a certain "magister Matheus F." (often identified as Matheus Ferrarius, 12th c. Salerno), who purports to possess independent knowledge about our author. According to this Master Matheus, Stephen was a Pisan who, after most of Constantine's translation of the practical part of the *Kitāb al-malakī* had been destroyed by water, went to Syria, learned Arabic and produced a new, complete translation of the same Arabic work.[33] Stephen's Pisan origin was soon noted and associated with the mercantile and logistic importance of the Pisan fleet for the principality of Antioch in the time after the First Crusade, and with the Pisans' presence in the city.[34]

Moreover, since Stephen has been identified as the 'treasurer Stephen at Antioch' of the Milan *Rhetorica ad Herennium*, he is probably also 'Stephen the treasurer of the church of St. Paul at Antioch', documented in a charter for the period from 1126 to 1130.[35] Another reference to Stephen describes him as the 'nephew of the Patriarch of Antioch.'[36] The testimonies suggest that Stephen's move to Syria, assuming that it was indeed motivated by academic interests, was at the very least supported by an involvement in clerical organisation and by influential family members in the East.

---

[30]Stephen's glossary is found only in some manuscripts of the *Regalis dispositio*. Steinschneider could therefore only speculate about the meaning of a reference from the thirteenth century to 'Stephen's *Synonyma*'; cf. Steinschneider, 'Donnolo', p. 313.

[31]Rose, *Verzeichniss*, p. 1059.

[32]Haskins, *Studies*, p. 131; Talbot, 'Stephen of Antioch', p. 38.

[33]Rose, *Verzeichniss*, p. 1060, gives a transcription of the passage from a manuscript in Erfurt.

[34]See Rose, *Verzeichniss*, pp. 1059 and 1062, Haskins, *Studies*, pp. 131 and 134, and Talbot, 'Stephen of Antioch', p. 38.

[35]Hunt, 'Stephen of Antioch', p. 173. An edition of the charter is published in Rozière, *Cartulaire*, p. 174.

[36]Talbot, 'Stephen of Antioch', p. 38, note 5, referring to MS London, BL, Sloane 2426.

A common feature of the historical testimonies of Stephen of Pisa and Antioch is that they either appear in medical works or, when referring to Stephen's scholarly activities, consider him only as a student of medicine. Yet, although later authors speak of him as a doctor, he does not present himself as such; not even in the *Regalis dispositio*, where he rather calls himself a 'disciple of philosophy'.[37] This corresponds to another statement of Stephen's in the *Regalis dispositio*, which mentioned that he had planned that work only to be his first translation, and that he intended to produce many more "of all the secrets of philosophy (philosophie archana) which lie hidden in the Arabic language" and which, compared to the medical treatment of the body, "are far more sublime, as they pertain to the excellence of the human mind (ad animi attinent excellentiam longe altiora)."[38] The *Liber Mamonis* reveals that at least one of the subjects that Stephen was thinking of after completing his medical work was the science of the heavens.[39]

Before Stephen eventually wrote the *Liber Mamonis*, it is known (from that work) that in the meantime he had published another treatise, entitled *Regule canonis*.[40] These *Rules of the Canon* are lost today, but Stephen refers to them on various occasions in the *Liber Mamonis*. They clearly contained a collection of astronomical tables with instructions for their use. Stephen thus published repeatedly on various areas of the science of the heavens, and there was a close relation between his works and ‹Abd al-Masīḥ Wittoniensis' translation of Ptolemy's *Almagest*. Besides Stephen's well-known expertise in medicine, therefore, he must have acquired a respectable level of competence in astronomy as well. Such proficiency in two equally demanding 'new' sciences corresponds to Stephen's repeated exhortations, in the *Regalis dispositio* and in the *Liber Mamonis*, to surmount dogmatic constraints and to be open to new thought, even though this may provoke criticism and enmity.

[37] Burnett, 'Antioch as a link', p. 9.

[38] Haskins, *Studies*, p. 134f., esp. note 26, containing a transcription of the passage based on the Lyon print.

[39] Stephen's turning to astronomy could give a very palpable meaning to the words 'longe altiora' and 'animi excellentia' in the cited passage, in the sense that the concern with astronomy trains and mobilises one's intellectual strength. Although this is hardly suggested from the present context of the *Regalis dispositio*, it would be in line with a later statement by Stephen in the *Liber Mamonis*; see below, p. 118:19f. Stephen uses the expression 'philosophie archana' in the *Liber Mamonis* as well; see below, pp. 112:2 and 154:18.

[40] See below, p. 88:2f. of the edition.

2.3 The Arabic Cosmography *On the Configuration of the World*

2.3.1 The Work

With Ibn al-Haytham's *On the Configuration of the World*, Stephen of Pisa and Antioch decided to translate a cosmography that drew on the long tradition of Aristotle's *On the Heavens* and *Metaphysics*.[41] In *On the Heavens*, Aristotle attributes the various observable motions in the heaven to a system of nested spheres to which the celestial bodies are fixed. Rotations of the spheres, at different speeds and with different orientations, move the stars and planets relative to one another around a resting earth at the centre. Although similar systems had been proposed before, as mathematical or metaphysical entities, Aristotle established the view that the spheres exist as real bodies with crystal-like characteristics.

Aristotle's theory found wide acceptance during the following centuries. At the same time, advances in astronomy were producing increasingly complex mathematical descriptions of the celestial motions, thus requiring corresponding adaptations of the spherical model as well, if the mathematical and the physical concepts were to remain consistent. In that process it became clear that certain elements of the astronomical theory were incompatible with axioms of celestial physics, in particular with the claim that the celestial spheres were in constant circular motion. The contradictions caused a separation between astronomical and physical descriptions of the universe. Whereas astronomical works concentrated on mathematical rules that could be derived from, and be applied to, observational data, cosmological reports provided explanations for the phenomena on the basis of physical concepts. Both approaches are manifest in two works by Ptolemy himself: in his astronomical compendium, the *Almagest*,

[41] An introduction to Ibn al-Haytham's cosmography, together with further references, is given by Langermann prefixed to his edition of the Arabic text. Langermann follows the traditional attribution of *On the Configuration of the World* to al-Ḥasan ibn al-Ḥasan ibn al-Haytham, the famous composer of mathematical commentaries on astronomy and optics, who lived from 965 till ca 1040 CE. This attribution has been seriously questioned in recent years; it has been forcefully argued that the author of *On the Configuration of the World* may be a different Ibn al-Haytham, named Muḥammad ibn al-Ḥasan ibn al-Haytham; Rashed, *Les Mathématiques infinitésimales du IXe au XIe siècle. Vol. 2. Ibn al-Haytham*, pp. 1–17, 490–91 and 511–38, and *idem*, 'The Configuration of the universe: A book by al-Ḥasan ibn al-Haytham?'. A critical reply to these arguments was made by A. Sabra, 'One Ibn al-Haytham or two?'. The question is not directly relevant to the present study, since the identity of the Arabic author does not seem to have influenced Stephen of Pisa and Antioch's reception of the text. Nevertheless, if we are looking for indications, Stephen's failure to name his Arabic source and his massive interventions in the content suggest that he may have been aware of a less authoritative provenance of his source. Here, unless otherwise clear from the context, references to Ibn al-Haytham generally relate to the Arabic author of *On the Configuration of the World*.

cosmographical elements are largely absent, but a description of the bodily structure of the universe in the Aristotelian tradition is found in the *Hypotheseis*, which he completed at a later stage of his life. The separation between the genres continued in the early scientific literature of the Arabs, where works on celestial science could once again be divided into primarily mathematical works on the one hand (e.g., al-Battānī's *Ṣābiʾ Zīj*) and cosmological treatises on the other (e.g., Yaʿqūb ibn Ṭāriq's *Tarkīb al-aflāk*), while reconciling the different approaches remained a driving motivation among Islamic scientists as well. In natural philosophy, a genre of texts developed which is commonly referred to as *hay'a* (Ar. 'configuration'). Ibn al-Haytham's *On the Configuration of the World* became an influential representative of that tradition especially outside the Arabic speaking world, where it became known mainly through translations into Hebrew and Latin.[42]

In *On the Configuration*, Ibn al-Haytham dedicates an individual chapter to the spheres of the Sun (ch. 9), the Moon (ch. 10), Mercury (ch. 11), Venus (ch. 12), the upper planets (ch. 13), the fixed stars (ch. 14), and to the highest sphere (ch. 15). In his discussion of the planets, he suggests that any constituent partial motion in Ptolemy's theory as laid out in the *Almagest* can in principle be produced by a rotating sphere, whereas combinations of circles in the astronomical model can be equated to corresponding combinations of spheres. Eccentricities and epicycles in the Ptolemaic theory are effectively accounted for by smaller spheres that are embedded in the bulk of larger spheres. As a general characteristic, *On the Configuration* contains a purely qualitative description of the spheres; the text contains no reference to astronomical procedures or parameters, and no information is given concerning the relative sizes of, or distances between, the celestial structures, as might be expected from a technical description. Moreover, a strong influence of the astronomical literature is visible, as Ibn al-Haytham discusses the heavenly spheres not in a strictly ascending or descending order, which would be appropriate in a structural depiction; instead, in accordance with the astronomical approach, he first discusses the sphere of the sun, then turns to the lowest, i.e., that of the moon, etc. In addition, the individual spheres are dealt with in the second half of *On the Configuration*, while in the preceding chapters (chs. 3 to 8) a discussion of fundamental astronomical terms and concepts is given.

The importance of astronomy in the conception of *On the Configuration* is reflected in a statement in the opening part, which indicates that Ibn al-Haytham intended the text to make astronomical concepts easier for the layman to understand. At the same time, though, relating mathematical astronomy with Aris-

[42] It has been argued that it was also through this treatise that the Aristotelian-Ptolemaic concept of the celestial spheres became known in Latin Europe during the late Middle Ages; Hartner, 'The Mercury horoscope of Marcantonio Michiel of Venice', and Schramm, *Ibn al-Haythams Weg zur Physik*.

totelian physics appears as one of the guiding principles of the work.[43] In the end, Ibn al-Haytham presents a cosmography in which the cosmological and the astronomical views appear to be in agreement. He achieves this by ignoring central problems of celestial physics and by concentrating instead on an analogy which connects the two concepts, based on the principle that each of the circles in Ptolemaic astronomy can be conceived as the intersection of a bodily sphere and an imaginary plane.[44] The non-technical character of *On the Configuration* and the concentration on a geometrical analogy between bodily spheres and abstract circles distinguishes Ibn al-Haytham's treatise from earlier cosmographies, especially from Ptolemy's *Hypotheseis*.[45]

2.3.2 Copies and Translations of *On the Configuration*

According to present knowledge, Stephen of Pisa and Antioch was the first translator of *On the Configuration of the World*. But he was not the only one: his translation was followed by various other Latin and Hebrew versions, which were produced mostly in Spain or elsewhere in the western Mediterranean. Including the preserved copies in Arabic, the following versions are known, in varying states of preservation.[46]

Three Arabic manuscripts of Ibn al-Haytham's cosmology are known today. The first announced, in 1883, was MS London, India Office, Loth 734 (Langermann's siglum **Y**).[47] The manuscript seems to date from the sixteenth or seventeenth century; it is well written but lacks the original diagrams and chapter titles. A second, incomplete copy, MS Kastamonu, Genel 2298 (Langermann's **K**), was catalogued in 1952 and made known to a wider public in 1972.[48] Based on information in the colophon, the manuscript has been dated to the late eleventh century,

[43]Regarding these diverse motivations, Ibn al-Haytham's main intention when writing *On the Configuration*, and what the text actually provides, have been matters of debate; see for example Schramm, *Ibn al-Haythams Weg zur Physik*, pp. 6f. and 65, and Langermann's introduction to Ibn al-Haytham (ed. Langermann), *Configuration*, p. 5.

[44]See, e.g., Langermann, 'Ibn al-Haytham', p. 556, and similarly Ibn al-Haytham (ed. Langermann), *Configuration*, p. 7.

[45]See particularly Langermann's introduction in Ibn al-Haytham (ed. Langermann), *Configuration*, pp. 11–25.

[46]A detailed list of the preserved Arabic, Hebrew and Latin manuscripts, except for MS Cambrai 930 of the *Liber Mamonis*, is given by Langermann in the introduction to his edition of the Arabic text; Ibn al-Haytham (ed. Langermann), *Configuration*, pp. 34–44, supplemented in Langermann's preface to the 2016 reprint. The first systematic report on the then known versions of Ibn al-Haytham's *On the Configuration* was given by Moritz Steinschneider in 1881, at a time when no Arabic manuscript of the text had yet been identified; Steinschneider, 'Notice sur un ouvrage astronomique inédit d'Ibn Haitham'.

[47]Steinschneider, 'Supplement'. Langermann's sigla will also be adhered to in this book.

[48]Ateş, 'Kastamonu Genel Kitaplığı', p. 33, and Sabra, 'Ibn al-Haytham', p. 205.

but a production at such an early date has recently been called into question.[49] The manuscript contains a single diagram at the end of Ibn al-Haytham's Chapter 8. In addition to the previous two copies, the existence of a third Arabic manuscript, MS Rabat, Malik 8691, copied in the late nineteenth century, was reported in 1978.[50]

An undated Latin version of *On the Configuration* was produced from a (lost) Castilian translation commissioned by King Alfonso X of Castile in the second half of the thirteenth century. The Latin text is preserved in MS Oxford, Bodleian Library, Canon misc. 45 (Langermann's **L2**). In a preface to this 'Alfonsine' version, the text is described as a very free paraphrase of the Arabic original, whose contents had been supplemented and purposefully rearranged[51]: Ibn al-Haytham's original division into fifteen chapters has been superseded by a division of the entire work into two books. Moreover, the discussion of the motions of the four sublunar elements, originally the subject of Ibn al-Haytham's Chapter 2, has been transferred to the second half of the work; the subject of Ibn al-Haytham's original Chapter 4, on climates and terrestrial coordinates, has been moved to the end of the first half to follow the content of Ibn al-Haytham's Chapter 8. The rearrangement caused an inconsistency in the development of the work, since an understanding of the terrestrial latitudes is already required in Ibn al-Haytham's Chapter 7, in the discussion of the ascensions. The preface in the Alfonsine version replaces Ibn al-Haytham's introductory Chapter 1, but still includes the announcement that a bodily equivalent to Ptolemy's imaginary circles will be presented. The preface also states that further illustrations had been added to the text to aid understanding. The number of eight figures mentioned in the Arabic original thus rose to a total of forty-five diagrams referred to in the Alfonsine text.[52]

Another medieval Latin version from the Iberian Peninsula which was translated from Arabic is preserved in a single copy, in MS Madrid, Biblioteca Nacional, 10059 (Langermann's **L1**).[53] The manuscript dates from the thirteenth or early fourteenth century and was originally in the possession of the cathedral of Toledo. Though very concise and lacking Ibn al-Haytham's first chapter, the text renders Ibn al-Haytham's concepts with notable accuracy.[54]

In addition to the poorly preserved Latin translations, five different Hebrew versions of *On the Configuration* have been identified, at least two of which had a

[49] Rashed, 'The Configuration', pp. 56–8.

[50] Sezgin, *Geschichte des arabischen Schrifttums*, vol. 6, p. 255. Rashed, 'The Configuration', p. 49f., provides further details. The Rabat manuscript was not used by Langermann for his edition of the Arabic text.

[51] An edition of the text was made by José Luis Mancha, 'La version alfonsi'.

[52] Samsó, 'El original arabe y la version alfonsi del Kitāb fī hay'at al-'ālam de Ibn al-Hayṯam'.

[53] An edition of the text is included in Millás Vallicrosa, *Las traducciones orientales*.

[54] Schramm, *Ibn al-Haythams Weg zur Physik*, p. 64. A dependency on the Alfonsine version has occasionally been proposed but is unlikely in view of the very different character of both texts; José Luis Mancha, 'La version alfonsi', pp. 134–7.

significant influence and survived in several, partly illustrated, copies. The earlier one was completed in southern France in the 1270s by Jacob ben Machir ibn Tibbon (extant in Langermann's **H2-5** and **H7** and in further copies), and the other by Shlomo ibn Paṭer in Burgos in 1322 (Langermann's **H6** and **H8** and further copies).[55] Regarding these two Hebrew translations, Jacob ibn Tibbon followed his source more closely, while Shlomo ibn Paṭer used a better Arabic text and gave more sense to technical concepts.[56] A revision of Jacob ibn Tibbon's translation based on an independent comparison with the Arabic is preserved in MS Vatican 399 (Langermann's **H1**). Jacob ibn Tibbon's text was also translated into Latin at the beginning of the sixteenth century; this late Latin derivative exists in a single copy, in MS Vatican 4566. A fourth Hebrew translation, of at least Chapters 9 and 10, is preserved in a fragment in MS Paris, BnF, Heb. 1045, fols. 219r-221v; a fifth version, which is lacking Chapter 1, has recently been found in a manuscript in St. Petersburg.[57]

A critical edition of the Arabic text together with an English translation was published in 1990 by Tzvi Langermann.[58] The edition uses the Arabic manuscripts Y and K and also considers the three Hebrew versions contained in H1-8 and both the medieval Latin translations, L1 and L2, which were known at the time.

As it was only discovered recently that the *Liber Mamonis* also contains a translation of Ibn al-Haytham's *On the Configuration*, MS Cambrai 930 is not considered in the available repertories of Arabic, Latin and Hebrew manuscripts pertaining to Ibn al-Haytham's work. Produced in Syria around the middle of the twelfth century, the *Liber Mamonis* is the earliest translation of *On the Configuration* and is independent of the other known traditions; what is more, the Cambrai manuscript, dating from the twelfth century, is the oldest physical evidence of Ibn al-Haytham's treatise.

[55] For further copies of both Hebrew versions see Langermann's account in Ibn al-Haytham (ed. Langermann), *Configuration*, pp. 34–44, and Langermann's preface to the 2016 reprint.

[56] Steinschneider, *Supplement*, p. 507, idem, *Die hebraeischen Übersetzungen des Mittelalters und die Juden als Dolmetscher*, pp. 559–61, and Langermann, *Configuration*, p. 36f.

[57] See Langermann's introduction to Ibn al-Haytham (ed. Langermann), *Configuration*, p. 38f., and Langermann's preface to the 2016 reprint.

[58] Ibn al-Haytham (ed. Langermann), *Configuration*.

Chapter 3

Stephen's Version of *On the Configuration*

3.1 Translation and Commentary

A comparison of the *Liber Mamonis* with Ibn al-Haytham's *On the Configuration* shows that Stephen translated almost all of the Arabic text. The Cambrai manuscript is lacking the final pages, but one can assume that the text originally also included the final part of *On the Configuration*, i.e., passages 365 to 385. However, in addition to the new title, Stephen's text shows many more deliberate changes. For instance, he omitted Ibn al-Haytham's introductory first chapter, and in its place presents his own programmatic introduction; he abandoned the original division of the *Configuration*, into fifteen chapters, in favour of a new division into four books, each one of which forms a thematic unit and is provided with an individual preface. Further, Stephen placed the contents of Ibn al-Haytham's Chapter 4, on geographical coordinates, after Chapters 5 and 6, thus moving this chapter from Book I to Book II. Ibn al-Haytham had arranged the contents of his treatise roughly in ascending order, starting from the sublunar world and ending with the highest sphere; so the discussion of geographical coordinates (Ch. 4) precedes that of the astronomical coordinate of the altitude (Ch. 7). In Stephen's rearrangement the coordinates are introduced together, just before they are incorporated in a discussion of how the observer's position on the earth influences certain astronomical occurrences (Ch. 8). Stephen resolves a single inconsistency caused by this modification by inserting a reference (see p. 192:15). His most

D. Grupe, *Stephen of Pisa and Antioch: Liber Mamonis*, Sources and Studies in the History of Mathematics and Physical Sciences,
https://doi.org/10.1007/978-3-030-19234-1_3

substantial modification, however, is the inclusion of a long commentary which appears neatly interwoven with the translation.[1]

In the *Liber Mamonis* passages of translation typically alternate with commentary. Nonetheless, the various parts have been elegantly merged into a coherent presentation. While the commentaries in the work are roughly equal in length to Ibn al-Haytham's original text, the individual comments may range from cursory details to lengthy digressions extending over several pages. Despite the protracted and often digressive character of Stephen's commentary, Ibn al-Haytham's *On the Configuration* clearly forms the foundation of the *Liber Mamonis*, both in terms of its content and in its general line of reasoning. Stephen also implicitly qualifies Ibn al-Haytham's work as his main source when, at the end of two longer insertions, he concedes that he has 'digressed lengthily' (p. 154:15) or proposes to return to what he calls 'our actual concern' (p. 140:25), meaning Ibn al-Haytham's arguments. The second of these examples is also symptomatic of his desire to spread knowledge in various fields of learning, an intention that he already stated in his first translation, the *Regalis dispositio*, and emphasises again at several points in the *Liber Mamonis*. For, despite the cited call *ad rem*, Stephen immediately abandons this plan and inserts another 'very useful' (non... inutile) digression, covering some five pages, on competing explanations of the heating effect of the sun (see pp. 140:29–154:23 and below, p. 68). In cases like this, Stephen seems to see Ibn al-Haytham's treatise merely as a source of inspiration; he takes every chance available to supplement his work with other Arabic knowledge, even from loosely related fields, if he deems these details worthy of being brought to Europe.[2] This is consistent with his repeated claims, in the *Regalis dispositio* and in the *Liber Mamonis*, that scientific knowledge in Europe has generally fallen behind the level of neighbouring cultures[3]; as a first step towards bridging the gap, Stephen wants to bring as much of the advanced foreign knowledge as possible to his European readers.[4]

Including only Stephen's longer insertions (i.e., those of over roughly half a manuscript page) and those that define the division of the work, the independent and translated parts in the *Liber Mamonis* are arranged as follows.

[1]Several of Stephen's changes to Ibn al-Haytham's work have a parallel in the later Alfonsine version, such as the division of the text into longer sections, moving Ch. 4 to a later place and replacing the original Ch. 1, and supplementing the work with a thorough commentary and additional illustrations. Unlike the Alfonsine author, whose repositioning of Ch. 4 caused a major inconsistency, Stephen wisely places the chapter before the discussion of the ascensions rather than after it; cf. also his comment on p. 200:6.

[2]Cf., e.g., p. 154:15: '... gratia eorum que nostratum auribus ignota erant.'

[3]See, e.g., p. 90:7f.

[4]See e.g. above, note 38, and p. 154:17: 'Nam honestare Latinitatem totius si posset fieri subtilitate philosophie cum desiderem'.

Added sections:	Translated passages of *On the Configuration*:
Book 1:	
Preface to Book 1	
	Ch. 2: The Whole World
	[13]–[22]
That the earth is resting at the centre of the world	
	[23]–[26]
On the inundation of the Nile and the existence of a land connection between the temperate zones in the northern and the southern hemispheres	
	[27]–[37]
	Ch. 3: The Orb
	[38]–[44]
On fixed, moved and unmoved poles	
	[45]–[76]
	Ch. 5: The Orb of the Ecliptic
	[107]–[116]
On solar heat	
	[117]–[119]
On Macrobius' inappropriate definition of the zodiac	
	[120]–[139]
Book 2:	
Preface to Book 2	
	Ch. 6: The Declination
	[140]–[148]
That any set of four ecliptical points having equal declinations form a rectangle	
	[150]–[159]
	Ch. 4: Longitudes and Latitudes
	[77]–[106]
	Ch. 7: The Altitude
	[160]–[167]
	Ch. 8: The Ascendant and the Ascensions
	[168]–[184]
Book 3:	
Preface to Book 3	
On the arrangement of the planetary spheres	
	[185]

	Ch. 9: The Orb of the Sun [186]–[202]
On the rectification of the Sun	[203]–[207]
On the zones of the Earth	[208]–[209]
	Ch. 10: The Orb of the Moon [210]–[248]
On the rectification of the Moon	[265]–[271], [249]–[257]
On the varying sizes of solar eclipses	[258]–[264], [272]
Book 4: Preface to Book 4	
	Ch. 11: The Orb of Mercury [273]–[308]
A new theory of Mercury	[309]–[314]
	Ch. 12: The Orb of Venus [322]–[337]
	Ch. 13: The Orbs of the Superior Stars [338]–[359]
On the rectification, the retrogradation and the latitudes of the planets	[319], [358 *rev.*]
	Ch. 14: The Orb of the Fixed Stars [360]–[366]

In addition to Stephen's prefaces and the extra subjects identified in the table above, most of his insertions in the *Liber Mamonis* consist of short remarks on Ibn al-Haytham's arguments in their translated form. An impression of this process of translating and commenting can be gained from a comparison of Ibn al-Haytham's discussion of the ascensions and Stephen's Latin version of it as reproduced in Table 3.2 (*On the Configuration*) and on p. 238:12–238:26 (*Liber Mamonis*). As can be seen, Stephen's translation is accurate in the technical concepts but its wording is very free, and he adds several interspersed comments. These comments often leave translated segments unchanged and follow them in an explanatory manner. Even where the content of Ibn al-Haytham's sentences is modified in the Latin, omissions are rare but additions are quite frequent. In the example given, Stephen extends the validity of Ibn al-Haytham's argument on ascensions also to descensions ('et occidentalia'), an analogy that is not found in the Arabic. Similarly, passage 177 is extended by the explanation that the phenomena at Libra and Virgo are contrary to those at Pisces and Aries, as stated in the Arabic, because these signs stand opposite the others on the ecliptic ('horum

Ibn al-Haytham, *On the Configuration*, passages 176–77

(ed. Langermann)	(tr. Langermann)
[١٧٦] فهذه حال المطالع في كلّ واحد من الافاق الّا انّ مطالع كلّ جزء من دائرة البروج في أفق من الافاق مخالفة لمطالع ذلك الجزء في غير ذلك الآفق واختلاف المطالع بحسب اختلاف العروض.	[176] This is the situation of the ascensions with regard to each one of the horizons. However, the ascensions of each part of the ecliptic circle in a given horizon are different from the ascensions of that part in a different horizon. The difference in the ascensions is according to the difference in latitude.
[١٧٧] ولمثال في ذلك مطالع الحمل اذا كان في أفق ما اجزاء ما فإنّ مطالعه في الآفق الّذي عرضه اكثر من ذلك العرض اجزاء ما أقلّ من تلك الأجزاء	[177] As an example of this, the ascension of Aries for a given horizon is a given number of parts. Its ascension in the horizon whose latitude is greater than that latitude is a certain number of parts less than those parts.
وكذلك الحوت وبلضدّ من ذلك السنبلة ولميزان.	It is likewise for Pisces; contrariwise for Virgo and Libra.
ومطالع النصف من دائرة البروج الّذي بين نقطتي الاعتدال أبدا نصف دائرة معدّل النهار.	The ascension of the semi-circle of the ecliptic which lies between the two equinoctial points is always half the circle of the celestial equator.

Table 3.2: Text example from Stephen's Arabic source for the *Liber Mamonis*; Ibn al-Haytham, *On the Configuration*, passages 176–77 (ed. and tr. Langermann)

oppositis'). So Stephen may include new factual information (see also the second longer insertion in passage 177), which in many cases aids understanding of Ibn al-Haytham's reasoning (see for example the comment that follows passage 176, and the last comment to passage 177). Furthermore, Stephen includes numerical or geometrical details which connect Ibn al-Haytham's description to the methods of mathematical astronomy; for instance, in the first major insertion to passage 177, numbers from Ptolemy's table of the ascensions (*Alm.* II, 8) are cited. In general the example shows that even when Stephen follows his Arabic source, he does not try to translate the Arabic literally, but a distinction between the translated and independent content is rarely suggested by the Latin text itself; in fact, in most cases only a word-by-word comparison with the Arabic allows one to identify Stephen's additions to the text.

The variety of subjects in Stephen's commentary bears witness to his ambitious plans for a large-scale transfer of knowledge from Arabic to Latin. However, the main subject in his additions to the text is astronomy; he includes quantifying parameters, astronomical procedures and diagrams. This focus on astronomy reflects what Stephen says were his main reasons for writing the *Liber Mamonis*: not only the importance of cosmological knowledge *per se*, but also its usefulness for teaching astronomy (p. 92:15f.). Stephen confirms this by several references to his earlier publication of astronomical tables, the *Regule canonis*. However, by introducing astronomical details Stephen breaks with Ibn al-Haytham's systematic restriction of his treatise to a non-mathematical depiction of the physical cosmos. In so doing, Stephen not only gives a new character to the *Liber Mamonis*; he also runs the risk of bringing to the surface the discrepancies between the astronomical and the cosmological approaches to celestial science, an issue on which Ibn al-Haytham intentionally remained silent.

Also, in view of Stephen's far-reaching plans for the transfer of Arabic sciences to Europe, as a work on celestial science the *Liber Mamonis* is a suitable follow-up to his earlier translation on medicine, the *Regalis dispositio*. However, the *Liber Mamonis* differs significantly from that earlier work in the translation standards applied. In the *Liber Mamonis* Stephen conceals his main source and does not feel obliged to follow it literally; he also rather nonchalantly mixes contents of different origins. All this recalls a practice which Stephen himself once criticised in his censure of Constantine the African, and indeed it was the supposed inappropriateness of this approach that had motivated Stephen's own first activities as a translator. A possible explanation, which will become clearer below, is that Stephen neither intended to provide a translation of a particular source when writing the *Liber Mamonis*, nor in fact considered the work to be a translation.

3.2 Language

3.2.1 Technical Terminology

The *Liber Mamonis* no longer shows the "slavishly literal" (Talbot) translation technique that characterised Stephen's earlier work. Rather, Stephen now allows himself considerable freedom to interpret his Arabic source and to render its content in a language that would conform to the tastes of an educated Latin readership. When it comes to technical terms, however, the majority of expressions in the *Liber Mamonis* are close translations from the Arabic (see Glossary). While admitting a few Greek expressions, Stephen avoids transliterations of Arabic terms throughout and instead offers suitable Latin equivalents—a practice that distinguishes his work from many of the later translations from Spain, in which at least some Arabic technical terms were adopted.[5] As will be seen, in some cases Stephen's terminology is also more precise than the Arabic counterpart. The distinctness of Stephen's terminology is occasionally undermined, however, by his habit of varying his expressions for stylistic reasons.

For the most part Stephen's terminology can be regarded as equivalent to the Arabic, although certain deviations in the terms he uses reveal a systematic reinterpretation of certain aspects of his source. Ibn al-Haytham brought together the physical and the astronomical descriptions of the universe by means of a particular analogy, namely that the circles used by the astronomers can be conceived of as intersections of spheres and planes, for example, corresponding to the planar curve that is produced by a point on a rotating sphere. To suggest the equivalence of the cosmological, spherical model and the planar concepts of the astronomers, Ibn al-Haytham makes use of a terminological ambiguity which blurs the distinction between the two images. Stephen also repeatedly emphasises the analogy between the two models, but, in contrast to Ibn al-Haytham, he strives to maintain a clear differentiation between them. Unlike the reality of the cosmic spheres, Stephen emphasises, the circles of the astronomers are merely imaginary abstractions of that reality which are produced by the human intellect (p. 124:9f.). In consequence, and in clear contrast to Ibn al-Haytham, throughout the *Liber Mamonis* Stephen distinguishes between the physical description of the 'real' heaven on the one hand and the astronomers' abstraction of it on the other.

The difference between the texts becomes most apparent in the expressions which each author uses in connection with the correspondence between spheres and circles. In Stephen's text the respective terms are 'sphere' (spera) and 'circle' (circulus). Although Ibn al-Haytham also makes frequent use of these terms (i.e., *kura* and *dāʾira*), the central term in his treatise is 'orb' (*falak*). Standing for any

[5] See, e.g., in the Glossary the terms 'aux' (apogee), 'hadid' (perigee) and 'cenith' (zenith) in the Madrid translation of *On the Configuration*.

circular or spherical object, the term does not allow any distinction between a planar and a bodily representation. Ibn al-Haytham in fact notes this ambiguity when he explains that *falak* can denote four different objects, namely the plane or the circumference of a circle as well as the body or the surface of a sphere (*Configuration*, passage 39). While the reader is thus made aware of the various possible meanings, Ibn al-Haytham does not specify which of the four meanings he intends in each case. By using the term *falak* ambivalently, he effectively suggests that the planar and the bodily images are equivalent.[6]

Ibn al-Haytham's suggestive method is not followed by Stephen, who seems unwilling to sacrifice terminological clarity on such a central question. Although the Latin word 'orbis' would provide for a suitable, and equally ambivalent, translation of the Arabic *falak*, Stephen refuses to take this obvious option. Instead, he translates Ibn al-Haytham's *falak* either as 'spera' or as 'circulus', depending on the image he considers more appropriate in each individual case. Stephen consequently also omits from his translation Ibn al-Haytham's passage 39, which includes the "all-important" (Langermann) definition of *falak* and its four possible meanings.

In an analogous case, Ibn al-Haytham points out that the Arabic scientific terminology does not distinguish between the epicycle and the associated epicyclic sphere, both of which were called *falak al-tadwīr* (*Configuration*, passage 332). As in the case of passage 39, Stephen omits this statement from his translation and instead introduces distinctive terms, by denoting the sphere of the epicycle as 'rotunditas' and the epicycle itself as 'rotundus circulus (rotunditatis)', that is, the circle associated with the epicyclic sphere. This derivation of the astronomical term, epicycle, from its physical correspondent, epicyclic sphere (contrary to modern usage) is consistent with Stephen's understanding of astronomical circles as abstractions of the bodily reality.

Similar changes in the terminology concern the equant. Ibn al-Haytham uses the same adjective, *mudīr*, to denote four objects: the equant point, the dirigent line from the equant point through the centre of the epicycle, the dirigent sphere, and the motion of the epicycle centre within that sphere. The term is thus applied equally to these two astronomical objects and to their physical correspondents. As in the previous cases, Stephen introduces a terminology that distinguishes between the astronomical concept of the equant on the one hand and the related physical sphere on the other, by translating *mudīr* in the first case as 'circinans' and in the second case as 'referens'; the first expression clearly has a geometrical connotation, and the second a physical one.

---

[6]Cf. Langermann, 'A note on the use of the term orbis (falak)', and Langermann's introduction to Ibn al-Haytham (ed. Langermann), *Configuration*, p. 4f.

While Stephen's terminology provides more clarity than the Arabic, it neutralises a key element of Ibn al-Haytham's argumentative strategy. Whatever Stephen's motives for these changes were, his terminology reveals considerations that went beyond the content of the Arabic text in front of him. Moreover, a detail in his terminology suggests that his studies in astronomy were not restricted to book learning. When introducing the highest sphere, which causes the daily rotation of the heaven, Stephen informs the reader of a casual term for that sphere: '...the highest sphere, which for its prominence we often simply call "the sphere" ' (See p. 234:20f.). Stephen does not use this simple expression later on in his text, and so its asserted 'frequent use' may have occurred in direct conversations among members of a learned community at Antioch. Stephen's astronomical terms also coincide broadly with those used by ‹Abd al-Masīḥ Wittoniensis in the translation of the *Almagest*; but it differs significantly from the terminology that would appear in later, western translations of astronomical works, as exemplified by the Madrid translation of *On the Configuration* and Gerard of Cremona's translation of the *Almagest* (see Glossary). This isolated state of the Antiochian terminology confirms the close relation between Stephen's and ‹Abd al-Masīḥ's works. It also shows that despite the sophisticated state of this Latin terminology, it did not catch on among scholars in Europe.

3.2.2 Style

Not bound by the ideal of literal translation, in the *Liber Mamonis* Stephen adapts his language to the stylistic preferences of a readership that was well acquainted with the prominent Latin authors of Antiquity. Stephen's good command of Latin and the elevated style of his prefaces have been noted earlier.[7] However, while skilfully arranged sentences are not unusual in the opening parts of scientific texts, Stephen does not refrain from stylistic ornamentation, even in the technical parts of his work; on occasions he seems to use artistic formulations to playfully demonstrate his mastery of the scientific subject matter.

One of the clearest characteristics of Stephen's style is variation, even in technical arguments, where variation for purely aesthetic reasons competes with clarity and consistency. Stephen's technical descriptions are typically plain and clear, and use a single preferred expression for each object. However, any immediate reoccurrence causes him to vary his terms—in stark contrast to the Arabic, where terminological consistency is never compromised; in fact, Ibn al-Haytham even seeks parallelism in the language to indicate analogies. In such cases, Stephen sacrifices intuitiveness for readability and beauty.

[7]Cf. above, note 20.

As an example, Ibn al-Haytham defines each of the four quadrants of the ecliptic by their terminating equinoctial and solstitial points and by the season that corresponds to each quadrant (passages 111–14). He maintains a close parallelism between his formulations for the different quadrants. What Stephen made of this account can be seen in Book I of the *Liber Mamonis* (see pp. 138:1–138:29). In order to avoid repetition, Stephen varies the formulations several times; and, although he extends the discussion of the topic considerably, he continues with an abundance of synonyms. A demonstration of Stephen's treasury of words is also found in a discussion of the Nile, which he inserted into the *Liber Mamonis* (see p. 112:3ff.). In that passage, synonyms for the inundation of the Nile comprise nominal expressions, 'exundatio', 'exuberatio', 'ubertas aquarum', which alternate with increasingly picturesque descriptions of the phenomenon, such as 'aquaticum humorem augmentant', 'exundantis largo benefitio Egipti plana irrigantur', 'aque laxius camporum plana pervagantur', 'alvei dedignans clausulas longe lateque totam perfundat terre viciniam', 'solito copiosior larga maximus ubertate circumfusa terrarum occupat', 'exuberantium larga copia irrigaret aquarum. . . quibus in superas prius funditur auras'.

Aware of the need for clarity, especially with regard to technical terms, Stephen normally restricts his variations in these cases in order to preserve the characteristic root of a term. In one passage (p. 234:1ff., corresponding to *Configuration*, passage 168), for example, Stephen uses four different expressions for the horizon, 'orizon', 'orizontis circulus', 'orizonteus circulus', 'circulus orizontalis', rather than repeating any of them. Nonetheless, each term still contains the significant root 'orizon(-)', which ensures terminological clarity. A single exception to this rule is found in Stephen's use of the word 'zodiac' (zodiacus). Perhaps as a concession to the habits of the reader, Stephen continues to use 'zodiac' as a synonym of 'circle of the signs' (circulus signorum), i.e., ecliptic, even after he has explicitly stated that the latter is his own technical term for the ecliptic in contrast to Macrobius' use of '(circulus) eclipticus' (p. 158:25).

As can be observed from Stephen's discussion of the Nile, with the increasing variation his language becomes more and more embellished. Traditional poetic attributes to the celestial bodies or to the zodiac signs in the *Liber Mamonis* may be perceived as attractive ornaments (e.g., p. 120:6f.: 'luna. . . mutuate mundo radios lucis refundit. [. . .] Quintum. . . Mars rutilus, sextum blandi luminis Iuppiter, septimum autem ceteris altior Saturnus vendicant', or p. 164:30f.: 'architenens Sagittarius', 'corniger Capricornus', 'Aquarius rectus' etc.). However, his personification of mathematical lines is slightly irritating; he makes them susceptible to happiness and indignation (p. 110:17: 'nulla <*scil.* linea> se maiorem minoremve letabitur aut queretur'). This flowery style, which characterises large parts of the *Liber Mamonis*, contrasts sharply with the plain technicality of the Arabic source. Stephen's evident delight in rhetoric also reminds us that the earliest text that we

can associate with him was the *Rhetorica ad Herennium*, of which he acquired a copy in 1121.

3.3 Technical Quality

In the example of passages 176–77 (see above, Table 3.2), Stephen was able to grasp the technical content of the Arabic text and to associate it with subjects from related disciplines. The insertions also revealed his knowledge of other scientific sources, of which he made equally intelligent use, such as Ptolemy's table of the ascensions. Moreover, his liberal linguistic approach allows him to make many of Ibn al-Haytham's arguments more precise or to render them more clearly than would have been possible by a close translation of the Arabic. By elaborating on Ibn al-Haytham in this manner, Stephen demonstrates a solid understanding of the subject matter while at the same time confirming the primacy of the didactic role that he intends the *Liber Mamonis* to have. Illustrative of the technical and didactic quality of the work is Stephen's treatment of particular passages of *On the Configuration* which allowed diverse interpretations and occasionally caused confusion among recipients of the work.

3.3.1 *On the Configuration*, Passages 146–48

The transmission of *Configuration*, passages 146–48, is partially contradictory in the Arabic manuscripts.[8] When ascribing zodiac signs to particular declinations, both texts (Y and K) refer to the point of a sign where the declination is largest. For signs in the first and the third quadrants of the ecliptic, this means their end points, whereas for signs in the second and the fourth quadrants it is their beginnings (see Fig. 3.1, left). However, the readings in the Arabic copies, Y and K, disagree as to whether the declination of any of these points should be attributed to the complete sequence of signs starting from the nearest equinox (outer arrows in Fig. 3.1, left) or to the corresponding sign only (inner arrows in Fig. 3.1, left). A question would thus be, for example, whether the declination of the head of Gemini, which coincides with the end of Taurus, should be defined as the 'declination of Taurus' (the reading in K) or as the 'declination of Aries and Taurus' (Y), or, correspondingly, whether the declination at the end of Cancer, which is identical with the head of Leo, should be defined as the 'declination of Leo and [i.e., being identical with that of] Aquarius' (K after emendation) or as the 'declination of Leo and Virgo' (Y, K prior to emendation).[9]

[8]See Langermann's notes in Ibn al-Haytham (ed. Langermann), *Configuration*, pp. 79f. and 109f.

[9]Cf. Ibn al-Haytham (ed. Langermann), *Configuration*, p. 80.

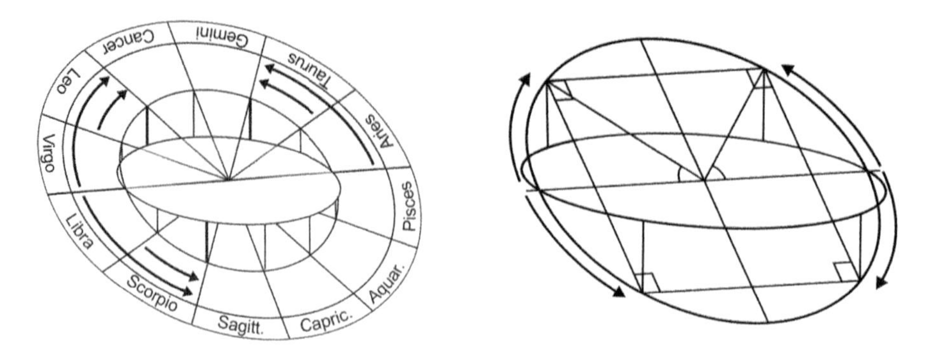

Fig. 3.1: *Left*: Alternative concepts of ascribing declinations to the zodiac signs, and *vice versa*, in the Arabic MSS Y (outer arrows) and K after emendation (inner arrows). *Right*: Any set of four points on the ecliptic that have equal distances from the equinoxes define a rectangle whose corners have equal declinations, as described by Stephen in the *Liber Mamonis*

Various indications favour the latter, accumulative concept as the authoritative reading. This interpretation is supported by the original text in both the Arabic manuscripts examined, and also seems more reasonable. It also corresponds to similar concepts, for example, as described in Chs 4 and 5 of al-Battānī's *Ṣābiʾ Zīj*.[10] In the tradition of K, however, in each series of two consecutive signs the name of the first sign was replaced with the sign opposite the second one. The 'declination of Aries and Taurus' was thus changed to the 'declination of Scorpio and [i.e., being identical with that of] Taurus', etc. Perhaps for consistency, in passage 148 on the declination of the head of Virgo, which in both interpretations would be ascribed only to the sign of Virgo, K again includes the detail that the opposite sign, Pisces, has the same declination. The modification described emphasises the symmetry in the declinations of opposite signs. The different interpretations also found their way into the Hebrew tradition.[11]

In view of the inconsistent transmission of the passage, Stephen's treatment of the subject is remarkable (cf. pp. 184:3–190:18). He propagates the accumulative concept found in parts of the Arabic tradition, but he also adds a commentary in which he emphasises the identical nature of declinations of opposite signs, as

[10] Al-Battānī (ed. Nallino), pt. III, pp. 19–21.

[11] See the notes in Ibn al-Haytham (ed. Langermann), *Configuration*, p. 109f. Langermann questions both traditions and suggests an emendation according to which the declination of every sign is determined by the sign's final point in the direction of the succession of the signs. For the first and the third quadrants this produces the same wording as the modified reading in K. For the second and the fourth quadrants, however, Langermann needs to make substantial changes to all the preserved instances; cf. Ibn al-Haytham (ed. Langermann), *Configuration*, p. 105f. and, correspondingly, p. 26, and the notes on pp. 80 and 110. Langermann's suggestion is followed by Samsó in his discussion of the Alfonsine translation; Samsó, *El original arabe y la version alfonsi del Kitāb fī hay'at al-'ālam de Ibn al-Hayṯam*, p. 126. However, the proposed changes to the text are unnecessary and not convincing.

found in the other parts of the Arabic tradition. Having comprehensively covered the subject in this way, Stephen extends the discussion further to the declination of the solstitial points, where symmetry allows one to ascribe the declination to both sequences of three signs on either side of the solstice. As an independent supplement to the discussion, Stephen then points out that a rectangle is formed by any set of four points on the ecliptic that have the same distance from the equinoxes (see Fig. 3.1, right; pp. 184:18–190:18). He presents this rectangle as a tool for easily identifying any set of four points on the ecliptic of equal declination. However, the concept also effectively conveys the significance of the equinoxes as reference points for the declinations (cf. the correspondence between the arrows in Fig. 3.1, right, and the outer arrows in Fig. 3.1, left).[12]

3.3.2 *On the Configuration*, Passages 178–79

Following a general introduction of the ascensions (cf. above, Table 3.2), Ibn al-Haytham discusses the special case of the ascensions for an observer at a far northern latitude of 'about 66 degrees', i.e., on the Arctic Circle (*Configuration*, passages 178–79). At this latitude the horizon coincides with the ecliptic once in a day. However, Ibn al-Haytham's formulation of this coincidence, '*fī kulli yawmin wa laylihi*', is ambiguous; it could be understood as '<once> in every succession of day and night', but also 'for an entire day and night'.[13] At the given latitude the ecliptic coincides with the horizon for an instant, exactly when the head of Capricorn culminates in the south from the observer's point of view. Due to the daily coincidence of the planes it is also a daily phenomenon that a certain half of the ecliptic does not have an ascension, because it rises instantaneously when, after coinciding, the ecliptic and the horizontal plane effectively tilt away from each other.

Stephen not only worked this out correctly from the ambiguous formulation in the Arabic, but he also carefully avoids misunderstandings by giving repeated and clear translations of the key phrase, 'semel in unoquoque die' (p. 240:6) and 'omni die fit semel' (p. 240:22). As in the previous example, he elaborates on the matter: he adds that the same half of the ecliptic that has a vanishingly short ascension will have, as a compensation, a descension of a full circle, whereas the other, autumnal, half of the ecliptic does not have a descension because it sets in an instant, but instead has an ascension of a full circle. Stephen also tries to explain the stated

[12]A similar construct is described at the end of Ch. 4 of al-Battānī's *Ṣābiʾ Zīj*; al-Battānī (ed. Nallino), pt. III, p. 19, with the corresponding diagram in pt. II, p. 58.

[13]The ambiguity affected Langermann's translation of the passage and his notes thereto as well as the proposed replacement of 'horizon' with 'instant' in passage 179, as found in parts of the Hebrew tradition, for which there is no necessity; Ibn al-Haytham (ed. Langermann), *Configuration*, pp. 119 and 124.

phenomena (pp. 240:1–240:24). Whereas Capricorn at the southern end of the instantaneously rising half of the ecliptic will soon set again, it will take Gemini, at the northern end, a full revolution to disappear completely from above the horizon. This half therefore has no ascension, due to its instantaneous appearance above the horizon, but a descension of a complete circle until the last point in that half has set. Analogously, Stephen explains that the autumnal half of the ecliptic needs the same full revolution to rise, but then sets in an instant. While the visible part of the ecliptic and its orientation relative to the observer are permanently changing, Stephen knows that for the instant after coplanarity with the horizon the visible part of the ecliptic will always be the half around the spring equinox, which at that moment will be oriented due east from the observer.

In the above cases the Arabic text was susceptible to misinterpretation. Stephen's confident treatment of these passages confirms him as an intelligent reader of Ibn al-Haytham's treatise and as someone who is also well informed about a range of related ideas. We have stressed that Stephen renders the technical content of *On the Configuration* accurately but without the intention of producing an exact Latin equivalent: his commented version is characterised rather by a didactic concern, aiming at completeness and ease of understanding for the untrained reader. To this end, Stephen makes notable efforts to prevent possible misunderstandings by providing detailed explanations and by discussing subjects from different angles; indeed, Stephen states that he conceived the *Liber Mamonis* for a European readership which was still largely unacquainted with Ptolemaic astronomy and cosmology.[14]

3.4 Ibn al-Haytham's Diagrams

Ibn al-Haytham refers to eight different diagrams in his treatise, only one of which is preserved in the Arabic tradition, in MS K. Although further drawings can be found in copies of the Hebrew and the Latin translations, the graphic evidence is highly inconsistent.[15] A reconstruction of Ibn al-Haytham's original set of illustrations on the basis of the scattered evidence has been suggested, but has not been performed to date.[16] Apart from the diagram in K, the most important testimonies seem to be two copies of Jacob ibn Tibbon's Hebrew translation, MSS Parma, Biblioteca Palatina, 2466 (De Rossi 568) and Paris, BnF, Heb. 1035, each of which contains six diagrams. However, there is little agreement between the drawings in the two manuscripts: the diagrams in the Paris copy are in general simpler and less accurate than the ones in the Parma copy.

[14] E.g., p. 242:17: 'introducendis scripsimus.'

[15] Ibn al-Haytham (ed. Langermann), *Configuration*, p. 35.

[16] Samsó, 'El original arabe', p. 118f.

The many illustrations in the *Liber Mamonis* provide valuable further evidence on the matter, especially because of the early production of the work and its independence from western translations. Since the Cambrai manuscript contains fifteen diagrams, obviously some of them are not taken from Ibn al-Haytham. Moreover, in view of Stephen's liberal approach to textual source material in the *Liber Mamonis*, one must expect that he also made changes to the diagrams which he took from his source. Nonetheless, several diagrams in the *Liber Mamonis* relate only to Stephen's commentary, and can therefore be ignored. Moreover, most of Stephen's modifications to Ibn al-Haytham's text pertain to particular types of information, such as the insertion of astronomical details, which Ibn al-Haytham avoided; where information of this kind appears in the drawings, it can therefore also be assumed to result from interventions by Stephen.

At the end of Book III of the *Liber Mamonis*, Stephen provides a diagram of the lunar sphere (see the reproduction on page 421; cf. also p. 322). A corresponding illustration, including largely equivalent captions, is found on fol. 41r of MS Parma, De Rossi 568 (see Fig. B.4 on p. 422).[17] Stephen's diagram differs from the one in Parma essentially in that its captions also quantify the daily motion of each sphere. Information of this kind, however, is typical of Stephen's additions to Ibn al-Haytham. The close agreement between the two diagrams in the Cambrai and Parma manuscripts makes it probable that Stephen found a corresponding diagram in his Arabic copy of *On the Configuration*; by contrast, the simpler diagram in the Paris manuscript Heb. 1035 does not contain the sun or the shadow of the earth, nor does it show the position of the moon at eclipses.

Similar observations can be made with regard to other diagrams in the *Liber Mamonis*, in which the agreement is generally better with the Parma manuscript than with the Paris copy. Images in Stephen's work which thus seem to be close adaptations of Ibn al-Haytham's are:

- Figure 1.2 (see p. 172): an incomplete and unlabelled illustration of Ibn al-Haytham's definition of the zodiac; cf. Fig. B.5 for a more complete equivalent in MS Parma, De Rossi 568, fol. 43r.

- Figure 3.8 (see p. 280): the plane of the ecliptic, shown as a cross-section through the celestial spheres and divided into the zodiac signs; cf. Fig. B.3 for the corresponding diagram in MS Parma, De Rossi 568, fol. 40v (fol. 40r shows a failed attempt). A similar illustration can be found in another Hebrew copy, in MS Paris Heb. 1031, fol. 90r, whereas MS Paris, Heb. 1035, fol. 18v, again has a much simpler variant.

[17] I am grateful to Josefina Rodríguez Arribas for her help in reading the Hebrew captions in the Parma manuscript.

- Figure A.11 (see p. 322): the sphere of the moon; cf. Fig. B.4 for the corresponding diagram in MS Parma, De Rossi 568, fol. 41r.
- Figure 4.13 (see p. 362): a corrupted illustration of the sphere of Mercury, with Stephen's modifications to Ibn al-Haytham's theory of Mercury appearing only in a caption for the annual sphere (cf. below, Sect. 4.4); cf. Fig. B.6 from MS Parma, De Rossi 568, fol. 41v, which, though unlabelled, shows the two eccentricities of Mercury. The shade in the right half of the drawing is caused by the lunar diagram shining through the page.
- Figure 4.14 (see p. 376): an incomplete illustration of the sphere of an outer planet. The diagram shows a similarity with the one in MS Paris, Heb. 1035, fol. 35r, but is obviously left incomplete. Analogous to the previous diagrams, it was probably intended to be labelled and also to show the dirigent line as well as the relative inclinations of the equator, the ecliptic and the deferent. Figure B.7, from MS Parma, De Rossi 568, fol. 42v, also contains these elements.

The *Liber Mamonis* does not contain a diagram of Venus. Stephen considers this to be obsolete due to the alleged similarity between Venus and Mercury after his modifications to the classical theory (cf. below, Sect. 4.4). For completeness, Fig. B.8 shows the sphere of Venus in MS Parma, De Rossi 568, fol. 42r. Furthermore, the single diagram which is preserved in Arabic K appears after Ibn al-Haytham's Ch. 8 and, according to a description, shows among other circles the equator, the ecliptic and the horizon.[18] It appears to be the diagram that Ibn al-Haytham announces in *Configuration*, passage 184. No comparison with Stephen's illustration at that place has yet been made, but there seem to be major differences. Also, at the end of his treatise (passages 382–85) Ibn al-Haytham announces an eighth diagram which, he says, shows a simplified view of the spheres of all the planets. This diagram seems to be lost in the known manuscripts.

[18] Ibn al-Haytham (ed. Langermann), *Configuration*, p. 42f.

Chapter 4

The Content and Purpose of Stephen's Commentary

Around half of the *Liber Mamonis* consists of commentary added by Stephen. The insertions that have been identified so far include explanations that help to expound Ibn al-Haytham's arguments. Stephen also adds genuine astronomical elements as well as digressions on subjects that Ibn al-Haytham does not address. Other comments in the *Liber Mamonis* can be summarised as programmatic or motivational statements. These are mostly, but not exclusively, found in Stephen's prefaces to the four books into which he divided the text. In the same category one can also place Stephen's criticism of outdated doctrines in European science. In the following, these and other elements of Stephen's commentary and their function for the composition of the *Liber Mamonis* will be discussed.

4.1 The Aim

Concerning the purpose, or the motivation, of Stephen's translations, two aspects can be highlighted from what he himself describes as the reasons for his efforts. In general, Stephen intends to import as much foreign knowledge as possible to Latin Europe, in order to compensate for the deficits he has noticed. And with regard to the *Liber Mamonis* in particular, Stephen considers a cosmography a useful supplement to the astronomical tables and procedures which he had published earlier, in his *Rules of the Canon*. It soon becomes clear that both aspects are embedded in a wider set of motivations, which can be reconstructed from various comments of his, some of them quite personal.

D. Grupe, *Stephen of Pisa and Antioch: Liber Mamonis*, Sources and Studies in the History of Mathematics and Physical Sciences,
https://doi.org/10.1007/978-3-030-19234-1_4

In the preface to Book I, Stephen emphasises the pedagogical importance of academic learning for the development of a person's character as early as childhood. It is the early encounter with knowledge, Stephen says, especially if such knowledge is beyond a boy's understanding, that will teach him modesty and foster in him a curious and open mind (p. 92:7f.). Stephen connects this educational benefit of learning with an ethical obligation to exert and develop one's mental abilities ('animus'); for the intellect is a divine gift that distinguishes man from animals and imbues him with human celsitude ('hominis celsitudo'): Stephen also says that to disdain this gift through passiveness would be nefarious ('gravissimum facinus').[1]

Stephen criticises Europeans for their profound and long-lasting neglect of learning. This has not only meant that European science is underdeveloped in comparison with that of neighbouring nations;[2] it has also caused a corruption of the moral, social and political constitutions of societies in Europe. Stephen elaborates further on this connection (See pp. 90:13–92:14). As scholarship in Europe declined, a jurisdiction which was based on codified law ('lex naturalis', 'lex posita'), and which therefore required professional studies of extensive legal and philosophical corpora ('litterali... exercicio adipiscamur'), lost ground; this opened up the way for primitive legal practices on the basis of outdated customs ('consuetudinum... inconsulta observantia'). In combination with the moral decline of an increasingly illiterate ruling class, this legal practice had led to despotism and unjust appropriations by those in power, which now threatened to tear apart the most solid bond of society ('societatis humane artissimum vinculum detrahunt').[3]

---

[1]See p. 246:8f.; see also p. 324:9. On Stephen's use of 'animus' in the sense of 'ratio', cf. Burnett, 'Antioch as a link', p. 11, note 38.

[2]See p. 90:7: 'Unde factum est ut que fere plenitudinem posset habere artium nunc ceteris gentibus Europa videatur humilior.'

[3]Stephen's concentration on the deficits of the customary law, in particular with regard to the protection of private property against arbitrary expropriation, and his advertising of the codified law reveals the influence of his northern Italian urban background. For it was mainly in the upcoming trading cities of northern Italy where in Stephen's day the *Corpus iuris civilis* remained in force at local level, though in a much reduced form, and where legal deficits were increasingly noticed; see e.g. Haskins, *Renaissance*, ch. 7, esp. pp. 194–96 and 220f. In the same environment the revival of the *Corpus iuris* (Bologna) began, as well as a new codified legislation, the *Constituta usus et legis* enacted in Pisa in 1160; Classen, 'Kodifikation im 12. Jahrhundert', see p. 313. Stephen's analysis that a functional jurisdiction on the basis of the *Corpus iuris* required an educated, professional elite is commonly shared today. The revival of the *Corpus iuris* and the advance of a new codified legislation is therefore considered an integral part of the scientific "renaissance" (Haskins) in twelfth century Europe; see e.g. Haskins, *Renaissance*, ch. 7, esp. pp. 220f., Classen, 'Die geistesgeschichtliche Lage', see p. 22, Otte, 'Die Rechtswissenschaft', pp. 127–131, Wolf, 'Gesetzgebung und Kodifikation', pp. 157f. Stephen's concern about the security of property also reminds us of his position as treasurer of the Benedictines in Antioch; and, his familiarity with judicial customs is also evident on another occasion, in the preface to Book III, where he asks for a fair trial in which he will present his new teachings against the defenders of traditional views (see p. 174:16ff.).

Stephen's exhortation for a greater public appreciation of scholarship may appear as timeless as they are commonplace. However, when criticising the situation in Europe, he exculpates the public and its rulers and blames no one other than the representatives of scholarship themselves. In his view, European scholars had abandoned their fruitful traditions and had adopted an abhorrent mentality of quarrelling over doctrines which had been inherited unquestioningly and passively. This led to a general disregard of scholarship in Europe and thus the downward spiral began (pp. 90:9–12). In Stephen's eyes, the rulers' 'unreflecting observation of consuetudes' is merely a consequence of the similarly unreflecting observation of equally outdated doctrines by the learned. Bereft of the spirit of critical and independent investigation, Latin science has fallen far behind the level of its neighbours, and throughout his commentary Stephen stresses the depths that Latin ignorance has reached (e.g., p. 388:3: '... in quo Latinitas diu caligat errore turpis ignorantie plurimum involuta').[4]

With the origin of the grievances thus identified, Stephen considers it to be the responsibility of the educated to halt the decline by returning to a culture of critical, creative and open-minded research and debate. In order to foster this scientific revival, and given the present advantage of the neighbours, Stephen decides to make the advanced foreign knowledge accessible to the Latins. Although he ambitiously aspires to provide Latin Europe with all the scientific and philosophical knowledge that it currently lacks,[5] he also sees the practical necessity of concentrating on the most useful aspects.[6] It becomes clear, however, that Stephen sees these imports of knowledge only as an initial aid and stimulus in regard of the more fundamental requirement of changing the scientific culture in Europe as a whole.[7]

Stephen demands no less from his fellow Europeans than to abandon traditional authorities and to establish a new investigative approach to the sciences, based on empiricism and reason.[8] He is aware that such advances will provoke

[4]Similar, p. 90:7, p. 154:15, p. 174:1, p. 174:29, and p. 324:14.

[5]See p. 154:17: 'Nam honestare Latinitatem totius si posset fieri subtilitate philosophie cum desiderem, ... '

[6]See p. 88:5: 'utilitatis ammonitione'; see also p. 154:20.

[7]It is noteworthy that in none of Stephen's statements the Arabic origin of his new knowledge seems to be of major importance. Stephen mentions this Arabic origin only in two casual remarks towards the end of his text; he does not use it either as an argument for advertising the scientific quality of his work, or to express his indignation about Europe's dependency on these foreign achievements. In combination with his careful creation of a Latin astronomical terminology, this gives credibility to Stephen's inward-looking motives, which concentrate primarily on a renovated and independent 'Latin' science.

[8]For Stephen's advertising of empiricism, see p. 112:22, p. 148:9, p. 120:1, and p. 96:6. The last example also includes one of Stephen's own empiric judgements, of which another can be found on p. 154:11. A scathing criticism of natural philosophy that claims to do without empiricism is found on p. 276:13.

protest. But, although he understands people's adherence to traditional views as a natural characteristic of man,[9] he is convinced of the necessity for a scientific renewal in Europe and of the value that his own work may have in this project. He thus encourages his readers in the spirit of a scientific avant-garde, while at the same time warning them of the enmity they may encounter (p. 92:11ff.). Seeing himself as a leading figure in the movement, Stephen compares his role with that of Solon, whose necessary but at the same time unwelcome reforms came to his countrymen as a blessing in disguise (p. 176:9f.).

Aside from the above reflections about scholarship in general, various remarks in the *Liber Mamonis* explain Stephen's decision to write a work specifically on cosmology. Firstly, he considers Ibn al-Haytham's theorems about the cosmos to be far superior to the traditional ideas that circulated in Europe at the time—especially Macrobius' *Commentary on Scipio's Dream*, whose weaknesses Stephen points out on numerous occasions.[10] At the same time, his chosen form of a commented translation allows him to include in his text details of Arabic natural philosophy which were unknown in Europe (p. 154:15: 'nostratum auribus ignota').

Important also is the connection with Stephen's previous publication, the astronomical *Regule canonis*. Stephen says in the opening part of the *Liber Mamonis* that he considers an illustrative description of the heavens to be an ideal text for illustrating to beginners the meaning of astronomical terms and methods (p. 92:15ff.). His perception of Ibn al-Haytham's *On the Configuration* thus broadly coincides with what medieval and modern recipients have seen as the main purpose of the latter's treatise, namely to explain basic concepts of astronomy to beginners.[11] Stephen's interest in the text also matches Ibn al-Haytham's interest in demonstrating that Ptolemy's circles can be conceived as abstractions of bodily spheres.[12] Apart from these parallels, however, Stephen expands the idea of using Ibn al-Haytham's cosmography for teaching purposes in astronomy. This involves important changes to the text, such as his terminological changes and the inclusion of astronomical details.

4.2 The Criticism of Macrobius

One of Stephen's proclaimed intentions in the *Liber Mamonis* is to eradicate the Latin world's misconceptions regarding the universe and to provide it with superior knowledge from outside. The errors Stephen is thinking of in particular are

[9]See p. 176:2f.: 'novitatis sepius comes nocet invidia'.
[10]See below, Sect. 4.2.
[11]See above, p. 27.
[12]Cf. above, p. 27, and Ibn al-Haytham (ed. Langermann), *Configuration*, pp. 3–7.

the cosmological doctrines of the neo-Platonist Macrobius (early 5th c.), whose *Commentary on Scipio's Dream* he considers to be the leading representative of the old learning. On numerous occasions, Stephen draws attention to deficits in Macrobius' theorems which, to his obvious annoyance, continued to flourish in Europe despite their obvious contradictions. Stephen then propagates Ptolemy's astronomy and the interpretation given to it by Arab philosophers. In practice, whenever Ibn al-Haytham's arguments offer an opportunity, Stephen points out their strengths with respect to the Macrobian theory.

Most of Stephen's criticisms of Macrobius are to be found in Book I of the *Liber Mamonis*. This may be incidental, as a result of the organisation of subjects in Macrobius' and Ibn al-Haytham's texts. But it also creates an effective argumentative order in the composition of Stephen's work: first to challenge the traditional learning, then to replace it with Ibn al-Haytham's superior concept, and finally, in Book IV of the *Liber Mamonis*, to make independent improvements in Arabic cosmology as well (see below, Sect. 4.4).

In his description of the sublunar world, Ibn al-Haytham states that the earth rests without motion at the centre of the universe. In his translation of this passage, Stephen adds an explanation saying that it was in agreement with the characteristics of heavy and light matter that the massless heaven should rotate with an extremely fast motion around a stationary earth rather than that the earth, being the agglomeration of the heaviest elements of the universe, should rotate with a much slower, yet still very fast motion under a stationary heaven (pp. 96:23–108:24, where Stephen's arguments concerning the heaviness of matter are found at the beginning and at the end). Macrobius, too, claims that the earth is motionless, but tries to prove it with mathematical arguments. He says that the earth, being at the centre of a rotating spherical universe, could not move on grounds of geometrical principles, because every rotating sphere revolved around its centre; its position thus remained unchanged (*Somn.* I, 19, 11). While this would apply to a motion of the earth from place to place, a generalisation to rotations would require that the earth be treated as a mathematical point. Macrobius accordingly cites the astronomers' common understanding that, when distances on the earth are related to the size of the heavens, one can ignore the size of the earth and regard it 'as a point' (*Somn.* II, 9, 9). Stephen refutes this argument with a reference to Macrobius' own division of the earth into climatic zones. Since this division implied that the earth is an extended body, the latter could not 'be' a point in the mathematical sense, because the definition of a point required indivisibility (pp. 100:12–106:4). Stephen also rejects Macrobius' attribution of the earth's unmoving permanence at the centre of the universe to its being blocked there by the homogeneous pressure of the surrounding light matter (*Somn.* I, 22, 7); he notes the ability of heavy bodies to easily traverse air when falling (p. 106:5 ff.).

Still in connection with Ibn al-Haytham's description of the earth, Stephen also rejects Macrobius' claim that an oceanic stream surrounds the earth along the equator (*Somn.* II, 9, 2 ff.). Stephen accuses Macrobius of spreading unproven hypotheses, since certainty could only be achieved by an exploration of the equatorial region which, if at all possible, had not yet been made (p. 110:20 ff.).[13] In an extended discussion Stephen then argues that, whereas Macrobius' hypothesis is purely speculative, the inundation of the Nile provides an argument against an equatorial ocean (see also below, p. 66).

Stephen's introduction to Ibn al-Haytham's definition of the zodiac contains another critique of Macrobius (pp. 156:18–162:22). Macrobius ascribes a width to the zodiac, saying that to either side of the ecliptic this width equalled the furthest distance of the moon from the ecliptic (*Somn.* I, 6, 53). On another occasion, Macrobius conceives the zodiac as a band which comprises all the zodiac signs (*Somn.* I, 15, 8–10). Stephen points out that, regardless of whether Macrobius conceived the borders of the zodiac as crooked due to the varying spread of the stars of different zodiac signs, or as straight because the moon, by the rotation of the dragon, will reach its greatest inclination at every point of the ecliptic at some moment, there are certain signs that extend further from the ecliptic than the inclination of the moon. Stephen concludes that 'Macrobius' theory is therefore ambiguous and contradictory in itself' (p. 160:14). He then solves the problem by giving Ibn al-Haytham's alternative definition of the zodiac.

Further Macrobian doctrines are cited, and rejected, in later parts of the *Liber Mamonis*. In connection with the order of the planetary spheres, Stephen once again shows that Macrobius' statements on the subject are contradictory (pp. 248:25–254:30). On one occasion Macrobius describes Mercury's sphere as lying above the sphere of the sun while being enclosed by the sphere of Venus (*Somn.* I, 19, 6–10), whereas later he places Venus beneath Mercury when he says that 'the distance from the earth to Mercury equals four times the distance from the earth to Venus' (*Somn.* II, 3, 14). The contradiction leads Stephen to launch a harsh polemic against Macrobius, which culminates in the diagnosis of mental insanity. Stephen finds not just Macrobius' order of the spheres to be inconsistent but also their sizes—and, by consequence, the velocities which Macrobius ascribes to their rotations. Macrobius says that the surface velocities on the equators of all spheres were equal and that it was only the spheres' different diameters that caused the different periods of the planets (*Somn.* I, 14, 27; I, 21, 6f.). But, since he had defined the radius of Mercury's sphere to be four times that of the sphere of Venus, this would be contradicted by

[13]In *Alm.* II, 6 Ptolemy gives a similar argument when refusing to make hypotheses on the equatorial zone.

the fact that the two planets have equal mean periods, which are both equal to the mean period of the sun.[14]

Stephen distances himself from Macrobius for the last time when he defines the terrestrial climates (p. 274:24 ff.). Stephen says that due to the position of the sun's perigee in the southern signs and in agreement with recent explorations, the uninhabitably hot zone on the earth could only lie south of the equator. The hot zone could therefore not extend symmetrically on both sides of the equator, as Macrobius stated (*Somn.* I, 15, 13).

In his criticism of Macrobius, Stephen makes many strong accusations. However, he is well aware that too much open hostility may discomfort European readers, among whom Macrobius was a highly respected authority: indeed, Macrobius' *Commentary on Scipio's Dream* at the time enjoyed the status of a standard work. After a series of anti-Macrobian arguments in Book I of the *Liber Mamonis*, Stephen uses the preface to Book II to explain his attitude towards the neo-Platonist.[15] He points out that his invectives were not aimed at Macrobius in person, whom he accuses of ignorance but not of dishonourable intentions (p. 174:11ff.): it was rather the wide circulation and the persistence of Macrobian errors in the West that led Stephen to challenge his influence (p. 174:25ff.). Aware that the Europeans would not easily abandon their traditional doctrines, Stephen compares his task of liberating unwilling compatriots from their deep-rooted errors with that of Solon (p. 176:9ff.). But whereas Solon could abolish harmful traditions by legal means, Stephen's efforts at disproving a tradition of learning with as yet unheard of ideas could only rely on a relentless rhetoric, in order to at least draw attention to the new arguments (p. 176:2ff.).

Indeed, despite his harangues against Macrobius, Stephen nonetheless praises his discussion of the falling rain (though not without qualifying it as a rare example of praiseworthiness (p. 94:22ff.)) and adopts Macrobian expressions, such as 'globositas' (p. 312:8, cf. *Somn.* I, 15, 16, I, 19, 23, II, 5, 10 etc.), 'iugabilis competentia' (p. 180:7, cf. *Somn.* I, 6, 24),[16] and possibly also one of Macrobius'

---

[14]Obviously, Stephen does not see Macrobius as a propagator of Heraclides' theory that Mercury and Venus rotate around the sun, as is held by several medieval and modern interpreters of Macrobius; see, e.g., Haskins, *Studies*, p. 101. Stephen therefore does not *deny* the 'Macrobian theory' of a rotation around the sun, as Haskins sees it, but does not even seem to know about this interpretation of Macrobius. As Stephen probably received his own first training in cosmology in the Macrobian tradition, an interpretation of Macrobius in the sense of Heraclides' theory may thus not have been common. Moreover, the contradictions in Macrobius' arguments would be resolved only partly by such an interpretation, while new inconsistencies would occur. In modern studies it has therefore been argued that Macrobius did not have a clear concept of the inferior planets; cf. Macrobius (tr. Stahl), *Commentary on the dream of Scipio*, Appendix A, p. 249f.

[15]See p. 174:1–176:30

[16]See also p. 104:13: 'ut quod ratio exequitur oculorum sensus comprobet'; cf. *Somn.* I, 18, 2: 'ut non solum mente concipi, sed oculis quoque ipsis possit probari.'

illustrations.[17] This would confirm the impersonal character of Stephen's objections and reflect that Stephen himself was probably introduced to cosmology via Macrobius' *Commentary on Scipio's Dream*. As Stephen includes various quotations from Macrobius' work in the *Liber Mamonis* (see, e.g., p. 158:19ff.), at least a partial copy of it must also have been available in Antioch.

4.3 Combining Cosmology with Astronomy

Another main objective of Stephen's in the *Liber Mamonis* was to provide an explanation for the astronomical tables and procedures which he had published in his *Regule canonis*. Ibn al-Haytham's *On the Configuration* was surely a suitable starting point, as it contains an illustrative bodily analogue to the abstract circles of astronomy. However, we have seen that Stephen's particular concern with numerical procedures and their geometric foundations is not among Ibn al-Haytham's priorities in that treatise. For Stephen, therefore, the original text required systematic reworking in its relation to astronomy.

Among the measures taken to adapt Ibn al-Haytham's text to the contents of the *Regule canonis*, Stephen includes numbers from his tables on occasions where Ibn al-Haytham gives only imprecise quantifying information, or none at all. For example, Ibn al-Haytham defines the obliquity of the ecliptic to be 'ca 24 degrees'. Stephen replaces this with the more precise parameter of 23;35 degrees (pp. 152:17 and 182:17).[18] And while Ibn al-Haytham is silent on the velocities of individual celestial motions, Stephen adds these details on various occasions.[19]

The above modifications construe a correspondence between the parameters in Stephen's publications. In addition, in this way Stephen relates particular parts of Ibn al-Haytham's cosmography to corresponding contents of the *Regule canonis*. One of his intentions in so doing becomes apparent from the previous discussion of passages 176–7, where he inserted numerical values for selected ascensions. Taken by themselves, these numbers, which are chosen arbitrarily from Ptolemy's extensive set of ascensions, would be of hardly any use to the reader; however, the situation is quite different if we assume that the reader also has the table in their hand, in a copy of Stephen's *Regule canonis*. In this case, the few quoted numbers in the *Liber Mamonis* are sufficient for the reader to associate the description with the entire table in the *Regule canonis*.

The same purpose must be assumed for numbers in the *Liber Mamonis* which would be irrelevant, and in fact unintelligible, unless the corresponding table is

---

[17] See Stephen's diagram on p. 244; cf. *Somn.* II, 7. See also below, Sect. 4.6.

[18] Precise values for the obliquity also appear in the other Latin translations of *On the Configuration*, 23;51° in the Madrid translation and 23;23° (amended) in the Alfonsine version.

[19] See, e.g., the velocities ascribed to the spheres in the diagram on p. 421.

studied simultaneously. This is the case of Stephen's discussion of the relations between ascensions and descensions at *sphera obliqua*, in a commentary on Ibn al-Haytham's passage 175 (p. 236:33f.). Stephen's numbers can be inferred to pertain to the fourth climate in Ptolemy's table of the ascensions (*Alm.* II, 8). Only here do the first 10 degrees of Aries have an ascension of 6;14° as stated by Stephen, which falls short of 9;10° at *sphera recta* by 2;56°, the second number in Stephen's description. This decrease is compensated for by a corresponding increase of the ascension at the autumnal equinox, i.e., for the last interval of Virgo, which has become larger than 9;10° by an identical amount of 2;56° to 12;6°, which is Stephen's third number. Notably, Stephen neither mentions the important value of 9;10° for the above considerations nor the validity of the numbers chosen only for the fourth climate. This means that Stephen's numbers are unsuited to support or visualise the present argument in the *Liber Mamonis*. Conversely, however, the numbers allow that argument to explain another aspect of the table of the ascensions in the *Regule canonis*, which Stephen seems to expect the reader to have at hand.

On occasions, the relation between astronomical procedures and Ibn al-Haytham's description is too complex to be sufficiently clear from the cursory insertion of numbers. In such cases Stephen adds more detailed explanations, as he does for the rectification tables. After translating each of Ibn al-Haytham's descriptions of the solar, lunar and planetary spheres, Stephen points out how the single columns of the tables in the *Regule canonis* can be derived from the described arrangement, or, rather, how the prescribed use of each table can be understood from the spherical model; see pp. 268:17–270:26 for the rectification of the sun, pp. 296:23–300:9 for the moon, and pp. 378:1–402:30 for the planets.

In the case of the moon, Stephen gives a simplified geometrical explanation, including a simplified diagram (cf. Fig. 3.9 on p. 300), of the first and second lunar anomalies. He ignores the *prosneusis* and the change in the epicyclic equation due to the varying distance from the earth. Stephen informs the reader of the deficient nature of the illustration and notes that it does not account adequately for all the details of the rectification procedure as described in the *Regule canonis*. A fuller description of a rectification is given later in the *Liber Mamonis*, in Stephen's discussion of the planets, which also includes a correction of the epicyclic equation. On the same occasion, Stephen also discusses the retrogradation based on a numerical example, gives parameters of the planetary latitudes, and includes astronomical diagrams (pp. 378:1–402:30, with Fig. 4.15 on p. 380 and Fig. 4.16 on p. 390). Stephen's discussion of the retrogradation is particularly detailed. The subject seems to be one of his main concerns in Book IV, as is also indicated by the only book title in the *Liber Mamonis*—'de retrogradatione'—which is given to this last book (cf. also Stephen's remark on p. 388:1ff.).

The discussion of rectification tables in the *Liber Mamonis* shows how much Stephen's use of *On the Configuration* diverges from Ibn al-Haytham's original intention. Stephen also introduces mathematical elements for purposes other than the above. Some of them merely supplement Ibn al-Haytham's account, such as the quadrangle for determining points of equal declination or the lines and arcs in Stephen's discussion of the phases of the moon; also, when refuting Macrobian views, Stephen often resorts to geometrical arguments. This mathematical reasoning, though generally alien to the Arabic source, adds a great deal to the astronomical character of the *Liber Mamonis*.

Stephen's clear distinction between astronomical and cosmological terms, as discussed in the previous chapter, can also be understood now as a part of his plan to illustrate the astronomical terms and concepts of the *Regule canonis* by relating the physical model to them as closely as possible. For this to be effective, Stephen had to use the same astronomical terms in the *Liber Mamonis* as he had before in the *Regule canonis* and associate these terms with their cosmological equivalents. If not, if Stephen had followed Ibn al-Haytham in using a different (for example, a more ambivalent) terminology, the *Liber Mamonis* would have failed in this particular purpose. Moreover, some of Stephen's astronomical terms, such as the epicycle, were found to be derived from their physical correspondents. Stephen's considerations about a comprehensive Latin terminology, through which astronomy and cosmology would become contrastingly associated with each other, must therefore have preceded the production of the *Regule canonis*, if terminological consistency with the *Liber Mamonis* is assumed. This is noteworthy, since, as far as we know, the *Regule canonis* dealt with mathematical astronomy only and, as such, would not have provided an incentive for any further-reaching thoughts about a universal terminology. A possible explanation would be that related ideas were already circulating among Arabic-speaking astronomers in Syria and were a timely inspiration for Stephen.

4.4 New Planetary Models

Complementary to the inclusion of astronomical elements, using an inverse approach Stephen also modifies parts of the cosmological theory to adapt it more closely to certain astronomical concepts. Such attempts, however, necessarily brought him face to face with some of the unsolved problems of celestial science: the physical implications of the oscillating deferent planes of Mercury and Venus, the moving planes of the epicycles, or the notorious equant. Among these problems, Stephen pays particular attention to the first two, which both relate to the planetary latitudes and for which Ibn al-Haytham made no serious effort to give physical explanations in *On the Configuration*. Stephen detects a first shortfall in this regard in Ibn al-Haytham's discussion of Mercury. Here,

Stephen expressly distances himself from his Arabic source and presents instead an alternative theory, which he claims to be his own.[20]

4.4.1 The Oscillating Deferents of Mercury and Venus

Ibn al-Haytham conceived his physical model of the planets to match the planetary motions as described by Ptolemy in the *Almagest*. As illustrated in Fig. 4.1, left, Ptolemy describes the centre of Mercury's epicycle, *C*, as moving on the deferent circle, while at the same time the centre of the deferent circle moves on a small circle in the opposite direction. The small circle lies shifted from the earth, E, towards the apogee, A. The motion of the deferent centre on the small circle is uniform with respect to the centre of that circle, whereas the motion of the epicycle centre on the deferent is uniform with respect to what is known as the equant point, Q (both motions are shown in Fig. 4.1 for a position α from the apogee). In addition, the apogee moves slowly eastwards with respect to the solstices.[21] Ibn al-Haytham accounts for these motions by a system of three great nested spheres; *Configuration*, passages 273–290. Figure 4.1, right, shows a cross-section through Ibn al-Haytham's arrangement. A first, slow sphere, *S*, which is concentric with the earth, produces the slow eastward motion of Mercury's apogee, while embedded in the shell of that sphere are two eccentrically nested spheres, *R* and *D*. The dirigent sphere, *R*, produces the motion of the deferent centre around its axis II, corresponding to Ptolemy's small circle. The deferent sphere, *D*, which is embedded eccentrically in the dirigent sphere, carries the epicyclic sphere around the axis III. In the same context Ibn al-Haytham also mentions the equant point and the equant, or dirigent, line Q-*C*, but he says nothing about the implications of the equant with regard to irregularities in the motions of the spheres; *Configuration*, passages 295 and 299.

Nor does Stephen comment on the equant problem when he translates the corresponding passages of Ibn al-Haytham's *On the Configuration*. Rather, he follows Ibn al-Haytham by quantifying the motions of the spheres simply by their mean speeds, in terms of revolutions per year. What prompts Stephen's criticism instead is Ibn al-Haytham's failure to present a physical cause for the changing inclination of Mercury's deferent plane. According to *Alm.* XIII, 2–3, the plane of Mercury's deferent oscillates around the centre of the world in such a manner that the centre of the epicycle is inclined south from the ecliptic when standing at the apogee or opposite to it, whereas it lies in the plane of the ecliptic every time it passes through one of the nodes. This implies an oscillating motion of the

[20]For Stephen's theory of the inferior planets as described below see also Grupe, 'Stephen of Pisa's theory of the oscillating deferents of the inner planets (1h. 12th C.)', *AHES* 71 (2017), pp. 379–407.

[21]For a detailed account of Mercury's various motions in the *Almagest* see Neugebauer, *HAMA*, and Pedersen, *Survey*.

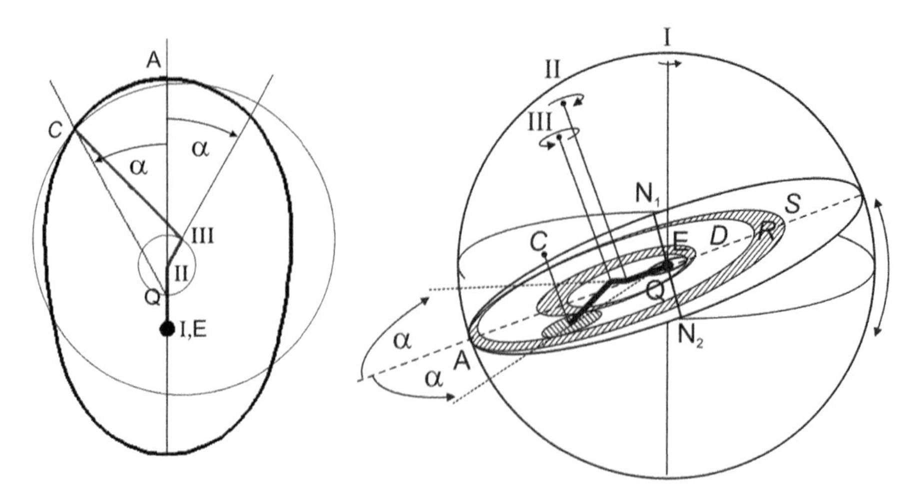

Fig. 4.1: The motion of Mercury's epicycle centre, *C*, according to Ptolemy's *Almagest* (left; radii and proportions not to scale), and a cross-sectional view of the oscillating deferent plane in Ibn al-Haytham's spherical model of Mercury (right)

deferent plane, which changes sides whenever the epicycle moves from the side of the apogee to the opposite side, and vice versa (indicated by the double arrows on the far right in Fig. 4.1, right). Although Ibn al-Haytham also reports these oscillations of the deferent plane, he abstains from suggesting any physical cause regarding the spheres involved.

When he comes to Ibn al-Haytham's discussion of Mercury, in the *Liber Mamonis* Stephen has already prepared the reader for his objections. In the preceding parts he stresses repeatedly that a rotating sphere cannot experience a change in the orientation of its rotational axis without the presence of an external motion acting on it. The poles of the rotating sphere thus need to be attached to another, enclosing sphere, whose own motion causes the change in orientation of the sphere it encloses (see p. 98:4ff. and, similarly, p. 120:27ff.). Nevertheless, Stephen first translates Ibn al-Haytham's description of Mercury's spheres and of the oscillating plane of its deferent without introducing any doubts; it is only at the end that he indicates the fundamental deficits (see p. 338:26ff.). He then explains that an arrangement in which the axes of the referent[22] sphere and the deferent sphere are parallel cannot produce a changing inclination of the deferent plane. However, if the axis of the deferent sphere were inclined from the axis of the referent sphere by the same angle as the referent's axis happened to be inclined from the axis of the slow sphere, the inclinations of the axes would alternately either add up or compensate, so as to produce an alternating inclination and coplanarity of the deferent relative to the ecliptic.

[22]The common term in the modern literature is 'dirigent' sphere, which is also used above. In connection with Stephen's model I follow the latter's terminology; cf. above, Sect. 3.2.

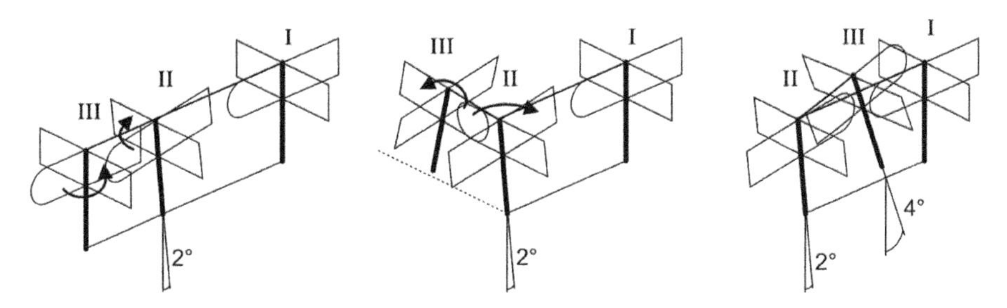

Fig. 4.2: Stephen's demonstration of the insufficiency of only three spheres for generating the motion of Mercury's deferent plane (Stephen's 'first' planes are drawn with one of their ends rounded). In the first position (left), the deferent's axis, III, is parallel to the axis of the slow sphere, I, but becomes inclined as it is moved by the rotation of the referent sphere, II (centre and right)

To visualise his argument, Stephen describes an arrangement shown in Fig. 4.2 (see p. 340:6ff.). Stephen points out that a corresponding model, though it provides for alternating instances of coplanarity and inclination, is still unsuitable for bringing about the required oscillations. The arrangement described produces neither an oscillating motion around the centre of the ecliptic nor an inclination from the ecliptic in opposite directions. Stephen refers to these aspects only vaguely, when he objects instead that every position during one circuit corresponds to a distinct orientation of the deferent. At the same time, Stephen remains silent about another apparent deficit: when the deferent's axis, III, stands furthest from the central axis, I, this would coincide with a position of Mercury's epicycle at the apogee. At that moment, however, contrary to the requirement, coplanarity of the deferent with the ecliptic, rather than inclination, is produced. As will be seen, Stephen's criticism of the model in Fig. 4.2 corresponds to problems that are solved (and to others that are not solved) by his own theory of the oscillating deferents. He now goes on to describe his theory.

To overcome the deficits identified, Stephen requires that an arrangement similar to the above rotates as a whole by one revolution per year. In that way, inclination in different directions (and in particular opposite directions) can be achieved. To do so, Stephen introduces a fourth great sphere, which he calls the 'annual' sphere, and which has its centre at the equant point. As the annual sphere produces an additional rotation eastwards, Stephen increases the westward motion of the referent sphere accordingly, making the latter perform two revolutions per year instead of only one. The longitudinal effects by the additional rotation of the annual sphere are thus largely compensated. At the same time, there results an arrangement in which the annual sphere produces one rotation per year of the entire inclination mechanism, which is embodied by the next two axes, each rotating in a direction opposite the other and at a speed twice that of the annual sphere. This double speed provides for the two cycles of inclination and coplanarity required per year.

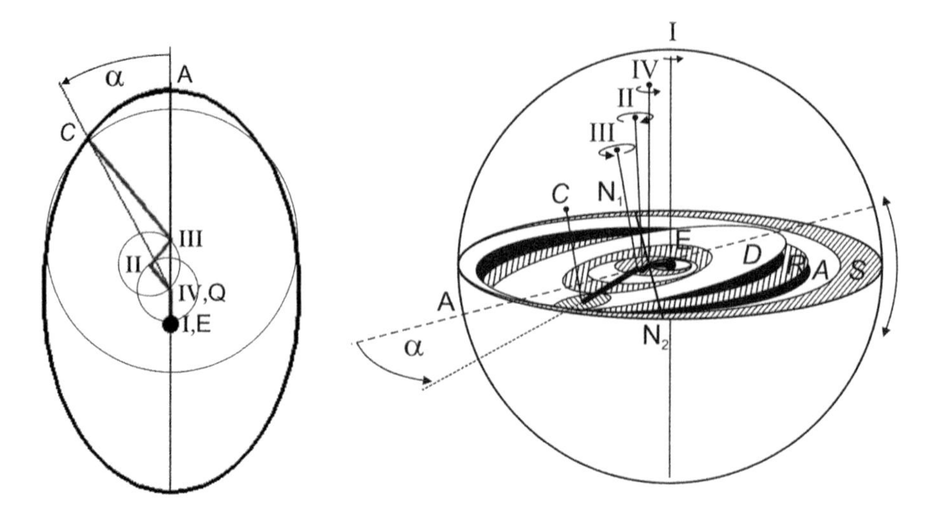

Fig. 4.3: The motion of Mercury's epicycle centre, *C*, after Stephen's insertion of a fourth great sphere, with axis IV, centred at the equant point, Q (left), and a cross-sectional view on the equatorial planes of the spheres in Stephen's model (right)

Figure 4.3 shows the above modifications. Stephen's newly introduced 'annual' sphere, *A*, rotates about its axis, IV, at the equant point, Q, parallel to the axis of the slow sphere, *S*. This makes the centre of the referent sphere, *R*, and its axis, II, which in Ptolemy's theory rests on the apsidal line, move along an additional small circle, around axis IV. Stephen makes one further modification relative to the arrangement in Fig. 4.2. Instead of having the referent's axis, II, incline in a plane with the axis of the ecliptic, I, he proposes that the referent's axis be inclined sideways, around the line through the centres of the annual sphere and the referent sphere. Thereby, according to Stephen's description, when the centres of all the spheres are aligned on the apsidal line, the north poles of the annual, the referent and the deferent spheres define an equilateral triangle (see p. 348:6ff.).

To describe his model in action, Stephen illustrates the changing positions and orientations of the axes by means of a diagram of trajectories. The diagram itself is not preserved in the manuscript. Figure 4.12, on p. 354, shows a reconstruction based on Stephen's description. In the diagram, the reader looks from the north onto the plane of the ecliptic, where the centres of the four spheres are aligned along the apsidal line, at points A to D. The centre of the world, identical to that of the earth, lies at A, the centre of the annual sphere at B, the centre of the referent sphere at C, and the centre of the deferent sphere at D. The axes of all spheres are assumed to stand vertical on the book page in the beginning, except for the axis of the referent sphere, which is inclined around the apsidal line such that its poles are seen at points Q (north pole) and H (south pole). By tracing the trajectories of the centre and each pole of the deferent sphere for one complete cycle, as projections

onto the plane of the ecliptic, Stephen shows that in the proposed arrangement the deferent's axis will stand vertical to the ecliptic when its centre reaches either of the extreme positions D and M, whereas it will be inclined on two occasions, and in opposite directions, when its centre passes through the geometric centre of the motion, at point B. To facilitate a demonstration of the effect which the combined motions of the annual sphere and the referent sphere have on the orientation of the deferent's axis, Stephen divides the cycle into four equal intervals; in each interval he decouples the motions of the spheres, first letting the annual sphere rotate by a quadrant and, in a later stage, the faster referent by twice that angle.

Stephen presents his theory in connection with Ibn al-Haytham's discussion of Mercury. Venus, being the second inner planet and the subject of Ibn al-Haytham's following chapter, also has an oscillating deferent. This aspect is again mentioned, but not physically accounted for, by Ibn al-Haytham. As in the case of Mercury, Stephen first translates Ibn al-Haytham's discussion of Venus completely and, in a concluding comment, remarks that for the same reasons as in the case of Mercury the number of spheres must be increased to be the same as that of Mercury (see p. 368:13ff.).

4.4.2 Nested Epicyclic Spheres

Stephen proposes further modifications to Ibn al-Haytham's theory, in connection with the epicycles of the five planets. Ibn al-Haytham claims a single sphere for the epicycle of each planet, whose rotation produces the planet's motion on the epicycle. However, as Stephen remarks, a planet's period on the epicycle does not coincide with the period of the orientation of the epicycle's plane relative to the deferent plane, which instead depends on the position of the epicycle centre on the deferent. Stephen accounts for that difference again by introducing further spheres (see p. 398:6ff.). He proposes one additional sphere for each epicycle, which encloses the first epicyclic sphere and which revolves at the same speed as the epicycle centre on the deferent, but in opposite direction. To compensate for the effect in anomaly of that additional rotation, Stephen increases the speed of the first epicyclic sphere accordingly.

Stephen gives no information as to how the axes in his epicycle couples are oriented. It is therefore uncertain whether he had a clear opinion about how his couples work. He may have introduced the second sphere of each epicycle simply as a general prerequisite for a changing orientation of the epicycle plane relative to the deferent. As a plausible alternative similar to the lunar model, Stephen may have considered the axis of the second epicyclic sphere to stand vertical on the deferent plane, while the axis of the inner sphere is inclined from the former by an angle corresponding to the planet's greatest latitude. In that way self-parallel epicycles with respect to their deferents could be produced. This would agree with another deviation from Ibn al-Haytham, as Stephen claims that both (and not

only one) of Ptolemy's small circles that are attached to the epicycle planes of the inner planets are fixed with regard to the deferent plane (cf. p. 360:1ff. and the note to passage 310). This would also correspond to a common simplification of the latitudes in astronomical tables.[23]

4.4.3 The Implications of Stephen's Planetary Models

In connection with the planetary latitudes, Stephen cleverly identifies shortcomings in Ibn al-Haytham's cosmography and presents physical explanations to fill these gaps. Stephen's theory of the oscillating deferents departs from Ptolemaic doctrines, and even contradicts Ptolemy, in proposing a moving centre of the referent sphere. By carrying out these modifications, oscillation of the deferent plane in opposite directions is obtained. Moreover, the deferent plane pivots about the geometric centre of the motion, as a result of the sideways inclination of the referent's axis.

However, Stephen's theory has various imperfections, which he leaves entirely undiscussed. Most notably, the changing orientation of the deferent's axis in his model for Mercury and Venus does not correspond to the simultaneous position of the epicycle. In accordance with the astronomical theory, Stephen describes the epicycle to be at the apogee in the beginning. At that moment, however, the deferent plane in his model is not inclined, as the astronomical theory requires, but coplanar with the ecliptic. Analogously, Stephen's deferent reaches its greatest inclination when the epicycle is at either of the nodes. When at the nodes, however, the epicycle would simply become slanted by an inclined deferent plane, contrary to the astronomical theory. This shift in the instances of coplanarity and inclination of the deferent recalls the case of the hypothesis rejected in Fig. 4.2, where Stephen already passed over the deficiency in silence. An explanation for this inaccuracy could be that the tables in Stephen's *Regule canonis* relied on a simpler concept of the latitudes, which did not consider oscillations of the deferents or at least did not allow their effect to be identified in isolation from others.[24] The same is also suggested by the parameters in the text, where Stephen defines the maximum inclination of Mercury's deferent to be 4°, which is identical to his value for Mercury's greatest latitude.[25]

Stephen was probably aware of certain flaws in his theory of the oscillating deferents, because he gives a very accurate Latin translation of Ibn al-Haytham's

[23]An interpretation of Stephen's concept of the planetary latitudes is made difficult by some confusing, and possibly corrupted, statements in his summary of the subject; cf. 400:16ff. and the notes to that passage.

[24]See below, Sect. 5.1.

[25]The former results from doubling the inclination angle of 2 degrees of the referent's axis, mentioned on pp. 340:21ff. and 348:2ff.; the latter is given on p. 402:24.

different description of the oscillations. Reservations regarding his own theory are apparent in the outcome of the argument, where Stephen restricts his conclusion to the necessity of a fourth great sphere—or that there must be more than three spheres, without excluding the possibility of further spheres. An awareness of remaining unsolved problems also emerges from Stephen's preface to Book I, where he says that his discussion of the number and the order of the heavenly spheres is restricted by his own limitations as well as those of the human intellect; cf. p. 92:21ff. For Ibn al-Haytham's *On the Configuration* it has been observed that the work does not solve any of the problems of celestial physics, but it gives a rather suggestive report in which the cosmological and the astronomical descriptions of the universe appear to be in agreement (see above, Sect. 2.3). In a similar manner, Stephen may have been content with making a plausible case that the astronomical concept of the oscillating deferents can in principle be given a physical analogue. This would also explain why in the case of Venus Stephen simply claims the same reasoning as for Mercury, ignoring that for Venus a circular, not an elliptical, motion for the epicycle had to be produced.

Already in the preface to Book IV, Stephen announces that he will present independent improvements on his Arabic source.[26] In each of these cases Stephen preserves the content of his source and contrasts it with his own views. Enabling the reader to compare and judge the opposing arguments, Stephen gives an example of the kind of open scientific discourse that he defends earlier in the *Liber Mamonis*. At the same time he demonstrates his ability to present a scientific argumentation at the level of the Arab scientists, thus giving a persuasive demonstration to the European reader of the *Liber Mamonis* that the achievements of foreign scientists can be equalled or surpassed. The way in which Stephen visualises his arguments also gives another example of his striving for an intuitive presentation. For his Mercury diagram, Stephen de-couples complex motions and divides them into alternating strokes; he further illustrates changing positions in space by projecting the trajectories of suitably chosen points onto the book page, and he describes the diagram in such detail that the survival of the argument will not depend on the skills of later manuscript illustrators. In spite of these accomplishments, Stephen apologises to the reader for the difficulties his discussion may still pose; he would have preferred to demonstrate the motions of the inclining axes by showing, in front of his audience, models of 'small sticks', which would make the argument immediately visible. But because he is limited to communicating in writing, as he puts it, abstract descriptions cannot be avoided (p. 346:17f.).

Stephen claims the theory of the inner planets as presented in the *Liber Mamonis* to be his own discovery and an advance on his Arabic source. However, the

[26]See p. 324:3: 'Verum cum in aliis Arabem quendam plurimum secuti sumus, in hoc quoque permultum sequemur licet quedam de sperarum numero et rotunditatum invenerimus et de circulis quidem et inclinationibus planetarum vera perstrinxit a quibus sperarum numerus dissonat.'

creation of an oscillating motion by a couple of nested spheres, which rotate in opposite directions around mutually inclined axes, had been known since Antiquity, in the form of the Eudoxan couple.[27] Despite essential differences, it is hard to believe that Stephen developed his criticism of Ibn al-Haytham, and that he arrived at the combination of inverse rotations about mutually inclined axes (for the oscillating deferents and, in a simpler form, also for the epicycles) without inspiration from elsewhere—especially bearing in mind that his theory involves aspects that were or would become important subjects of debate among Arab astronomers. When he placed the centre of the annual sphere onto the equant point, not only was the pivot of the oscillations brought closer to the earth than in the rejected case of Fig. 4.2, but the conflict between geometric and dynamic centres of Mercury's motion, as a consequence of the equant hypothesis, was mitigated. Furthermore, the centre of Stephen's deferent sphere, and with it the entire deferent sphere, oscillates in a quasi-linear motion on the apsidal line; this follows from the eccentricities and motions of the spheres involved that would become central in the development of the Ṭūsī-couple.[28] With the deferent sphere's own rotation finally superimposed onto its linear oscillations, the quasi-elliptical shape of the resulting trajectory of Mercury's epicycle is also much more evident than in the case of Ptolemy's mechanism.[29] Although Stephen does not mention any of these implications of his theory, notions of them may have circulated in Antioch and thus affected his thinking. This suggests once more that Stephen was influenced by the presence of like-minded experts in Antioch, where speculating about alternative cosmological theories and using advanced techniques to visualise them may have been common practice.

It should also be clear by now why Stephen did not continue to apply the very strict translation standards when writing the *Liber Mamonis*. He could hardly consider that work a translation: although Ibn al-Haytham's *On the Configuration of the World* appears almost complete in the *Liber Mamonis*, it merely forms a

[27] Ibn al-Haytham himself applied the Eudoxan couple in his later *Treatise on the Movement of Iltifāf* (now lost) to explain the motion of the epicycles. Stephen's concept differs from the Eudoxan device, essentially in that it uses eccentrically, instead of homocentrically, nested spheres. Also, in contrast to Ibn al-Haytham's use of the Eudoxan couple, Stephen introduces his mechanism to explain the oscillating deferents, not the moving epicycles, whereas for the epicycles he gives a simpler explanation. The earliest appearance of the Eudoxan couple in Latin writings has been dated to the fourteenth century; J. L. Mancha, 'Ibn al-Haytham's homocentric epicycles in Latin astronomical texts of the XIVth and XVth centuries', in *Centaurus* 33, 1990, pp. 70–89.

[28] A recent account on the different versions of the Ṭūsī-couple and their respective development is given in F. J. Ragep, 'From Tūn to Turun: The twists and turns of the Ṭūsī-couple', preprint, Berlin, 2014.

[29] With the particular parameters given in the *Almagest*, the shape also comes close to an ellipse; Hartner, 'The Mercury horoscope of Marcantonio Michiel of Venice'. The first notion of Mercury's deferent as an ellipse has been identified in the work of the eleventh century Andalusian astronomer Ibn al-Zarqālluh (Azarquiel); cf. Samsó, Mielgo, 'Ibn al-Zarqālluh on Mercury'.

basis for what Stephen actually intends to transmit to his European readers. At the same time, translating the various Arabic source materials literally would have hindered their consistent combination, let alone Stephen's own innovations, for the particular purpose of the *Liber Mamonis*. The importance of Ibn al-Haytham and of his work may have been unduly downplayed by Stephen when he refers to him only in the final part and merely as a 'certain Arab' whom he followed 'in most of his arguments'; but given the extent of Stephen's reinterpretation and reworking of the source, an attribution to Ibn al-Haytham would have been equally misleading. In view of the innovative quality of the *Liber Mamonis* compared with its individual sources, Stephen was justified in considering his work as an independent account and in presenting it as such.

4.5 Non-Astronomical Teaching

According to Stephen's own testimony, the introduction of Arabic medicine and astronomy to Europe was only one part of his attempt to close the gap in knowledge between the Arabs and the Latins. The fact that his interest was indeed not restricted to these two disciplines becomes clear from his frequent treatment of non-astronomical questions in the *Liber Mamonis*.

4.5.1 Aristotelianism

Haskins noted already that Stephen calls himself a Peripatetic, but disagrees frequently with the Aristotelians.[30] This provides an adequate summary of Stephen's philosophical and methodological commitment in the *Liber Mamonis*, for he accuses Plato of missing the truth very often, and also firmly rejects the Epicureans.[31] He also ridicules Sophism and any natural philosophy that disdains empiricism.[32] However, there is no corresponding disrespect towards Aristotle, and it is only in individual doctrines that Stephen disagrees with Aristotelian beliefs: for example, the question of God's existence before the creation of the world, Aristotle's ignorance of the ninth celestial sphere, and an explanation for solar heat, which Stephen also attributes to the Aristotelians.[33] In the latter case Stephen does not in fact reject the Aristotelian argument as such but combines it with a competing idea.

Stephen proposes reasoning applied to objects of sense-perception as the most reliable way to achieve scientific knowledge (p. 112:22ff.):

[30] Haskins, *Studies*, p. 101, referring to the passages on p. 246:3f. and p. 140:29f.

[31] See p. 88:13: 'Plato in multis a veritate dissonat', and p. 246:3: 'neque... Epicureum aliquando dogma audivimus'.

[32] See p. 112:22 and p. 276:13.

[33] See p. 88:13f. and pp. 140:29–154:23.

"[I]t is better to pursue a theory which nature itself suggests, by any kind of experience ('experimentum') to the judgement of human senses ('sub humani sensus arbitrio') through cogent reasoning ('certa cum ratione')."

In accordance with this conviction, Stephen emphasises the heuristic value of sense-perception, including his use of diagrams, and includes empirical reasoning in some of his own arguments.[34] Moreover, on occasions he can be found attempting to perform logical reasoning in a noticeably formalist manner.[35] While he thus commits himself to Aristotelianism with respect to its methods, he does not uncritically adopt what he believes to be Aristotelian views. The following cases show how Stephen puts this attitude into practice.

4.5.2 The Flooding of the Nile

One of Stephen's first digressions in the *Liber Mamonis*, embedded in an anti-Macrobian argument, is devoted to the annual flooding of the Nile (pp. 110:20–114:21). Stephen connects the discussion to Ibn al-Haytham's brief remarks about water in the chapter on the sublunar world. To refute Macrobius' theory of an equatorial ocean, Stephen refers to the inundation of the Nile as proof that there must be a continental connection between the northern and the southern hemispheres of the earth.

Stephen starts the discussion with Macrobius' statement of a closed oceanic stream along the equator. Stephen criticises the statement for its lack of empirical evidence and logical cogency, and presents an argument that purports to avoid such mistakes, while also leading to a different result. The empirical foundation of Stephen's argument is the flooding of the Nile, which occurs in late summer—six months before or after the changing water-levels of all the other rivers in the northern part of the earth, where floods mostly occur in late winter as a result of rain and snowmelt. From that six-month shift in seasonal effects, Stephen argues, one must assume that the source of the Nile is located in the southern hemisphere. Before explaining this conclusion, he first refutes other theories about the inundation, in particular all those that explain it, directly or indirectly, via reference to the extreme heat near the northern tropic during summer. Stephen does this by noting the sun's generally adversary nature with respect to water. He also cites Lucan's historical report in the *Pharsalia* (*Phars.* X, 305ff.) as further proof that the source of the Nile lies inaccessibly far south from the borders of the northern temperate region. This agrees with the assumption that the Nile originates south of the tropical zone.

Elaborating further on the half-year shift of the Nile's water-level, now also considering the Nile's equally inverted dryness in winter, Stephen concludes that

[34]See e.g. p. 96:6, p. 120:1, p. 148:9, p. 154:11 and p. 242:19.

[35]See p. 98:5, p. 100:13, p. 254:24, etc.

a location of the source in the southern temperate region is the most plausible explanation of the phenomena observed. For whereas rain and dryness are the causes of a river's changing level, the temperate zone of the southern hemisphere is the only place where these climatic conditions can, or even must, occur inverse to the seasons in the north. With the Nile's crossing of the equator thus proved, Stephen finally refutes Macrobius by pointing out that the Nile could not possibly pass through an equatorial ocean, since in that case it would mix with sea-water—while in fact the Nile's water is fresh when it arrives in the north.

The digression on the Nile provides an example of Stephen's assiduous and reasoned polemics against Macrobius and of his commenting technique, which is always ready to interrupt Ibn al-Haytham even on incidental matters. Stephen's factual knowledge about the Nile is also noteworthy. The (pseudo-)Aristotelian treatise *On the Inundation of the Nile* mentions, among various theories, the hypothesis that the floodings were an effect of the inverted seasons in the southern hemisphere, and it attributes the idea to Nikagoras of Cyprus.[36] Plutarch (*Plac. phil.* IV 1, 7) later ascribed the same hypothesis to Eudoxus, while the only Latin author to mention it is the first century geographer Pomponius Mela (I 54).[37] The theory did not find wide acceptance, because it was believed that a river could not possibly cross the vast burnt zone of the earth.[38] Pomponius Mela therefore modified the hypothesis, allowing the Nile to avoid the greatest heat by passing the equator through underground channels. With this modification, however, the theory was of no use to Stephen, since an underground passage could also pass underneath an equatorial ocean.

As far as we know, Nikagoras' theory is not mentioned in the Latin literature of Antiquity; other theories appear quite frequently, such as tropical rainfalls, north winds that partly dam up the Nile, or an absorbing effect of the hot air in summer, through which more ground water is drawn to the surface.[39] As most of these explanations in some way or other involve the hot climate in the Nile's upper course, Stephen effectively refutes them by pointing to the adversary nature of sun and water. Clearly, then, Stephen was aware of other theories; but it is not apparent how he gained knowledge of Nikagoras' hypothesis and came to favour it over the more influential theories.

The only author on the Nile who is mentioned in the *Liber Mamonis* is Lucan, to whom Stephen refers for the historical evidence that the spring of the Nile is inaccessible to people from the north. In Lucan's epic this anecdote is embedded in a longer *excursus* on the mysteries of the Nile (*Phars.* X.188–332), where sev-

[36] Aristoteles Pseudepigraphus (ed. Rose), p. 637.

[37] An account of the various theories of the inundation of the Nile and their reception by Greek and Latin authors is given in Postl, *Die Bedeutung des Nil in der römischen Literatur*.

[38] See Aristoteles Pseudepigraphus (ed. Rose), p. 637.

[39] Postl, *Die Bedeutung des Nil in der römischen Literatur*, pp. 71–89.

eral of the classical theories about the inundation are also listed. But as Lucan does not mention the inverted seasons in the southern hemisphere as a possible explanation, he cannot have inspired Stephen's argument. Lucan himself appears to prefer the Aristotelian theory of Ethiopean rains, and in fact none of the theories in his report would serve Stephen's purpose by rejecting the presence of an equatorial ocean. Although Lucan later states that the Nile traverses the equator from south to north, which would provide an immediate argument against an equatorial ocean, Stephen does not make use of this remark. Put into the mouth of a Memphis priest who gained his knowledge through inspiration, the information must have failed to meet Stephen's peripatetic standards. Stephen also knows of the characteristic freshness, or sweetness, of the Nile's water (p. 114:18). This topic had been discussed since Diodorus (I, 40, 5–8), but again it does not appear in Lucan's epic. Also, Stephen's reference to Lucan is somewhat inaccurate, as the latter treats the flooding and the source of the Nile as two separate subjects, whereas Stephen combines the two saying that the expeditions to the source had been made for the sole purpose of discovering the cause of the flooding. This connection, in turn, is suggested in Book IV of Seneca's *Naturales Quaestiones*, whose similarity to Stephen's formulation may not be incidental.[40] Seneca also mentioned the particular sweetness of the Nile's water, just before the transmission of his text breaks off.[41] However, although large parts of Seneca's discussion of the Nile are lost, he probably did not mention Nikagoras' theory either. This is again suggested by Lucan, who follows Seneca's catalogue closely.

Stephen may have learned of Nikagoras' hypothesis from a later Latin tradition. More probably, though, he received his information from a local source in Antioch. In this case, he would have considered the hypothesis as another example of foreign learning that was worthy of being brought to Europe. This is also suggested by the manner in which Stephen presents the *excursus* on the Nile; although it is connected with an anti-Macrobian argument, the extent of the digression, its loose relation to Ibn al-Haytham's cosmography, and the self-contained character of its argumentation make it appear as a lesson in itself.

4.5.3 A New Theory of the Sun's Heat

When Ibn al-Haytham discusses the astronomical seasons as quadrants of the ecliptic, Stephen takes the opportunity to add a lengthy discussion of alternative theories about the warming effect of the sun (pp. 140:29–154:23). Although this will inevitably divert the reader from Ibn al-Haytham's discussion of astronomical

[40] Sen. *Quaest.* 4a, II, 3; 1f.: 'Unde crescere incipiat si comprehendi posset, causae quoque incrementi inuenirentur'; cf. Stephen's similar formulation on p. 112:26: 'quod... possent invenire originem, et sic... intellectu caperent causam'.

[41] Sen. *Quaest.* 4a, II, 30; 10: 'Nec enim ulli flumini dulcior gustus [...]'.

circles, Stephen considers this an acceptable price to pay, because it is a chance to discuss another highly important subject of which the Europeans know nothing (see p. 154:15f.).

Stephen first presents what he considers to be the older theory, held by 'pre-Aristotelians and still by many people after Aristotle who did not belong to his followers' (p. 142:1f.). According to this theory, the sun's heating was caused by the hot and dry nature of its body, whereas the greater heat during summer was caused by the longer presence of the sun above the horizon during those parts of the year. Stephen then contrasts this opinion with an idea which, he says, was held by many Aristotelians, namely that the warming was caused by the impact of the solar rays onto the earth. Using the analogy of the impact of a falling stone, they said, radiation which falls more vertically, like sunlight in summer, had a stronger impact and experienced stronger reflection than obliquely incident radiation, such as the sunlight during winter. The greater heat in summer was therefore caused by the steep incident angle of the sunlight rather than by the allegedly hot nature of the sun's body.

After presenting the conflicting views, Stephen rejects the two theories, though conceding that they both include elements of truth (see p. 144:18ff.). Before presenting his own explanation, he points out the deficiencies in both traditional theories. Against the first, which ascribes the changing temperature on the earth only to the duration of the sun above the horizon, he argues that in this case the equator would need to have the most temperate climate, since every day of the year lasts exactly 12 h there; conversely, the polar region during summer would have to be the hottest place on the earth, because for six months the sun does not set there but stays permanently above the horizon. Stephen then turns against the Aristotelian position, pointing out that only the sun causes warmth on the earth, whereas other shining bodies, like the moon, whose light should have a similar warming impact, do not. In fact, Stephen himself has observed that nights with a full moon high in the sky tended to be colder and more humid than nights without moonlight. A steep impact of light, therefore, could not be the only reason for the warmth caused by the sun.

Stephen sees the true reason in a combination of the two theories, that is, the hot and dry nature of the solar body and of its rays, and the incident angle of the radiation.[42] This combined explanation resolves the inherent contradictions of the

[42]See p. 150:23ff.: 'At nobis quidem videtur ex utrorumque rationibus temperanda veritas, et eorum scilicet qui calidum aiunt solem et siccum, et illorum qui ex eius propinquitate puncto capitis fieri aeris calorem dixerunt.' Haskins, *Studies*, p. 100, misinterpreted the term 'propinquitas puncto capitis' when he says that according to Stephen "the greater heat of the sun in summer is due to its nearness, not, as the Aristotelians think, to the angle of its rays". In fact, in the discussion of the solar apogee Stephen says that the sun is not nearest but furthest from the earth in summer; cf. p. 272:1ff.

two theories. Moreover, it allows Stephen to account for the formation of clouds mainly in the temperate regions of the earth; whereas in the tropical region the hot and vertically incident sunlight is reflected from the earth to a high altitude and the air is thus heated by the double effect of direct and reflected heat so that clouds are prevented from forming, in the temperate region the oblique radiation of the sun is not reflected equally strongly so that clouds may form in the cooler air above a certain height from the ground (cf. p. 150:29f.). This argument involves both the effect of impact and reflection as well as the warmth of the rays themselves.

4.5.4 Historical Considerations

Stephen shows detailed knowledge of certain historical episodes. He cites Lucan's description of ancient expeditions to the upper course of the Nile (p. 112:25f.), and he reports the circumstances of Solon's reforms in great detail (p. 176:9ff.).

Aside from particular events, Stephen tells the reader of a noble European past, when scholarship flourished and public life prospered. That former world, Stephen laments, has suffered a terrible decline up until his own day (see above, Sect. 4.1). A similar view is expressed in the preface to Book III, where Stephen cites Cicero,[43] saying that the order of the Roman Empire was shaken by Caesar's temerity (cf. p. 246:23ff.). These events were followed by a replacement of the former imperial order by the 'higher and more extended, truliest Empire of Christ', whose foundations were laid by Constantine and his early successors. After that period of most splendid rise, however, Stephen says, the present generation is witnessing another, much worse decline, with 'all the virtues expelled' (p. 246:25ff.).

At first glance, Stephen's view of past splendour and present decline appears highly pessimistic. However, as his call for more scholarly efforts in Europe shows, he considers that flourishing and decline are not ordained by fate but lie to a large extent in our hands; with notable enthusiasm, he encourages the reader to work with him towards a reversal of the current trend (cf. above, Sect. 4.1).

4.6 Added Diagrams

Several diagrams in the *Liber Mamonis* were found to be taken, and in some cases adapted, from Ibn al-Haytham's *On the Configuration*. Other diagrams seem inspired by other sources. The following illustrations in the *Liber Mamonis* do not relate to Ibn al-Haytham's original arguments, but have close correspondences in other texts:

[43] Apparently *De officiis* I,26.

- Figure 2.4 (see p. 244): the correspondence between the climatic circles in the sky and on the earth. Although Ibn al-Haytham announces a corresponding diagram, the one given by Stephen does not contain the information mentioned in the Arabic; in fact, it resembles a design found in Macrobius; cf. Macrobius (ed. Willis), p. 209.

- Figure 3.6 (see p. 272): the unequal motion of the sun; cf. e.g. Ptolemy's diagrams in *Alm.* III, 4–5.

- Figure 3.7 (see p. 278): the borders of the northern inhabitable zone. The image resembles Macrobius' diagram, adapted to Ibn al-Haytham's divergent report; cf. Macrobius (tr. Stahl), p. 202. As the illustration is oriented south up, it might be taken from an Arabic source.

- Figure 3.9 (see p. 300): a simplified illustration of the lunar anomaly; cf. e.g. Ptolemy's diagrams in *Alm.* V, 4–5.

- Figure 3.10 (see p. 308): the phases of the moon; cf. e.g. al-Battānī (ed. Nallino), *Ṣābiʾ Zīj*, pt. I, p. 62. Stephen's diagram combines different perspectives: a first one onto the plane of the moon's deferent, and a second one at the phased moon as it would appear to an observer on the earth. A combination of the same perspectives is also found in the Alfonsine paraphrase of *On the Configuration*, in MS Oxford Canon. misc. 45, fol. 53a–v, and in various Arabic writings.

- Figure 4.15 (see p. 380): anomalies of the planets; cf. e.g. Ptolemy's diagram in *Alm.* V, 4.

- Figure 4.16 (see p. 390): retrogradation; cf. e.g. Ptolemy's diagrams in *Alm.* XII, 1.

As a third group, some designs in the *Liber Mamonis* seem to have been conceived by Stephen himself:

- Figure 1.1 (see p. 106): a geometric argument against Macrobius.

- Figure 2.3 (see p. 204): the correspondence between celestial and terrestrial coordinates; possibly a commonplace illustration.

- Figure 3.5 (see p. 252): different parallaxes of different planets. *Alm.* V, 13 contains similar illustrations for the parallax of the moon only.

- Figure 4.12 (missing, see p. 354 for a reconstruction): trajectories of Mercury's axes in connection with the oscillating deferent plane.

Figure 4.12, though announced in the text, is missing in the manuscript. Stephen apparently anticipated the difficulties the diagram would pose to an ignorant illustrator, as he gives a detailed description of the diagram in the text. Other very figurative descriptions by Stephen can be found in connection with the axes of Fig. 4.2 (see p. 59) and the rectangle of Fig. 3.1 (see p. 42).

The illustrations in the *Liber Mamonis* show that Stephen primarily follows Ibn al-Haytham in the diagrams as well. Nevertheless he modifies and supplements these illustrations whenever his own understanding of the subject requires such changes. In this process, Stephen also combines illustrations from different origins and, if necessary, is able to conceive functional designs by himself. The diagrams in the *Liber Mamonis* thus reveal essentially the same creative pragmatism on Stephen's part as we find in his use of textual source material.

Chapter 5

Stephen's Astronomical Sources and his Lost *Regule Canonis*

5.1 Astronomical Traditions in the *Liber Mamonis*

Many of Stephen's modifications of Ibn al-Haytham's cosmography were designed to adapt the text more closely to astronomical works and methods. Many, but not all, of these efforts focused on the astronomical tables contained in Stephen's *Regule canonis*, which in all likelihood he also encountered in Antioch. Stephen also refers to Ptolemy in certain questions, but disagrees with several of his views and prefers more recent observations. Although Stephen does not tell us which astronomical source(s) he eventually follows, the *Liber Mamonis* contains indications of various astronomical traditions that influenced the work.

5.1.1 Ptolemy's *Almagest*

Ptolemy's three major works on astronomy, the *Almagest*, the *Handy Tables* and the *Hypotheseis*, do not present a uniform theory of the heavens. Regarding the differences between them, Ibn al-Haytham conceived his *On the Configuration* to match particularly the astronomy laid out in the *Almagest*. Therefore, the *Almagest* necessarily influenced the *Liber Mamonis* whenever Stephen translated indicative passages from Ibn al-Haytham. In addition to this implicit dependency,

D. Grupe, *Stephen of Pisa and Antioch: Liber Mamonis*, Sources and Studies in the History of Mathematics and Physical Sciences,
https://doi.org/10.1007/978-3-030-19234-1_5

Stephen's commentary also contained astronomical values which are in agreement with those in the *Almagest*: for example, some of his ascensions.[1]

In view of the prominent role of the *Almagest* in medieval astronomy, it would hardly be surprising if the work also had an immediate influence on Stephen's studies. In his preface to the *Liber Mamonis*, Stephen says that he wants to give an illustration of the 'circles which Ptolemy presents in his *Almagest*' (p. 92:21f.). The close relation between Stephen's writings and ʿAbd al-Masīḥ Wittoniensis' Arabic-Latin translation of the *Almagest* indicates that Ptolemy's work was indeed available, and was studied, in Stephen's milieu. Ptolemy is the only astronomer whom Stephen mentions by name in the *Liber Mamonis* and indeed Stephen shows great respect for him on these occasions. In a strong expression of appreciation, he says that Ptolemy's acknowledged expertise in many fields means that he sets greater store by some of Ptolemy's bolder statements than by the more direct knowledge of later observers (p. 220:5ff.).

Stephen refers to Ptolemy in connection with the definition of the seven climates (p. 218:11f.), the parallaxes of the planets (p. 250:20f.), the shift of the sun's apogee (p. 264:25f.), the speed of the precession (p. 332:2f.), correspondences between the three outer planets (p. 370:1f.), the latitudes of the planets (p. 402:24f.), and the fact that the precession is around the poles of the ecliptic (p. 406:3f.). Traces of a possible influence of Ptolemy's work can also be identified in Stephen's discussion of the solar anomaly and in some other parameters that he cites.[2] Further similarities exist in Stephen's formulation that 'there is no dispute among the ancient philosophers about the order of the outer planets' (p. 248:23; cf. *Alm.* IX, 1) and in his criticism of intuitive judgement if applied to cosmic scales (p. 96:16; cf. *Alm.* I, C7).

Although there are thus numerous references to Ptolemy and the *Almagest* in the *Liber Mamonis*, none of them provides conclusive proof that Stephen actually used Ptolemy's text. In particular, no direct quotations from the *Almagest* are to be found. Moreover, Stephen refers to Ptolemy to support the statement that, at least up to Mars, the order of the planets can be proved by differences between their parallaxes. While this might be inferred from Ptolemy's *Hypotheseis*, in the *Almagest* Ptolemy denies a noticeable parallax of the planets (cf. p. 250:20f. and *Alm.* IX, 1). Similarly, when Stephen defends Ptolemy's mistake of stating a stationary solar apogee, he displays a thorough knowledge of Ptolemy's methods and arguments (cf. p. 264:25ff.). But, while some of these details cannot be read from the *Almagest*, Stephen does not seem to know of Hipparchus' important role in this context, despite the fact that the latter is named in the *Almagest*. In the same context, Stephen's esteem for Ptolemy does not prevent him from supporting the view

---

[1]Cf. above, pp. 36 and 54.

[2]Burnett, 'Transmission', p. 31.

of later astronomers that say that the sun's apogee is moving; but, in this case of open disagreement with Ptolemy, Stephen points out that the erroneous statement in the *Almagest* was not Ptolemy's fault, but a result of inaccurate observations by a certain 'earlier astronomer' ('precessor astronomus', i.e., Hipparchus) whom Ptolemy had had to rely on.[3] Obviously, this conclusion cannot be drawn on the basis of the *Almagest* alone, but requires further assumptions about the motion of the solar apogee. Clearly, then, Stephen had further sources on the subject at his disposal. Following a common practice, it seems, Stephen learnt of Ptolemy's doctrines not by reading the *Almagest* directly, but through later commentaries or from a teacher in Antioch.

5.1.2 Al-Battānī's *Ṣābiʾ Zīj*

Richard Lemay proposed the *Ṣābiʾ Zīj* of the Arabic astronomer al-Battānī (ca. 858–929 CE) as another source that Stephen may have used.[4] Indeed, Stephen's discussion of the lunar phases, including his diagram, has a close correspondence in al-Battānī's work.[5] A possible influence has also been noted in Stephen's discussion of the declinations,[6] and, in a commenting note to the ascensions, Stephen informs the reader that the descension of any degree on the ecliptic equals the ascension of the opposite degree relative to the nearest solstice (cf. above, p. 54). This information is not given by Ptolemy or in *On the Configuration*, but is included in al-Battānī's discussion of the subject.[7]

Stephen is also convinced of the presence of a moving solar apogee, as stated by al-Battānī, and he gives a motion of 1 degree in 106 years as its speed. This may be a corruption of al-Battānī's parameter for the precession, 1 degree in 66 years. Another value in Stephen's text that agrees with al-Battānī's while differing from Ptolemy's is the obliquity of the ecliptic, which Stephen defines as 23;35° instead of Ptolemy's 23;51,20°.[8] Moreover, Stephen determines the maximum latitude of Venus to be 9 degrees (cf. p. 160:27f.), which, he says, was stated by 'other astronomers' in preference to the 6;22 degrees given in the *Almagest* (cf. p. 402:24ff.). Since in the *Liber Mamonis* some of the planetary latitudes appear as rounded variants of traditional values (see below, Table 5.1), Stephen could have obtained his value by rounding up the 8;56 degrees mentioned by al-Battānī.

[3]The discussion of Ptolemy's negligence regarding the solar apogee continued in modern times, e.g., in Manitius' reproach and Toomer's defence of Ptolemy; cf. *Almagest* (tr. Manitius), vol. I, pp. 428–9, and *Almagest* (tr. Toomer), p. 153, note 46.

[4]Richard Lemay, unpublished edition. Lemay assumes that al-Battānī was also the source for Stephen's astronomical tables in the *Regule canonis*.

[5]Cf. p. 304:4f. and al-Battānī (ed. Nallino), pt. III, pp. 92–94.

[6]Cf. above, p. 41.

[7]Al-Battānī (ed. Nallino), pt. III, p. 39.

[8]See pp. 152:17 and 182:17.

In the cases mentioned, Stephen rejects information which is given in the *Almagest* and clearly follows a different source. It is debatable, nonetheless, that this source was al-Battānī. The value of the obliquity as used by Stephen is not originally al-Battānī's, but dates back to the Mumtaḥan astronomers of the early ninth century; since that time, it had been in widespread use. What is more, most of Stephen's values of the ascensions are in agreement with those in the *Almagest* but differ from those of al-Battānī, who recalculated the ascensions based on his different value for the obliquity. Moreover, while al-Battānī also attributes the alternative value of Venus' latitude to the Ptolemaic tradition, Stephen does not seem to know of the Ptolemaic origin of both values.[9] Not only is al-Battānī not mentioned at any point in the *Liber Mamonis*, but Stephen does not even seem to be thinking of a particular person when he refers to the origin of his data; he rather attributes it to an anonymous group ('alii... astronomi/-nomici'; cf. p. 264:22ff., p. 332:2ff., and p. 402:24ff.).

5.1.3 Astronomical Parameters in the *Liber Mamonis*

Table 5.1 displays most of the astronomical parameters that appear in the *Liber Mamonis*, and the corresponding information from Ptolemy's *Almagest*.[10] Numbers in parentheses are not given explicitly in the respective works, but can be determined from related data.

Regarding the obliquity, Stephen follows the Mumtaḥan tradition and al-Battānī. At the same time, his ascensions of 10 and 30 degrees of Aries and the last 10 degrees of Virgo in various climates equal Ptolemy's values in *Alm.* II, 8. This combination of the Mumtaḥan value for the obliquity and Ptolemy's ascensions entails an inconsistency in Stephen's data, as the ascensions depend directly on the assumed obliquity. In addition, Stephen's value of the ascension of 15° of Aries in the fourth climate, 9;24°, stands out, because the corresponding information is found neither in the *Almagest* nor in al-Battānī's *Zīj*. The value agrees precisely with an obliquity of 23;35°, for which it may have been calculated, and not 23;51,20°, like Stephen's remaining ascensions. The value may nevertheless result from Ptolemy's obliquity, through an imprecise interpolation of Ptolemy's ascensions for 10 and 20 degrees of Aries; it appears in the tradition of Ptolemy's *Handy Tables*, where the ascensions from the *Almagest* were interpolated for single degrees.[11] From the *Handy Tables*, the alternative

[9] Cf. Al-Battānī (ed. Nallino), pt. III, p. 175.

[10] Cf. Pedersen, *Survey*, and Neugebauer, *HAMA*.

[11] Here and in the following, references to the *Handy Tables* will be made based on the transliteration of MS Vat. gr. 1291 made by W. Stahlman, *Astronomical Tables*, here pp. 41 and 223.

	Liber Mamonis	*Almagest*
Obliquity	23;35°	23;51,20°
Ascensions		
30° ♈ at 3rd cl.	20;53°	20;53°
30° ♈ at 4th cl.	19;12°	19;12°
30° ♈ at 5th cl.	17;32°	17;32°
10° ♈ at 4th cl.	6;14°	6;14°
last 10° ♍ at 4th cl.	12;6°	12;6°
15° ♈ at 4th cl.	9;24°	(9;22,54°)
Mean solar motion	0;59,8$^{\circ/d}$	0;59,8,17,13,12,31$^{\circ/d}$
Solar apogee -Pos.	28+° ♊	5;24° ♊
-Shift	'some say 1°/106 years'	–
Lunar motions		
1st motion (dragon)	0;3,10$^{\circ/d}$;	0;3$^{\circ/d}$
	18 years 263 days 1 h	(0;3,10,41,6,46,49$^{\circ/d}$)
2nd motion (referent)	11;9,15$^{\circ/d}$;	11;9$^{\circ/d}$
	28 days 16 h 30 min	(11;9,7,43,0,18,39$^{\circ/d}$)
3rd motion (deferent)	24;23$^{\circ/d}$;	24;23$^{\circ/d}$
	14 days 18 h 20 min	(24;22,53,22,40,35,58$^{\circ/d}$)
Lon. (3rd-2nd-1st mo.)	13;10,35$^{\circ/d}$	13;10,34,58,33,30,30$^{\circ/d}$
Elo. (Lon. -mean sol.)	12;11,27$^{\circ/d}$	12;11,26,41,20,17,59$^{\circ/d}$
Epicycle	13;3,54$^{\circ/d}$	13;3,53,56,17,51,59$^{\circ/d}$
Planetary motions		
Mercury (anomaly)	3;6,24$^{\circ/d}$	3;6,24,6,59,35,50$^{\circ/d}$
Venus (anomaly)	0;36,59$^{\circ/d}$	0;36,59,25,53,11,28$^{\circ/d}$
Mars (longitude)	0;31,26$^{\circ/d}$;	0;31,26,36,53,51,33$^{\circ/d}$
	1 year 322 days [4 h?]	
Jupiter (longitude)	0;4,59$^{\circ/d}$;	0;4,59,14,26,46,31$^{\circ/d}$
	11 year 315 days 14 h 29 min	
Saturn (longitude)	0;2,[1?]$^{\circ/d}$;	0;2,0,33,31,28,51$^{\circ/d}$
	29 years *'ed'*d 15 h 24 min	
Latitudes		
Moon	4;46°	4;58° N, 5;0° S
Mercury	4;0°	4;5°
Venus	9°	6;22°
Mars	4;20° N, 7° S	4;21° N, 7;7° S
Jupiter	2°	2;4° N, 2;8° S
Saturn	3°	3;2° N, 3;5° S
Apogee to northpoint		
Mercury	180°	180°
Venus	0°	0°
Mars	0°	0°
Jupiter	−20°	−20°
Saturn	50°	50°

Table 5.1: Astronomical parameters in the *Liber Mamonis* and their equivalents in Ptolemy's *Almagest*

Ptolemaic value of 8;56° for the maximum latitude of Venus, as reported by al-Battānī, can also be obtained, from which Stephen's value of 9° possibly resulted through rounding.[12]

Several other parameters in the *Liber Mamonis* appear as approximations of Ptolemaic numbers. These include the mean solar motion, the planetary motions as well as most of the lunar motions, which are largely rounded or cut-off variants of Ptolemy's values. Stephen's latitudes of the planets also correspond to the ones in the *Handy Tables*, if the latter are rounded to 1/3 of a degree, i.e., 20 arc-minutes. As a consequence, Ptolemy's distinction between different northward and southward inclinations for Jupiter and Saturn is eliminated. However, in this scheme of simplified Ptolemaic values in the *Liber Mamonis* there are some parameters that do not fit: besides the obliquity, these parameters include the position and the shift of the solar apogee, the second lunar motion, and the latitude of the moon, whose appearance among Stephen's data requires a different explanation.

5.1.4 The Mumtaḥan Tradition

The most distinctive of Stephen's parameters is the value of 4;46° for the greatest latitude of the moon. The value is attributed to the Mumtaḥan astronomers in early ninth century Baghdad, but it did not find much acceptance among later astronomers.[13] Apart from Stephen's value for the obliquity, the lunar latitude thus indicates another connection between Stephen's tables and this early Arabic tradition.

A distant dependency of Stephen's tables on the Mumtaḥan tradition would be passably reconcilable with the above findings. The inconsistency in Stephen's ascensions has a parallel in the Arabic manuscripts that contain parts of the Mumtaḥan *Zīj*, i.e., MSS Escorial árabe 927 and Leipzig Vollers 821. In each of these manuscripts an obliquity other than Ptolemy's is assumed, although the ascensions at *sphera recta* correspond to Ptolemy's. Moreover, ascensions at *sphera obliqua*, including in both cases the compiler's latitude, have been calculated for the assumed obliquity: 23;35° in MS Escorial and 24° in MS Leipzig.[14] Similarly, Stephen's ascension for 15° of Aries at the fourth climate may not stem from an interpolation of Ptolemy's values, as in the *Handy Tables*, but from an accurate calculation based on the assumed obliquity of 23;35°, possibly for a user near the fourth climate. Except for the latitude of the moon, the Mumtaḥan astronomers also adopted planetary latitudes from Ptolemy's *Handy Tables*, rounded variants of which seem to underlie Stephen's values; the

[12] Cf. Stahlman, *Astronomical Tables*, pp. 153 and 331f.

[13] Cf. Yaḥyā ibn Abī Manṣūr, *Verified Astronomical Tables*, p. 41. I am grateful to Dr. van Dalen, who drew my attention to the Mumtaḥan tradition concerning this parameter.

[14] van Dalen, 'A Second Manuscript of the Mumtaḥan Zīj', p. 19.

same rounding of these values, i.e. to a third of a degree, is known from the work of al-Farghānī (9th c.), whom Stephen may have followed on this point (cf. p. 402:24ff.).[15]

The *Liber Mamonis* contains a little more information on the arrangement of Stephen's tables, which is compatible with the above assumption. When Stephen discusses the retrogradation of the planets, he gives an example based on Saturn's anomaly at two consecutive positions, 5;55° at 114° and 5;53° at 115° (cf. p. 392:1f.). The same information is found as consecutive entries in the *Handy Tables*, but also in the Escorial manuscript of the Mumtaḥan *Zīj*.[16] Furthermore, Stephen's instructions on the use of his rectification tables match with the arrangement in the Escorial manuscript. The five columns of Stephen's planetary tables as described in the *Liber Mamonis* (cf. p. 378:1f.) agree in their order and purpose with columns 1 to 5 in MS Escorial.[17] Similarly, in the case of the lunar tables, the four columns mentioned by Stephen (cf. p. 296:23f.) correspond in the same order to the first four columns in the Escorial manuscript, whereas different arrangements are found in other traditions.[18]

It is conceivable that Stephen's *Regule canonis* derived from a revised, possibly regional, part of the Mumtaḥan tradition.[19] Stephen may have had this origin in mind when he referred to a group of unnamed astronomers as the source of his information (alii... astronomi; cf. pp. 264:22ff., 332:2ff., and 402:24ff.). This awareness may also have determined his choice of title for the *Liber Mamonis*. The 'Book of al-Maʾmūn' may thus refer to the *Verified* (Ar. 'mumtaḥan') *Zīj for al-Maʾmūn*, whose rules and tables it is meant to illustrate.

---

[15] Yaḥyā ibn Abī Manṣūr, *Verified Astronomical Tables*, pp. 110 and 114; Stahlman, *Astronomical Tables*, pp. 153 and 325–334; Viladrich, 'The planetary latitude tables in the Mumtaḥan Zīj', p. 265; Mozaffari, 'Planetary latitudes in medieval Islamic astronomy', p. 517, note 3. I am grateful to Dr. Mozaffari for having drawn my attention to al-Farghānī.

[16] Cf. Stahlman, *Astronomical Tables*, p. 298, and Yaḥyā ibn Abī Manṣūr, *Verified Astronomical Tables*, p. 52.

[17] Cf. Yaḥyā ibn Abī Manṣūr, *Verified Astronomical Tables*, pp. 49–54. The same correspondence exists with columns 3 to 7 in the *Handy Tables*, and al-Battānī also arranged the columns of his planetary tables in that order; cf. Stahlman, *Astronomical Tables*, pp. 295–300, and, accordingly, al-Battānī (ed. Nallino), pt. II, pp. 108–113.

[18] Yaḥyā ibn Abī Manṣūr, *Verified Astronomical Tables*, pp. 39–44. In *Alm.* V,8, the order of corresponding columns is 1, 4, 2, 3; in al-Battānī (ed. Nallino), pt. II, pp. 78–83, it is 3, 1, 2, 4; in the *Handy Tables* (trl. Stahlman), pp. 249–254, an additional column for the double elongation is prefixed.

[19] A connection between Abraham ibn Ezra's *Pisan Tables* and Stephen of Antioch's astronomical work has occasionally been assumed; cf. Burnett, 'Transmission', p. 36, and Samsó, '«Dixit Abraham Iudaeus»'; see also Mercier, 'The lost zij of al-Sufi in the twelfth century tables for London and Pisa'. However, the astronomical parameters in the different works do not indicate a direct relation.

5.1.5 Situational Influences

Stephen locates the apogee of the sun in his time at 28 degrees and a few arc-minutes of Gemini (see p. 264:22ff.). While errors of up to several degrees in calculated apogee positions were not uncommon, his value corresponds closely with the true position of the sun's apogee during the early second quarter of the twelfth century. This accuracy contrasts with a very inaccurate value for the motion of the apogee, 1 degree in 106 years, which Stephen proposes on the same occasion. Even if one assumes a scribal corruption of an original figure of '66' years in the manuscript, a calculation from the known epochs would still not come close to the position that Stephen mentions.[20] In addition, the apogee position is the only value in Stephen's text which he admits to be not precise ('... and a few arc-minutes'). An explanation for this could be that he had access to a recent determination of the solar apogee which—either incidentally or systematically—was extremely accurate and which he accepted without knowing its foundation. He therefore may not have made an attempt to adjust the apogee position precisely to his own days, but rather "fudged" the value slightly in order to accommodate the short time interval that had lapsed.[21]

While the solar apogee position is specific to Stephen's own time, he also shows a geographic preference, for the fourth climate, at 36;22° of northern latitude. Stephen repeatedly refers to this latitude: for example, in his commentary to passage 177 (see above, p. 36) a value for the fourth climate is placed centrally, while the results for the third and the fifth climates are used to demonstrate the effect of deviating to either side from it. Similarly, in Stephen's commentary to passage 175 (see p. 54), numbers that are distinctive of the fourth climate are presented as if they were of general validity; and, in the case of the ascension of 15° of Aries at *sphera obliqua*, Stephen gives a value which is calculated specifically for the fourth climate, while corresponding values for other latitudes do not seem to be available. In contrast to Stephen's later examples of the ascensions, whose amounts and intervals agree with those in the *Almagest*, the last one seems to have been taken from a different set of ascensions, with a finer step width. Additional sets of this kind are known, for example, from the manuscripts related to the Mumtaḥan tradition, for the composer's own latitude. Since Antioch lies near the fourth climate, Stephen's set of ascensions may have been prepared for an observer in that place.

[20] See e.g. Mozaffari, 'An analysis of medieval solar theories'.

[21] Stephen has no doubt about the position of the apogee in his time, and he also considers Ptolemy's determination of the solar apogee to be accurate (cf. p. 266:4ff.). So it is surprising that he does not use both positions to discuss the apogee's motion independently. Also, the speed of the apogee as given by Stephen does not support his statement that Ptolemy determined the apogee position correctly. This adds to the impression that Stephen did not have a clear opinion about the apogee's motion and did not calculate apogee positions by himself, but rather collected information from different sources.

5.2 Independent Calculations

For most of the values of lunar motions in Table 5.1, Stephen was able to rely on astronomical sources. However, in astronomical works constituent motions are typically determined only as far as necessary for calculating a resulting position; in the *Almagest*, for example, Ptolemy gives only rough approximations of several constituent motions of the moon, although he could have determined them more precisely from his other data (cf. the values in parentheses in Table 5.1). In contrast, Stephen starts from a 'synthetic' depiction of nested spheres and so needs to quantify the motions of all of them. Furthermore, this must be done with a sufficient degree of accuracy to ensure that, after combination, consistency with the observable motions is obtained. To achieve this, as well as other aims, Stephen had to make some calculations of his own.

Ptolemy's astronomical theory for the moon demands symmetry between the moon's second anomaly and the elongation with respect to the sun. The rotation of the deferent's centre, i.e., Stephen's 'second motion', must therefore equal the mean lunar elongation from the sun minus the combination of the mean solar motion and the motion of the dragon. At the same time, the moon's elongation from the sun must be half of the deferent's motion. To meet these requirements, Stephen provides values for the first and the second lunar motions which are more precise than, for example, those mentioned by Ptolemy, who had no need to determine all the constituent motions with equal precision. Nonetheless, Stephen finally admits that even with his numbers the required symmetry of the motions is not fully achieved; there remains a discrepancy of three arc-seconds between the sum of the claimed first and second lunar motions and the mean motion of the sun on one side, which add up to 12;11,33$^{\circ/d}$, and half the third lunar speed on the other side, which would be 12;11,30$^{\circ/d}$. Subtracting the former, combined, motion from the third motion of 24;23$^{\circ/d}$ would thus yield an elongation that is not equal to the combined motion, but slightly lower: 12;11,27$^{\circ/d}$. Stephen tolerates this minor discrepancy, claiming that it is compensated for by 'certain subtleties in the motions' (p. 292:13f.). However, while he describes the deviating value for the elongation as a result of a discrepancy, it is rather this particular value which agrees with the astronomical data (cf. Ptolemy's value in Table 5.1) and which is directly observable. Moreover, the discrepancy in Stephen's numbers results from a systematic mistake, for he quantifies the mean elongation up to arc-seconds, 12;11,27$^{\circ/d}$, whereas the deferent's motion is rounded to arc-minutes, 24;23$^{\circ/d}$, as in Ptolemy's approximation. Although the latter number is not exactly twice the former one, Stephen accepts it as equally accurate. This inconsistency cannot be resolved. Moreover, Stephen seems unaware that in the astronomical theory the deferent's motion is derived from the observable mean elongation, and is exactly twice its size.

Another group of numbers which Stephen seems to have calculated are planetary periods which he provides in correspondence to motions. However, few of these conversions match well. The first lunar motion, 0;3,10° per day, would correspond to a period of 18 years, 251 days and 1 h, but Stephen gives a different period, which would result from a daily motion of ca 0;3,9,40°. There is an even larger discrepancy between the second lunar motion and the given period, while for the third lunar motion Stephen's values are in fair agreement. Similar inconsistencies are also found for the outer planets. In the manuscript the period of Mars is given as '1 year, 322 days and 24 h'. If we assume the number of the hours to be corrupted, as they would make another full day, but read them as '4 h', this would match with the daily motion as given, i.e. cut off after the arc-seconds. This, however, would contrast with Stephen's calculations for Jupiter and Saturn, which suggest that he considered further hexagesimal places: in the case of Jupiter the period in the *Liber Mamonis* corresponds closely to Ptolemy's ca 0;4;59,14$^{\circ/d}$, and Stephen's period of Saturn, though evidently corrupted, contains fractions of a day. This means that he did not use a daily motion of 0;2° for his calculations, as this would equal 1° in 30 days, hence producing an integer amount of days per revolution. A suitable rounding to 0;2,1$^{\circ/d}$, in turn, would mean a period of 29 years, 125 days, 17 h and 50 min, which would still deviate significantly from Stephen's number.[22]

Some of the above problems may result from transmission errors. It is also possible that Stephen did not calculate the periods by himself but took them from another source. However, as in the case of his determining of constituent motions of the moon, his conversions from daily motions to periods provide an indication of the lengths to which he went to familiarise himself with the mathematical side of celestial science. In the discussion of Mercury, Stephen gives an impressive demonstration of his skills in practical geometry. He also intelligently uses the definition of the point in an argument against Macrobius (p. 102:2f.), and, when translating from Ibn al-Haytham the definition of the sphere, Stephen adds that the same definition is well established in the mathematical tradition (p. 94:2f.). When dealing with the definitions of chords and sines in *On the Configuration*, Stephen also expresses his intention to write more on this subject in the future (p. 230:27f.). Regarding his arithmetical skills, the evidence in the *Liber Mamonis* is less consistent. Stephen shows a good understanding of astronomical tables and their use, and he would have been well practised in numerical calculations due to

[22] A rounded daily motion of 0;2,1$^{\circ/d}$ for Saturn would agree with Stephen's habit of giving the planetary motions to a precision of arc-seconds. Moreover, the notation of Saturn's motion in the *Liber Mamonis* does not always end with the arc-minutes. It is sometimes followed by the phrase 'et sec', which in the absence of a number can be understood as 'one' arc-second.

his work as treasurer of the Benedictines in Antioch. Nevertheless, he rarely gives exact results when he associates periods to the mean daily motions of spheres. Although there may be other reasons for this, it would not be surprising that Stephen had difficulties with the less customary operation of dividing in the hexagesimal system.

II Edition and Translation

About the Edition

The edition is based on MS Cambrai, Médiathèque d'Agglomération, A 930, the only known copy of the *Liber Mamonis*. Editorial additions to the Latin text normally appear in angle brackets < >; further emendations are indicated in the apparatus. Typical occurrences of *-ti-* instead of classical *-ci-* (as in 'superfities') have been amended without indication. Punctuation has been inserted as deemed helpful for the understanding of the text. Emendations, descriptions and comments which have been adopted from Richard Lemay's unpublished edition of the same manuscript are indicated by a following '(L)'.

The text in the manuscript shows numerous diligent corrections, which apparently were made by the copyist himself. For describing these corrections, the following abbreviations are used in the apparatus: *add.* (*addidit*)—added; *corr. ex/in* (*correxit ex/in*)—corrected from/into; *del.* (*delevit*)—expunged; *in marg. add.* (*in margine addidit*)—added in the margin.

Hindu-Arabic numerals of the eastern type are transcribed as the letters to which they became assimilated in the manuscript; the numerals of 'four' and 'six', which became assimilated to Tyronian notes, are transcribed as their modern equivalents, '4' and '6'. Computer drawings of the diagrams have been produced and placed in accordance with their relevance to the text rather than to reflect their sometimes far-scattered placing in the manuscript. The diagrams are partly reconstructed and some were given hatching for an easier understanding.

To facilitate a comparison of the *Liber Mamonis* with Ibn al-Haytham's *On the Configuration*, Langermann's numbering of passages in the Arabic text has

D. Grupe, *Stephen of Pisa and Antioch: Liber Mamonis*, Sources and Studies in the History of Mathematics and Physical Sciences,
https://doi.org/10.1007/978-3-030-19234-1

been applied. Moreover, translated passages in Stephen's text that show close agreement with the content of *On the Configuration* are distinguished from his additions to the text by the use of italic type. Short pieces of the Arabic that have no correspondence in the *Liber Mamonis* are added in footnotes to the English translation, in the wording of Langermann's translation. Langermann's sigla have also been used for references to other Latin (L1), Hebrew (H1-H8) and Arabic (Y, K) manuscripts of *On the Configuration*; cf. above, pp. 27ff. Structuring elements in the Cambrai manuscript, especially highlighted initials and paraphs, which occasionally differ from the chapter breaks in the Arabic, are indicated in the edition. Chapter titles have been added in the English translation in imitation of the Arabic, adapted where necessary to Stephen's changes to the contents.

The English translation intends to give an impression of Stephen's particular approach to the subject matter, but also to render his arguments in a readable form. Compromises between these two purposes were unavoidable. No attempt has been made, for example, to provide an English equivalent of Stephen's variations in technical terms when such changes are considered to serve only stylistic purposes; for example, his alternating use of the terms 'orizontis circulus', 'orizonteus circulus', 'circulus orizontalis', etc., which have all been translated as 'circle of the horizon'. Where Stephen's use of synonyms was considered to be possibly meaningful to the subject, for example in his use of 'eclipsis' (mostly in connection with lunar eclipses) and 'defectus' (mostly with solar eclipses), a corresponding distinction in the English has been sought by using 'synonyms' such as 'eclipse' and 'evanescence'. However, this example already shows the limitations of this practice, which tends to obfuscate the simplicity of the original, therefore, it has been used only sparingly.

Conversely, attempts have been made in the English translation to avoid distinctions which are not present in the Latin, for example, in the case of the declination, the obliquity and other forms of inclination or tilt, most of which Stephen denotes with the same term, 'inclinacio'. Nonetheless, several exceptions to this rule have been made, especially where the meaning of an argument was believed to be otherwise critically obscured. This is sometimes the case, for example, with Stephen's indiscriminate use of 'tangere' with any of the meanings 'being tangential to', 'terminating at' or 'intersecting with', even where the argument depends on a distinction between these different meanings. On the other hand, as Stephen uses 'discordia' as an astronomical technical term for the anomaly, the same English translation (anomaly) has been used where Stephen refers by 'discordia' to the planets' periodic variations in latitude. The reader is asked to pardon inconsistencies of this kind in the translation, which result from competing priorities.

Book I

D. Grupe, *Stephen of Pisa and Antioch: Liber Mamonis*, Sources and Studies in the History of Mathematics and Physical Sciences,
https://doi.org/10.1007/978-3-030-19234-1

f. 2r |Incipit liber Mamonis in astronomia a Stephano philosopho translatus.

Quoniam in canonem astronomie, quas proposueramus, regularum exsequto tractatu promissum exsolvimus, secundum hoc opus, licet arduum et subtilissimo ac multiplici nature celatum archano, non inconsulta aut impudenti temeritate sed 5 frequenti et animi et utilitatis ammonitione aggredior. Sit enim licet magnorum super his gravissimorumque disputatio philosophorum, tamen mediocres persepe maxima quemadmodum maiores curant minora. Illud quoque attendendum est plurimum, quod cum omnis a Deo sit sapientia, ea autem verior et sine scrupulo fallatie concessa sit nemo noverit. Unde et qui graves habentur philosophi sepe 10 errasse maximis in rebus, eorundemque verius et perspicatius alios, qui nec philosophiam adepti essent nec ad eam aliquando posse pertingere existimarent, de divini muneris larga benignitate hausisse noticiam comperimur. Testes sunt Plato et Aristotiles, quos omnium liberalium artium fere magistros habemus. Quorum Plato in multis a veritate dissonat, Aristotiles mundum non esse a Deo conditum 15 de nichilo, sed cum eo, sicut nunc est, tamquam cum corpore umbram processisse et condidit et argumentis fallacibus conatur asserere, eo nimirum in loco intellectus et animi et oculorum privatus offitio, quod fidelium simplicitati divina nascitur misericordia. Idem ipse in hac de qua proposita est disputatio questione, cum de celestibus speris dissereret, octo positis de nona non, ut quidam arbitran- 20 tur, consulto tacuit, sed se ad eius noticiam nequaquam pervenisse manifestum nobis reliquit testimonium.

Quod nullatenus arroganter dictum cuipiam videri velim, et quod tante gravitatis et sciencie et ex eisdem auctoritatis adepte philosophus ignorasse dicatur, me non latuerit. Nam etsi inter maximos locum non obtineam, ad eosdem tamen aspi- 25 rans mediocrium invasi disciplinam. Habet enim ille sua quibus plurima consumpta opera perpetuitatis, dum phillosophantes vixerint, nomen adeptus est, quorum tamen pleraque a maioribus, omnia autem a Deo preter obfuscata falsitatis errore accepit.

---

2 exsequto] *corr. ex* exequto (L) **10** errasse] extra se **11** existimarent] existimaverent (L) **14** Aristotiles] Aristotilis **17** quod] qui **24** eosdem] *corr. ex* eos (L)

Here begins the Book of Mamon on astronomy, translated by Stephen the Philosopher.

1.1 <Preface to Book I>

As I have fulfilled my promise by completing the treatise of the announced *Rules for the Canon of Astronomy*, I now begin this second work; and although it is demanding and concealed by the most abstract and complex secret laws of nature, I will not do this in a state of inconsiderate or flagrant temerity but with a conscious mind that stays well aware of the primacy of usefulness. Though this subject may usually be discussed by the greatest and most profound philosophers, nevertheless, very often meaningful contributions are made by humbler minds, just as the great thinkers often also deal with things of minor importance. Above all we have to keep in mind that while all wisdom comes from God, one never knows whether it is given fully true and without a flaw of error. We therefore often see those who are regarded as important philosophers to have failed at the most fundamental questions, whereas the same problems have been discussed with much more understanding and truth by others who were not taught in philosophy nor ever saw a chance to learn it, but who received their knowledge from the great benignity of God's gift. As witnesses take Plato and Aristotle, whom we regard as masters of almost all the liberal arts. Of these men, Plato is wrong in many of his views, whereas Aristotle denies that the earth had been created by God from nothing but states that, as it exists today, it was created simultaneously with God, like a shadow appears together with an object. He tries to prove this by fallacious arguments, but in so doing he is obviously bereft of the sense of intellect, mind and sight, which is given to the believers in their simplicity by God's mercy. On the subject to which this treatise is dedicated, when Aristotle discusses the heavenly spheres and presents eight of them, he did not deliberately remain quiet about the ninth, as some believe; he left us clear evidence that he did not even come to think of its existence.

This shall not appear to be said arrogantly, and I am well aware that my words imply that a philosopher of such seriousness and knowledge and, resulting from these virtues, of such authority was ignorant. For even though I cannot claim a place among the greatest philosophers, nonetheless, aspiring to them I have reached a mediocre learning. He, however, has produced many and much-read works that have given him eternal fame as long as there will be people philosophising; nevertheless, many of his ideas came from his predecessors and everything

Quare nobis quoque, qui nichil aliis derogamus, si quidem idem omnium ditissimus Deus annuat, invideri dedecet, cum ab eo accepta alios docere quam ignavie silentio tegere malumus. Hec autem ideo, quia nisi tanta foret obtrectantium multitudo, ferociores habuisset Latinitas auctores, fertiliorque apud nos philosophie 5 seges pullularet. Cum etenim plurimi essent exercitus detrahentium, pauci qui benigne susciperent, pauciores certe artium scriptores magis exterrebantur multitudinis immanitate, quam adunarentur aliquorum benigno studio. Unde factum est, ut que fere plenitudinem posset habere artium, nunc ceteris gentibus Europa videatur humilior, quippe que quos educat contra fontem scientie, sepius oblatrantes 10 sentit sibi ipsis rebelles, nunc hec, nunc illa, numquam consona ruminantes. Que res tantum attulerit litteralis scientie odium, ut a quibus summe venerari debuerat, rerum publicarum rectoribus, summe odiretur.

Qua ex re illud quoque malum ortum non dubito, quoniam cum equitatis observande causa reges ceterique bonorum ordines, qua et leges constitute sunt, eius 15 autem quedam pars naturalis, quedam sit posita, quarum alterius omnem fere scientiam, alterius maiorem partem, litterali plus exercicio adipiscamur, ignorata semita veritatis qui reges esse debuerant nomen tantum retinentes tirannos se operibus exibebant. Ad consuetudinum enim inconsultam observantiam delapsi f. 2v eadem populum lege iudicant, qua qui induxerunt in usum | tiranni. His in con- 20 suetudinibus et eorum qui eas pepererunt observantia magis avaritie ardor exprimi quam hominum societatis communis utilitas, ut ita dicam, sonniari potest. Earum qui adepti sunt noticiam, apud improbos sapientis nomen impudenter arrogant, dumque eas ubicumque possunt irritant, iuste se iudicasse falso gloriantur. Quapropter veritatis inscii pars avaritie laborant, pars sui profusi inconsiderate 25 sua dilapidant, atque ut habeant que sociis sui furoris largiantur, alios falso iudicio condempnant, aliis vi sua palam extorquent existimantes se in suos liberales videri, si eos quacumque locupletent ratione.

Qua in re duplex malum esse prospicio, quoniam et hii qui eripiunt aliis quo

3 silentio] silentie (L) **16** plus] plurius exercicio] *corr. ex* exercio (L) **20** observantia] observantiam **25–26** iudicio] iudici (L) **26** vi] in

ultimately from God, except what is darkened by error.

Therefore, as we do not deny anyone his achievements which God, most rich in everything, has granted to him, it shall not arouse jealousy when we ourselves prefer spreading to others what we have received from Him rather than concealing it in lethargic silence; especially, as there would have been more courageous writers in the Latin world and the seed of philosophy would flourish more copiously among us, if there was not that huge crowd of critics. There are great armies of antagonists, while only a few people give benevolent support, and so the scientific authors, even fewer in number, get intimidated by the threatening masses rather than uniting in the benign effort of the others. As a result, Europe, who could be in the possession of all scientific knowledge and skills, meanwhile seems inferior to any other peoples. And those men whom she once guided towards the font of knowledge she sees as yapping rivals today, who are at odds with one another and inconsistently repeat different views but never produce anything coherent. This has discredited scholarship so much that those who had to venerate it most, namely the leaders of the states, despise it utterly.

I have no doubt that this was also the origin of that other well-known grievance. As kings and the other respectable classes have been installed for the same reason for which the laws have been enacted, namely to protect a just balance, and one part of the law is the natural law and another part the positive law, but the knowledge of one of these parts is gained almost entirely, and that of the other part to a great extent, through intensive studies of written works, those who should be kings unlearned the way that leads to the truth and preserved merely the name of kings while acting as tyrants. Passively bound to inconsiderate 'observance of consuetudes', they still judge the people by the same law that was brought into use by tyrants. And in these consuetudes, as well as in their observance by those who engendered them, burning greed can be expressed rather than deliberation about the common benefit, so to say, of people's society. And those who did study the consuetudes claim for themselves shamelessly the name of wisemen among the ruthless, and by reciting them at every occasion they erroneously boast that they have judged justly. Ignorant of the truth, some of them suffer from avarice while others squander all their property and, in order to regain capital which they may give to the companions of their fury, condemn some by false judgement while from others they extort it for their own profit violently and in public. In that manner, they think, they will appear generous to their henchmen by making presents to them for any conceivable reason.

I see a twofold mischief in this. Those who raptorially snatch from some what

alios gratificent, rapine vitio societatis humane artissimum vinculum detrahunt, et hii ipsi quibus conferunt aliena turpiter occupant. Que quantum a iusticie liminibus arceantur, facile cuiusvis sane ponderantis omnia nec se ipsum nimium amantis animus deliberat. Nec hoc quidem mirum, si is qui adolescentia, que etas
5 vitia maxime suadet, turpiter transacta turpitudinis usum quippe incorreptus iuventuti intulerit, firmus iam his que consuevit ad naturalem germaneque iusticie veritatem non revertitur. Qui cum a pueritia in philosophie cunabulis enutritus, factus iuvenis puer sibi ipse videatur, tanta est huius virtutis cautio. Quanto enim quis plura de eo novit, tanto plures difficilioresque occurrunt questiones, quarum
10 scrupulositate non animus inquirentis deterretur, sed discendi cupiditas quedam cum iocunditate augetur. Sed de his hactenus. Alterius enim sunt negotii. Verumtamen locus hic paululum attingendus fuit, ut detractionis venenum quantas pariat incommoditates manifestum sit, atque sic corrupti mores detrahere quiescant sive a bonis et simplicibus caveantur.

15 Quoniam autem in canonis regulis multa tetigimus que in hoc opere explicari desiderant, promissum preterire consilium non fuit, ut quod illic dubietatis scrupulus fastidium generaverit, huius operis benefitio sopiatur. Atque hec est ratio que me maxime ad hoc opus coegit, ne aut anxium lectorem a studio repulsum iri paterer, nostratumque utilitati quoad posse consulerem, neve quod pollicitus
20 fueram aut ignorasse aut inertia neglexisse arguerer.

Placet igitur celestium sperarum circulos, numerum, ordinem quo verius potero quantumque humana patitur ratio aperire, ut qui a Ptholomeo in sua sinthasi disponuntur circuli, in speris etiam quomodo possint inveniri, laborantibus in hac arte via teratur. In quo—nichil enim perfectum mihi vel cuiquam ad explicandum
25 concessum arbitror—si quid pretermissum superflueve positum fuerit, sapientium arbitrio corrigendum relinquo.

---

3 facile] *add.* (L) **13–14** sive a bonis] sui ab oris **17** huius] hoc **18** anxium] anexium (L) **25** quid] quidem

they need for alluring others, by this wrongdoing, tear the most solid band of human society, while those who are given take somebody else's property into disgraceful possession. Everyone whose mind is free of immoderate narcissism and who thoughtfully takes into account all implications will see how incompatible this is with the foundations of justice. And it is not surprising that a man who spent his youth, which age is most affectionate to vice, in a dishonourable way, and who unhamperedly retained the disgraceful habits in his early years of manhood, will not return to the natural truth of undepraved justice once he has become strong in the manners that had become his habit; whereas he who was brought up in the cradle of philosophy from boyhood will even as a youth still feel like a boy to himself, such is his decent cautiousness. For the more a person knows about it, the more abstract questions will arise, whose complexity, however, will not frighten off the investigating mind, but a particular eagerness and even delight in learning will be enhanced. But let us not talk about this any longer, as we have other things to discuss. However, some words needed to be said here to make clear what incommodity the poison of criticism can cause, so that corrupted manners shall no longer distract us or may at least be avoided by those who are well-disposed though unexperienced.

Since in the *Rules of the Canon* we have touched upon many aspects that wait for an explanation in this book, I did not intend to ignore the promise, so that whatever displeasure might have come up from obscurity in the other work shall be appeased by the benefit provided by this book. For it was my main concern when writing this book that I shall not accept that a timid reader might get discouraged from studying, but that I give our people a tool as useful as I can, and also that I would not be accused of having forgotten or languidly neglected an earlier promise.

Hence it is our intention to unveil the circles of the heavenly spheres, their number and order, to the best of my own ability and the limitation of the human mind, so that for those who exert themselves in this field a way shall be paved to an understanding of how the circles that Ptolemy presents in his *Almagest* can also be identified on spheres. But in the belief that I, like anyone else, cannot give a perfect explanation of anything, I leave it to the judgement of wise men to correct whatever I have left out in this work or demonstrated too abundantly.

[13] ***Mundus** nomen est ad placitum, per quod omnia fere que condita sunt designantur. Forma eius rotunda atque speralis est. Spera autem*, ut a maioribus accepimus, *corporalis forma est, quam circuit superficies, cuius in medio punctus est, a quo in diversa tendentes si egrediantur plures linee que terminantur in eiusdem spere superficie,*
5 *equalem eandemque omnes longitudinem sortiuntur. Huius puncti nomen* sapientes *centrum spere posuerunt.* [14] *Mundus igitur est solidum corpus, forma eius figura spere, quem circuit una* equalis rotundaque *superficies. In medio enim eius punctus, de quo ad superficiem usque quocumque imaginarias duxeris lineas, equales eas advertes longitudine. Punctus autem hic centrum mundi esse dicitur.* Apparet igitur mundi superficiem
10 equaliter rotunditatis planitie circunductam. *Solidum vero ideo mundum esse posui, quoniam in eo nichil est vacuum, sed corporea plenus est substantia.*

[15] *Ea autem quibus impletur mundus corpora discordia sunt specie, que dividuntur*
f. 3r *in tria,* | *leve, grave et medium.* [16] *Grave est id, quod ab omni parte mundi in centrum eiusdem suo pondere refertur. Leve illud esse dicimus, quod a centri puncto* vel ei proxi-
15 mis locis sua levitate *in exteriora,* scilicet in partem circularem, *defluit.* Circularis autem pars, *que et medium intelligitur, illa est, que nec in centrum nec a centro fugit, sed circa ipsum continuo motu quietis inscia volvitur.*

[17] *Locus igitur gravis partis medius est locus mundi*, medie vero circularis et exterior, horum medium levis pars vendicat. Et prime quidem ratio est, quoniam
20 si quam gravium partem motum habere contigerit, sua vi et pondere *ab omni parte mundi in centrum refertur, ut in suo naturalique loco quietis proprie sedem adipiscatur.* Huius rei Macrobius testis est eo in loco, quo de pluviarum diffluxu in terram disserit, cuius licet pleraque absurde dicta sint, hoc tamen natura duce positum est. *Grave autem inter duas partes distribuitur, quarum altera terra, aqua altera est.*
25 [18] *Et terra quidem atque omnes eius partes, metalla scilicet, animalia et plante, gravia sunt natura. Horum proprius locus est is, qui medius in mundo propinquiorque centro mundi habeatur.* [19] *Si enim horum quelibet pars a naturali loco sublata in levis partis sede, que aer est, relinquatur, proprie nature gravitate in centrum terre tendit. Hoc autem ita est, nisi aliquod* eiusdem nature *obstiterit;* [20] *sed et si adversans res de medio*

---

6 solidum] solidus (L) **15** scilicet] solis (L) **22** diffluxu] *corr. ex* difflusu (L) **26** proprius] proprium (L) is] *corr. ex* ei (L) propinquiorque] *corr. ex* propinquior (L)

1.2 <The Whole World>

Config. 2: The Whole World

[13] *The term 'world' is commonly used to designate generally the entirety of what has been created. Its shape is round and spherical; whereas, a sphere,* as we learn from our ancestors, *is a bodily shape that is enclosed by a surface in whose middle there is a point from which multiple lines that originate and extend in any direction and end at the surface of the sphere happen to be of equal length.* The learned *named this point the 'centre' of the sphere.* [14] *The world, therefore, is a solid body of spherical shape, which is surrounded by a single* and equally round *surface. For there is a point in its centre from which imaginary lines drawn until the surface will all be found to be of equal lengths. And this point is called the 'centre of the world'.* It becomes apparent, therefore, that the surface of the world is shaped in even roundness, equally all around. *And I have said that the world is solid, because there is nothing empty in it, but it is filled completely with corporeal substance.*

[15] *The bodies that fill the world are of unequal type and divided into three sorts; light, heavy and medium.*[1] [16] *The heavy is that which, by its weight, is taken from any place in the world towards the centre of the world. We say that the light is that which, from the centre* or its proximity, *volatilises* by its light weight *towards the periphery*, that is, the circular part. The circular part, however, *which is also considered the medium, is that which tends neither towards the centre nor away from it but rotates around it in an eternal motion, not knowing rest.*

[17] *Hence the place of the heavy part is the central region of the world*, that of the medium is the outer and circular region, whereas the light part takes the place between them. The reason for the first is that whenever any part of the heavy happens to have a motion, *it is driven from any place of the world towards the centre* by its own force and weight *to reach its natural position of rest.* This has been proved by Macrobius where he discusses the falling of the rain to the earth[2]; and even though most of his arguments are absurd, in this case his reasoning follows nature. *The heavy, however, is divided into two parts, one of which is earth and the other one water.* [18] *The earth and all its parts like metals, animals, and plants are heavy by nature, thus their natural position is in the middle of the world and in the proximity of its centre.* [19] *For if any of those parts has been lifted from its natural place into the place of the light part, which is the air, once it is no longer suspended it will by the heaviness of its nature*[3]

---

[1] 'medium': Arab., 'neither heavy nor light.'

[2] *Somn.* I,22,4–8 (L).

[3] 'by the heaviness of its nature': Arab., 'by the natural power which it has'; L1: 'virtute naturali'.

auferatur, grave statim in sua relabitur.

Ea autem que partes terre posita sunt habent differentiam. Graviora enim metallica, que et ipsa ab invicem distant; minus gravia plante; medium quorum animalia obtinent. His quippe super aquam depositis metalla citissime, animalia quo-
5 que, nisi artificio detineantur, sed non eodem pondere in profundum ferantur, plante desuper natant. Hec non solum ratione sed oculorum visu manuumque tactu, quibus facilius animus assensum prebet, probantur. Eadem ratio est aque omniumque ipsius partium. Et mihi quidem omnis aqua eiusdem videtur esse ponderis natura, sed si que sint aliis graviores, hoc aliquibus fieri accidentibus, quibus infecte
10 in terra qua oriuntur aut discurrunt parum alterantur.

[21] *Est autem forma terre similis spere, cuius superficies tametsi rotunda sit non tamen plana. Quibusdam nempe locis altior iuxta eosdem se videt inferiorem, que res nichil nos conturbare debet formam terre non esse speralem suspicari.* Montes etenim quos sublimes et valles quas aimus profundas, *ad magnitudinem terre comparatis*
15 *asperitatem quandam terre spere non sublime vel profundum quid esse deprehendes.* Homo namque ad sue magnitudinis mensuram cetera iudicans, que sui quantitatem excedunt magna, que infra sunt parva habet. Philosophorum ratio alia est, quorum iuditio terreum corpus centri magnitudinem parum excedit ad quantitatem tocius mundi comparatam. Nichil igitur cuiquam mirum videri decet, si montium quan-
20 titatem non magnum quid sed asperitatem terre dixerim supra, que unum est IIII^{or} elementorum, parvissima dicitur. [22] *Terra itaque forma speralis est, cuius centrum idem est quod et mundi, in medio eius posita, durum corpus et immobile.*

Cui rei quidam extasim patientes contrariam senserunt. Celum enim quasi, quod omnia clauderet et choerceret, fixum et immobilem ponentes, terram circum-
25 volvi certo motu putabant. Quibus tametsi ponderum natura resistat, sunt tamen

---

24 fixum] fixit (L)

move towards the centre of the earth. This happens, unless something of the same nature *comes in its way*[4]; [20] *and if the obstructing object is put out of the way, the heavy will immediately fall to its proper place.*[5]

But there is variety among the above mentioned parts of the earth. For metals are heavier, showing diversity also among themselves, whereas plants are less heavy, and the animals stand in between. Thus, if one puts each of these on water, the metals will sink to the ground very fast, and also animals will sink, though with less weight, unless they prevent it artificially, whereas plants will swim on the surface. This is not only proved by reasoning but also by our senses of vision and touch, through which the human mind is convinced most readily. The same reasoning holds for water and all its parts, although I have the impression that every water has the same natural weight and, if there are some waters heavier than others, that this is due to certain influences which nevertheless have little effect on the springing or flowing of the affected waters on earth.

[21] *The shape of the earth is similar to a sphere whose surface, though round, is not perfectly even. For in some places it is elevated, while immediately next to these places there are depressions*[6]*; but this should not keep us from assuming the earth to be spherical.* For although we call mountains 'high' and valleys 'deep', *you will not consider them as such once you compare them to the size of the earth, which makes them a mere roughness on the sphere of the earth.*[7] For man judges the magnitude of objects in relation to his own size, as big if they are bigger than him, and small if they are smaller.[8] But the philosophers have a different understanding, according to which the body of the earth exceeds its centre by a small amount compared to the size of the world. Nobody should therefore be surprised that I said above that the size of the mountains is nothing great but a surface roughness of the earth, which is one of the four elements and said to be the smallest. [22] *The earth is thus spherical in shape and its centre is the same as that of the world, and it is a firm and unmoved body.*

But some people, apparently suffering from lunacy, thought it the other way round. Assuming that the heaven, which encircles and encloses everything, was fixed and unmoved, they believed that the earth was revolving with a certain mo-

[4]Arab. add.: 'It then stops since there is no path for the motion.'

[5]Arab. add.: 'If it were possible to do this forever, it would continue to move until it reached [the place] where the centre of the world would be in its middle. So it is for every particle of water.' Stephen refers to this scenario at the end of his longer commentary following passage 22.

[6]Arab. add.: 'which come about due to the effect of the motions of the heavenly bodies.'

[7]Arab. add.: 'like the roughness which occurs in the surface of some small spheres.'

[8]Ptolemy makes a similar critique of intuitive judgement, in *Alm.* I,7.

alie plures firmissimeque rationes que eorum falsissimam sentenciam comprobant. Harum multitudine, cum breves esse proposuerimus, longissimi videamur supersedentes, communis iste error conteratur.

Omne quod certum habet circularem nec fortuitum motum, volvitur in duobus 5 fixis punctis, quos polos appellant. Hoc ita consequens est, quod si tollas postremum, perit primum. Si enim quelibet spera citissime nunc in aliis, nunc in diversis polis volvatur, quelibet eius partes in affinium partium circulos transeunt, nec earum aliqua certum habet in quo semper volvatur circulum. Incertus igitur spere f. 3v motus est. Constat ergo, si | certus sit in circulum motus, ipsum in duobus fixis 10 fieri polis. Terra autem, si movetur, certum habet circularem motum. Nichil enim a Deo conditum fortuitum, sed omnia certa lege cohercentur. Quod si fortuitu volvi dicatur, illud iam inconveniens oriri videbimus, quod eadem terre pars nunc circulo recto, nunc polis subiecta videretur, quod nequaquam fieri omnes sentimus. Certo igitur cum circunferatur motu, duos habet in quibus volvitur fixos polos. 15 Hii, si non movetur celum, ut illi sonniant, semper eidem parti celi suppositi sunt. Omnes autem poncti in spera positi, que in duobus fixis polis circumvolvitur, motu spere faciunt paralellos circulos, a quibus nunquam in quamlibet partem inclinantur. Punctis itaque quibuslibet terre hoc idem concedi necesse est. Quemvis igitur circulum horum, si imaginatione crescentem duxeris, totum ut dividat mundum, 20 faciet in celo circulum, in quo positas stellas semper videt in eandem subiectionis reversa motu terre lineam pars, que circulum circunduxit. Hoc non ita esse in sequentibus lucidissime apparebit, cum de motu celestium sperarum disputabitur.

Nunc susceptam controversiam ob simplicitatem introeuntium quam levius poterimus exsolvemus. Dies omnes non esse equales vel simplicibus patet, tantoque 25 eos augmentari, quanto vicinior sit sol septemtrionali polo, quem esse organum diei omnis scriptura testatur. Solem autem in celi spera fixum esse accepimus celumque non moveri. Cum igitur et celum fixum sit et sol in celo fixus, in eodem circulo semper invenietur quem ei supposita pars terre motu sui corporis facere positum est. Quod si ita est, equales in eodem loco omnes dies sunt, atque minor

4 certum] centrum (L) **5** polos] palos (L) **14** polos] palos (L) **18** Quemvis] Quamvis (L) **25** quem] quam **28** quem] quam motu] motum

tion. Although the nature of the weights would refute them, there are also many other strong reasons that prove the faultiness of their judgement. The multitude of these reasons, which we skip because they are so many that against our intention to be concise we would appear tedious, eradicates that widespread error.

All that has a regular circular and non-accidental motion rotates about two fixed points called 'poles'. This consequence is necessary such that, if you deny the latter, the former will fall. For if a sphere rotated about always changing sets of poles, every part of it would cross the circular tracks of neighbouring parts and none of them would have a particular circle on which it always revolves; hence the motion of the sphere is irregular. It is therefore clear that whenever there is a certain circular motion, it takes place about two fixed poles.[9] But the earth, if it moves, has a certain circular motion. For God created nothing accidental, but everything is bound by a certain rule. If one said that the earth was moved randomly, we would witness the absurd phenomenon that the same place of the earth would in one moment appear on the equator and in another moment under the poles. But we all perceive that this does not occur. Therefore, as it revolves with a certain motion, it has two fixed poles about which it revolves. And if the heaven does not move, like those people fantasise, the poles are always placed under the same part of the heaven. But all the points on a sphere that rotates around two fixed poles describe parallel circles by the sphere's motion, from which they never incline in any direction. Hence we must say the same for every point on the earth. Consequently, each one of these circles, if you imagine it extended such that it divides the entire world, will produce a circle in the heaven; and when the motion of the earth makes that point which has defined that circle return to the same position under a straight line, it would thereby always face the stars that are located on that circle. However, that this is not the case will become totally clear later, when the motion of the heavenly spheres will be discussed.

For the inexperience of beginners, we will now solve the started controversy as simply as possible. Even to the untaught it is evident that not all days are equal, but that their length increases the closer the sun comes to the north pole, which [*scil.* the sun] in all writings is unanimously shown to be the organ of the day. But we have assumed that the sun was fixed in the heavenly sphere and that the heaven was unmoved. Then, as the heaven is fixed and as the sun is fixed in the heaven, the sun would always be found on the same circle, which we have said will be

[9]The argument will also play a role in Book IV, in Stephen's criticism of Ibn al-Haytham's model of Mercury, starting on p. 340:6.

maiorve in diversis per latum est locis. Hoc autem non sic est. Nam cum terra, ut illi aiunt, certo cursu volvatur celo fixo, et sol in celo fixus sit, aliquando propinquior alteri polo in alto parallello, aliquando medius invenitur. Quamobrem maiores modo, modo minores dies fiunt, aut fixum non esse solem in celo respondeant necesse est aut celum moveri concedant. Solem non esse fixum si velint dicere, id quam puerile sit quamque absurdum non intelligunt celum esse liquidum, in quo tamquam in aqua alterius materiei corpus huc atque illuc defluat. Quod cum absonum sit, in celo fixum esse solem confiteantur oportet. Moveri igitur celum necesse est. Quod autem terra non moveatur, ponderum ratione facile probabitur, que superius posita est. Motus enim gravium in centrum, levium a centro non circa, mediorum non in centrum nec a centro, sed circa centrum est.

Importune enim Macrobius et terram centrum et, quia centrum sit, non moveri dixit; nam si constaret primum, et secundum. Verum est enim in omni certum habente spera motum centrum non moveri. Igitur si terra centrum est, verissime quia centrum est, non movetur. Quod si centrum non est, falsum est, quia centrum est, immobilem esse. Satis itaque nobis erit terram non esse centrum ostendisse, quia hoc ostenso fallacem videre licet Macrobii consequentiam 'quia centrum est terra, non movetur.'

¶Terram esse centrum mundi multis contenditur rationibus, quibus nulla obici potest calumpnia. Earum partem non modicam ex ipsius Macrobii disputationis serie hauriri licet, pars vero ex spere diffinitione colligitur. Prius ergo que in se ipse Macrobius argumentorum semina proferat videamus; post quemadmodum per diffinitionem comprobentur inducemus.

Terram globum esse et omnium philosophantium sentencia est, et Macrobius testatur. Globus autem omnis aut rotundus est aut angulatus. Si rotundus, spera est. Terra vero rotundus est globus, spera igitur est, non ergo centrum. Neque fieri

---

1 per latum] pellatum **13** certum] centro (L) **21** serie] ferie (L)

produced by the earth's motion by a point on the earth that is placed below the sun. If this was true, all days would be equal for one and the same place, while being longer or shorter at different places only according to the latitude. But this is not the case. For if the earth, as they say, revolves with a certain course while the heaven is fixed, and if the sun is fixed in the heaven, we nevertheless see it sometimes on a parallel that is closer to one of the poles and sometimes midway between the poles. While this causes the periodical increase and decrease of the day-length, those people either have to admit that the sun is not fixed in the heaven or that the heaven is in motion. If they prefer to say that the sun is not fixed, they do not understand how immature and absurd it is to imagine the heaven like a liquid in which a body of a different substance could float around like in water. Since this is absurd, they need to admit that the sun is fixed in the heaven. Therefore, the heaven must move. But that the earth is unmoved will be easily demonstrated from the reasoning about the weights which has been given above. For the motion of the heavy bodies is towards the centre, that of the light bodies away from it but not around it, whereas that of the medium bodies is neither towards the centre nor away from the centre but around the centre.

When Macrobius said that the earth was the centre and that, because it is the centre, it did not move, he did so inappropriately.[10] For if the first statement was true, then necessarily also the second one; because it is a fact that in every sphere with a constant [rotating] motion the centre does not move. Thus, if the earth is the centre, it is true that 'because it is the centre, it does not move'; whereas, if it is not the centre, it is wrong to say that 'because it is the centre, it is unmoved.' Hence we only need to show that the earth is not the centre, for then Macrobius' conclusion that 'because the earth is the centre, it does not move' will be refuted.

That the earth might be the centre of the world can be objected to with many arguments, none of which can be blamed as polemic. A great number of them can be taken from Macrobius' own line of arguments and others from the definition of a sphere. First we will consider the clues for arguments that Macrobius himself brings forward, and then we will show how they are supported by the definition.

All those who philosophise agree, and it is also affirmed by Macrobius, that the earth is an aggregated body.[11] Meanwhile, every aggregated body is either round or angular. If it is round, it is a sphere. And since the earth is a round body, it is a

[10] *Somn.* I,19,11: 'quia in sphaera quae volvitur nihil manet immobile praeter centrum, mundanae autem spherae terra centrum est, ideo sola immobilis perseverat.' Similar, *Somn.* I,22,3–4 (L).

[11] *Somn.* II,5,5.

potest, ut ex sui densitate corporis sit et alterius centrum. Amplius centrum omne
f. 4r punctus est. Punctus | omnis simplex quoddam est et incompositum, in superficie minima linee principalis pars. In terra autem multe sunt linee et superficies, compositum namque corpus est. Globus enim est, non igitur centrum. Amplius
5 quoque omne centrum indivisibile est; punctus enim est. Macrobius vero terram in V[que] zonas quibus et celum dividi confitetur. Non igitur terra centrum est.

Sed sunt quidam qui fallaci propulsant a Macrobio iniuriam calumpnia. Aiunt namque ipsum non centrum mundi sed tamquam centrum mundi in medio locatam dixisse. Quod nostre cavillationis approbationi quemadmodum et eorum defensio-
10 ni accomodari potest, quasi neglectum transimus, eadem illis verba respondentes quibus Macrobius terram non moveri comprobat dicens: 'non movetur, ait, est enim centrum.' Quibus in verbis non terram in medio mundi locatam, sed manifestissima predicatione centrum esse pronuntiat. Nam de eius locatione statim subiungit: 'est et infima, recteque hoc modo; nam quod centrum est, medium est.'
15 Quibus ex verbis manifeste datur intelligi eum non terram ad similitudinem centri, sed quia centrum sit, in medio mundi locatam sensisse. Ad hec autem solis orbem magnitudine terram quantitate excellere argumentans, solem certam partem in quo decurrit circuli obtinere cum dixisset, terram ad eiusdem circuli comparatam latitudinem punctum, 'quia pars esse non potest' subdidit. Scito hoc quidem,
20 quasi illud aliter probari non possit, nisi suam terre deroget quantitatem. Quam etenim frivolum sit, facili patet argumento. Circulus quippe, qui solarem semper novit motum, non in latiori est, sed infra quinque speras continetur. Terram autem quandam partem totius mundi esse omnes qui sano sunt cerebro predicant, nec falluntur, est etenim. Multo igitur magis partis mundi pars quedam esse poterit.
25 Insanum est enim dicere, si toto celo in duo media diviso, quia eorum nemini terra equalis esse poterit, punctum esse. Atque hec sunt que de Macrobiana oratione

18 circuli] circulus **22** speras] *corr. ex* peras (L) **23** esse] *add.* (L)

sphere and, therefore, not the centre. It is also impossible that the centre consists of its own as well as another body's density.[12] Furthermore, every centre is a point. Yet, every point is something simple and non-composite; in a plane it is the smallest elementary part of a line. But on earth there are many lines and planes, because it is a composite body. For it is an aggregated body and, therefore, not the centre. Furthermore, as every centre is a point, it is indivisible. But Macrobius himself admits that the earth is 'divided into the same five zones as the heaven.'[13] The earth is therefore not a centre.

However, there are some who defend Macrobius with fallacious arguments, saying that he had not claimed that the earth was *the* centre of the world, but that it was located in the middle *like* the centre of the world.[14] We pass over the fact that this can be used just as well in support of our criticism as in their defence and rather recite for them the very words by which Macrobius tries to prove that the earth does not move, saying: «'It does not move,' he [*scil.* Cicero] says, because it is the centre.»[15] With these words Macrobius does not say 'the earth is located in the middle of the world' but states in clearest speech that the earth 'is the centre.' Moreover, concerning its location he immediately adds: «'and it is the lowermost,' which is also true, because a centre is always the middle.»[16] From these words we clearly understand that Macrobius did not think of the earth as a mere analogue of the centre but as being located in the middle of the world 'because it is its centre.' And when he argues that the sun's disc exceeds the earth in size by saying that the sun occupies a measurable arc of the circle on which it revolves, whereas the earth compared to that circle's width is merely a point, he adds «as it cannot even be a fraction of it.»[17] Just as if the former [*scil.* the larger size of the sun] could not be proved, unless he denies the earth any extension at all. The ridiculousness of this is shown by a simple argument. The circle on which the sun is perpetually moved is not on the widest sphere but encircled by further five spheres. But all who have a healthy brain correctly affirm that the earth is a certain part of the world, because

[12]Alternative translations: '...that the earth consists of its own body's density and is also the centre of another one,' or 'that the earth exists outside its own body's density to be the centre of another one,' or 'that the centre exists outside its own as well as another body's density.'

[13]*Somn.* II,5,7 (L). See also *Somn.* II,7,1–9.

[14]Cf. *Somn.* I,16,12: 'terra ipsa in punctum quasi vere iam postrema deficiat;' *Somn.* II,9,9: 'omnis terra [...] ad quemvis caelestem circulum quasi centron puncti obtinet locum;' and, *Somn.* II,5,10: 'huius <*scil.* terrae> igitur ad caelum brevitas, cui punctum est, ad nos vero immensa globositas.'

[15]*Somn.* I,22,3 (L).

[16]*Somn.* I,22,4 (L), with a slightly different wording.

[17]*Somn.* I,16,11 (L).

pauca de multis adversum ipsum hausimus. Sufficiunt enim istec nostrorum que sequentur validissimo subnixa testimonio, que sunt huiusmodi.

Speram esse rotundum corpus in medioque eius centrum, a quo in diversa ducte usque speralem superficiem linee equales essent longitudine, superius positum est. Que si, ut est vera diffinitio spere, esse ponitur, hoc ideo in mundo, cum sperale corpus sit, provenire necesse est. Habet ergo centrum, a quo ducte usque ad eius superficiem linee pari protrahuntur quantitate.

Hic si terram ut apud Macrobium esse dixerimus, dispares lineas ad superficiem usque ducemus ut, verbi gratia, diamethros terre a capite ad superficiem alte spere linea ducta brevior; longiores conspicit lineas, que ab alterius diametri capitibus super hoc rectis angulis positi egresse punctum, quam prior in superficie celi tetigerint. Habet enim teste Macrobio terra diametrum, lineam scilicet. Quod ut manifestius fit, descriptionis formulam subitio, ut quod ratio exequitur oculorum sensus comprobet.

Sit enim alte spere circulus ABCD, in medio huius quam minima loco centri terra posita, ubi tamen diametrum habere possit, et super eam EFGH. Ducaturque diametrum terre ab E usque G, et recto superpositum angulo alterum ab F usque H. Ducatur igitur linea ab F usque B, <altera ab E usque B,> et tertia ex altero latere ab G usque B. Eodemque modo in reliquis fiat, ut ab unaquaque diametri terre littera IIIes egrediantur linee, quarum media in oppositam litteram alte spere circuli, due relique in altrinsecus positas finiantur. Statim ergo, quod ratio premonstraverat, oculis verum deprehendes longiorem esse lineam EB sive GB linea FB. Idemque in ceteris facile erit videri. Quantum | itaque vis terram brevitate arces, ita tamen ut diametro insigniri possit, celumque quantumvis spaciositatis immensitate dilates, eadem semper occurret linearum alteratio, quod docet subiecta descriptio.

8 dixerimus] *corr. ex* disserimus (L) **17** H] B (L) **18** B] H (L)

it is so. Hence it can even more be a part of [what itself is only] a part of the world. It is therefore nonsense to say that, if the entire heaven is divided in two halves, as the earth cannot be equal to either of them, it is a point. And these are only a few arguments which we have chosen from among many in Macrobius' work to be used against him. For sufficient are our own arguments below, based on hard and solid proof, which are as follows.

It has been said above that a sphere is a round body with a centre in its middle, from which lines that are drawn until the surface in different directions are of equal length. If we accept this as the correct definition of the sphere, the same must also apply to the world, as it is a spherical body. The world therefore has a centre, from which lines that are drawn until its surface are of equal lengths.

If we now assume the earth like Macrobius did, we will draw unequal lines to the surface [of the heaven] so that the line that is drawn from the end of a *diametros*, so to say, of the earth until the surface of the high sphere will be shorter, while those lines that touch the same point on the surface of the heaven like the first but originate from the ends of another diameter which lies at a right angle to the first one will seem longer. For Macrobius testifies that the earth has a diameter, which means a line.[18] The following illustration shall make this clearer, so the sense of the eyes may confirm what reason has revealed.

[Fig. 1.1] Let ABCD be the circle of the high sphere; in its middle let the earth be located, at the centre and as small as possible but still with a certain diameter, and on it EFGH. Let us draw a diameter of the earth from E until G and another one, crossing the former at a right angle, from F until H. Then let us draw a line from F until B, another one from E until B and, on the other side, a third one from G until B. Let the same be done for the other points such that from each letter on a diameter through the earth three lines will emanate, with the one in the middle running until the letter opposite on the circle of the high sphere and the other two until those at the sides. Then you can grasp with your eyes what reason had already revealed, that the lines EB and GB are longer than the line FB. The same is easily seen for the other lines. Therefore, as long as a diameter can be ascribed to the earth, you can compress the earth and increase the heaven by an unmeasurable expanse as much as you want, there will always occur that difference between the lines, as is demonstrated by the diagram below.

[18] *Somn.* I,20,20.

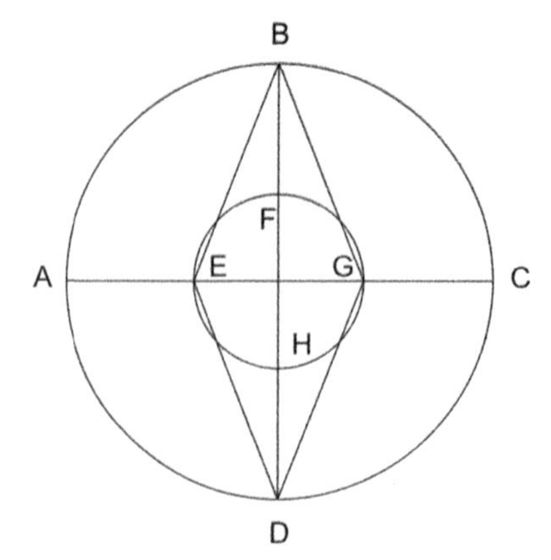

Fig. 1.1: From MS Cambrai 930, fol. 4v. According to the description in the text, four further lines should be drawn, two from each of the points F and H to A and C

Aut igitur falsum est terram centrum mundi esse, aut diffinitio fallax est. Sed diffinitio vera est, neque enim verior poterit inveniri. Falsum est ergo terram esse centrum mundi. Quod si centrum non est, nec 'quia centrum est, non movetur,' verum est; non enim centrum est.

5 Alia igitur immobilitatis eius ratio querenda est, sed neque illa probanda est, cui ipsum quoque Macrobium assentire video, terram undique et aeris et spiritus densitate fulciri, eaque lege cohibitam moveri non posse. Elementorum enim tria sunt liquida, aer, aqua et ignis; quartum terra durum est. Omne autem durum grave, liquidumque aut leve aut minus grave. Leve liquidum: ignis, aer. Aer igitur 10 liquidus est et levis, terra dura et gravis. Durum autem omne sui gravitate ponderis in quavis parte aeris si relinquatur, liquidi partes desecat et in ima defertur, nec coherceri durum aliquando poterit. Nec aer igitur, cum terre circumfusus sit, sua densitate terram a motu cohibere poterit. Neque enim aeris densitas terream densitatem equare poterit, cum hec ipsa aeris densitas, si qua est, terre fumo et 15 aque fiat.

¶Verissima autem et nichil fallens immobilis constantie terre ratio de ponderum natura, que superius posita est, facili poterit elimari argumento. Omnium corporum que mundi implent magnitudinem trifaria facta est divisio in levia, gra-

17 elimari] eliminari (L)

Thus, it is either wrong to say that the earth is the centre of the world or the definition [of the sphere] is false. But the definition is correct and there cannot be found a more adequate. Hence it is wrong to say that the earth is the centre of the world; and if it is not the centre, it is also untrue that 'because it is the centre, it does not move'; because it is not the centre.

We therefore need to find another explanation for its immobility; but one also cannot approve of that explanation which again Macrobius is found to affirm, namely that the earth is supported from every direction by the density of air and breeze and, thus enclosed, cannot move.[19] For there are three liquid elements, air, water and fire, whereas earth, being the fourth one, is hard. But everything hard is heavy, while everything liquid is either light or less heavy. The light liquids are fire and air. Air is thus liquid and light, whereas earth is hard and heavy. But everything hard, when released at any place in the air, will by its heaviness cut through the parts of the liquid and fall to the depth, and the hard can never be bound. Hence also the air that is flowing around the earth cannot 'by its density' prevent the earth from moving. For the density of air can never be equal to the density of earth, as that density of air, given there is any, stems from fog of earth and water.

However, the truest and flawless reason of the earth's unmovable constancy can be easily derived from the above described nature of the weights. There has been made a three-fold division of all bodies that fill up the wide space of the

[19] *Somn.* I,22,7.

via et media. Grave illud esse positum est, quod ab omni parte mundi in ima eius, id est centrum, descendit, levia, que a centro fugientia, quod locus est gravium, in exteriora et altiora feruntur, media, que equali semper a centro distantia spatio perpetuo motu circumvolvuntur. Idcirco enim gravia intelliguntur esse, quia in
5 ima, levia, quia ab imis, media, quia nec in ima nec ab imis, sed circa equaliter feruntur. Cum igitur tria sint eaque III[es] habeant non permixtos sed particulares motus, gravia in centro, levia a centro, media circa, sola gravia leviaque non moveri sed in propriis posita sint sedibus, manifestissima colligitur ratione, media autem nunquam motu carere. Quo enim moverentur gravia, si in centro vel pro-
10 ximis posita sint locis eorumque motus in centrum sit, levia vero in levium sede eorumque motus ad alta contendat, non video, nisi propriam alterius vie iubeantur variare naturam. At vero gravium leviumque media, quorum motus circa est, in propria sede locata nunquam a motu quiescunt, circa centrum etenim moveri habent. Terra autem et gravissima est et in medio mundi locata, qui locus est
15 gravium. Cum igitur gravis ad gravium locum usque descenderit, motusque eius in ima et non ab imis nec circa sit, qua ratione quove moveatur motu, si motum habet, non video. Neque enim descendere amplius quam in centrum potest, cum media eius pars in centro sit, reliqua eius ipsi tamquam duro corpori ipsisque alie usque superficiem terre superponantur. Fixa igitur est, nec aliquo movetur terra.

20 Atque hec eadem ponderum ratio, ut proxime innuimus, celum, quod medium est levium graviumque, incessanti continuoque motu circumvolvi vel pertinaciter negante comprobat. Quare et terra immobilis est et celum semper volvitur, falsaque est eorum opinio qui terram circumvolvi, celum non moveri sed fixum esse arbitrabantur. Quibus quoniam satis responsum est, iam nunc ad alia pergamus.

f. 5r 25 [23] | ¶*Gravium* minus grave *elementum aqua est, que circa terram diffunditur quanto propinquioribus potest centro mundi locis. Cum enim natura sint gravia ea que in centrum ferantur, terra* gravius et solidum corpus cui, ut diximus, proprius est locus centro propinquior, minus gravi *aque, ne in centrum descendat, obsistit.* [24] *Quia vero terra alias altior, alias est profunda, elementum aque suo pondere in centrum de-*

---

6 particulares] circulares **11** eorumque] eorum **16** nec] sed

world, into light, heavy and medium. We have said that the heavy is defined as that which falls from any place in the world towards its depth, i.e. the centre, and the light bodies, as those which strive away from the centre, which is the place of the heavy ones, and rise to the outer and heigher regions, and the medium bodies, as those which revolve by a perennial motion at an always equal distance from the centre. Accordingly they are identified, as heavy if they tend towards the depth, as light if they tend away from the depth, and as medium if they tend neither towards nor away from the depth but equally around it. Hence, as there are three sorts and as they have not mixed but particular motions, that is, the heavy towards the centre, the light away from the centre, and the medium around it, it is most clearly understood that only the heavy and the light do not move but have their proper stationary places whereas the medium never come to rest. For I do not see where heavy bodies that are located at the centre or its proximity, while their motion is towards the centre, and light bodies that are located in the natural place of the light bodies, while their motion is towards the height, should move unless they are forced to change their own nature to that of another sort. But the medium between heavy and light, whose motion is around, once put in their proper place, will never stop moving, because they have a motion around the centre. Now, the earth is extremely heavy and located in the middle of the world, which is the place of the heavy bodies. And as something heavy always descends to the place of the heavy bodies and as its motion is towards the depth and not away from it or around it, I cannot see why it should move, nor in which direction if it has a motion. For it cannot move further down than to the centre; and when its mean part is at the centre, its remaining parts will be laid upon it like on a solid body and also upon one another, until these parts make up the surface of the earth. Therefore, the earth is fixed and does not move anywhere.

That same reason of the weights, as we have just indicated, proves even against stubborn objections that the heaven, being the medium between the light and the heavy, revolves in a tireless and steady motion; and why the earth is unmoved whereas the heaven is always revolving, and that those are wrong who believe that the earth revolves while the heaven is unmoved and fixed. As we have given them enough in response, we shall now proceed to other things.

[23] *Among the heavy elements, water* is the less heavy one, *washing around the earth at those places where it comes as close as possible to the centre of the world. For, as those elements are heavy by their nature which are driven towards the centre and as the earth* is a heavier and a solid body which takes as its place the region closest to the centre, *it obstructs the way for the* less heavy *water to descend to the centre.*

scendens in imis terre consedit. Sicque factum est, ut altitudo terrei corporis vacuam aqua non minimam eius partem efficiens certis aquaticum liquorem finibus cohiberet. Quod mirabili conditoris providentia factum esse non dubito, qui cum creaturum se animalium naturas presciret, quorum maxima ex parte aeris temperie nutriretur vita, locumque quieti eorum solidum oporteret esse corpus quadam in parte terre speram exaltans, in alia humilians, in humilem aquam undique precepit confluere, ut animantibus, quorum spiritus aqua suffocatur, locus esset in quo et viverent et vite alimenta invenirent. Licet eadem nobis de montium valliumque productione sana mente interpretari. Ad usus enim animalium viteque oportunitatem conditi sunt. [25] *Manifestum autem ex his est omnem, tam hominum quam animalium reliquorum,* piscium excepto genere, *habitacionem insulam esse,* quod nobis tamen propter sui magnitudinem possibile minime videtur.

[26] *Superficies vero aque est superficies cuiusdam partis spere, cuius rotunditas plana est, nichil habens asperum.* Liquidum enim corpus in inconcinnas cumulari altitudines non potest. Unde et, si qua eius fiat vi ventorum asperitas, illis quiescentibus perplanatur. *Huius speralis superficiei aque centrum idem est quod et mundi, a quo si diversas imaginarias proteles usque superficiem aque lineas, nulla se maiorem minoremve letabitur aut queretur. Eadem invenit rationis prompta inquisitio, in quibus aque partibus, que separate a mari seorsum posite sunt, ut lacus, stagna et paludes.*

¶Et de mari quidem, quod quo ambitu quibusve locis terram circumfluat, incertum habeo preter id quod septemtrionales norunt habitatores, de quibus quoniam apud illos sepe dictum est taceamus. Nam de aliis quidem maris alveis apud Macrobii commentarios in sonnium Scipionis falsa quedam inserta esse comperior. Ait enim veriorem occeani alveum perustam terre ambire zonam imitantem equinoctialis circunductum circuli. Que res apud tanti mentem philosophi, que causa tantum obtinuerit fidei, ut relatu quoque inter philosophica digna videretur, non mediocriter ammiror, cum et nullus ad perustam possit fieri hominum accessus,

---

2 aqua] aquam **10** est] *add.* (L) **14** inconcinnas] concinnas **25** equinoctialis] equinocialis **27** ammiror] *corr. ex* mmiror (L) perustam] perusiam (L)

[24] *But, as the earth is higher at some places and deep at others, the element water, tending towards the centre due to its weight, collects in the depressions of the earth. As a result, the elevations on the body of the earth let a great part of it remain free of water*[20] while setting well-defined limits to the watery liquid. I have no doubt that this was arranged by the admirable providence of the Creator; anticipating the nature of the animals which he was about to create, whose life would to a great extent be nourished by the moderating air and for whom a solid structure should give a place to rest, he decided to have a part of the earth elevated and to have the water flow together from everywhere into the lower region, so that animals who would suffocate in water had a place to live and to find their nourishment. And with good reason we can assume the same for the creation of mountains and valleys, as they were made for the benefit of the animals and as a favourable condition for life. [25] *It is evident from this that the entire habitat of mankind and all animals*—except the species of the fishes—*forms an island;* which appears implausible to us nonetheless due to its vast extension.

[26] *The surface of the water is the surface of a certain section of a sphere whose roundness is even, without anything rough.* For, a liquid body cannot pile up to unaesthetic heights. And even if some roughness is produced by the force of the winds, it will flatten again as soon as the winds calm down. *The centre of this spherical surface of the water is the same as that of the world, from which if you draw different imaginary lines until the surface of the water, none of them could boast or lament any excess length or shortfall. Enquiring reasoning easily finds the same for parts of the water that are located separate from the sea, such as lakes, ponds and swamps.*[21]

I have no certain knowledge along which coastline or at what places the sea flows around the earth, except what is known by the inhabitants of the northern part of the world, who have written a lot about it, so we will remain silent about this. In Macrobius' *Commentary on Scipio's Dream* I find erroneous remarks on some parts of the sea. For he says that the ocean actually flows around the burnt zone of the earth, following the course of the equator.[22] I am quite surprised that this statement and cause could find so much approval of a philosopher so talented that it became an accepted philosopheme. For there is neither a way for a human

[20] Arab. add.: 'like the islands in the middle of the sea.'

[21] The inclusion of inland waters makes clear that different concentric spheres can be defined by different surfaces of water. L1 omits the passage, whereas the Hebrew translations extend the theorem also to rivers, although in that case the water would not be in its 'natural', i.e. its resting, place.

[22] *Somn.* II,9,2–5 (L).

quorum auctoritate certiores super hac re certi fierent, et nulla nec geometricali aut quavis alia philosophie archano digna sententia comprobari possit.

Nam quod totam quidem perustam non ambiat, hoc me credere compulit ratio, que pressius etiam intuentes in idem fidei argumentum reducet; Nili flumen, 5 cuius exundantis largo benefitio Egipti plana irrigantur, ut in fecunditatis ubertatem terre semina turgescant. Cum omnis frigore torpet septemtrionalis habitatio, suo totum continetur, nec ipsum improbet alveo, quo tempore ceterorum omnes septemtrionalis habitationis fluminum aque suis dedignantur cohiberi laxiusque camporum plana pervagantur. At vero cum estatis sub ardore reliqui fluvii ad con- 10 sumptionis fere parvitatem redacti facilem vadantibus transitum prebent, solus hic f. 5v aquas augmentat in tantamque exuberat magnitudinem, | ut sole in Leonis signo estatis nobis maximum ardorem vibrante alvei dedignans clausulas longe lateque totam perfundat terre viciniam. Quare eius cum in septemtrione non inveniatur originis principium, ultra perustam in temperata australi esse credendum est. Si 15 enim non sic eius in estatis ustu exuberationis causa, nulla poterit inveniri ratione. Nam illud quidem quod plerique arbitrantur, hoc per solem fieri, quanta quamque puerili involutum sit ignorantia, facile diligentius secreta nature rimantibus patet. Cum etenim solis corpus calidum sit et siccum radiusque eius identidem, frivolum sit esse liquet per adversans aque organum tantam ubertatem aquarum fieri. Calor 20 namque et siccitas quanto humorem fugiunt et frigiditatem, tanto contrarietatem proprie nature aquaticum humorem potius minuunt quam augmentant.

Quamobrem naturam imitantibus magis oportet credi argumentis quam quibusdam fictis et phisice subtilitatem modum excedentibus Sophistarum inventionibus. Placet igitur hanc potius sequi sententiam cui aliquod natura sub humani 25 sensus arbitrio certa cum ratione prebet experimentum. Hec huiusmodi est. Constat Egiptiorum, ut apud Lucanum est, maiores plurima enixos intencione, quod fontis Niliaci teste hominis visu possent invenire originem, et sic tandem fontis locum scientes exundationis quoque certissimam intellectu caperent causam. Quorum laboris spes sepius frustrata a proposito decidit. Quos etenim ad inve- 30 stiganda desiderata legaverant, multa defatigatos immense vastitatis via tandem peruste zone incendium a persequendo ultra fluminis accessum arcebat, sicque congressi itineris impotes vix sese suis reddebant.

---

15 ustu] usu **27** visu] usu **28** causam] creavi (L) **30** tandem] *add.* (L)

to get to the burnt zone, from whose witness those who think to know the answer would then really know it, nor is there a geometrical or any other method adequate for philosophical questions to prove that statement.

There is a reason that forces me to believe that the ocean does not embrace the entire burnt zone, and it will convince everyone who looks harder into the matter. It is the Nile, by whose rich gift of floodings the plains of Egypt become irrigated to make the seed on the fields grow in plentiful fecundity. When the entire northern inhabitable region is struck with cold, the Nile is well bound and does not exceed its banks, at a time when all other rivers of the northern region refuse limits and spread over the plains. And when under summer's heat all other rivers are almost dried out and travellers can easily wade through them, only the Nile increases its waters to such an extent that, when the sun standing in the sign of Leo brings us torrid heat, the Nile rejects its borders and widely floods the surrounding soil. For that reason, as its spring is not found in the northern part, one must believe that it lies beyond the burnt zone, in the temperate southern part. For if this was not the cause of its flooding during the heat of summer, the cause could not be found in any way. Especially that widespread belief that the sun caused the flooding stems from deep and infantile ignorance, as becomes clear to those who attentively study the secrets of nature. For as the sun is a hot and dry body and as also its rays are such, it is clearly unreasonable that the great excess of water could be caused by an organ that is an adversary to water. For just as heat and dryness avoid wetness and cold, they would diminish the watery wetness as something contrary to their own nature rather than increase it.

Students of nature should therefore rather believe in arguments than in fictional ideas of Sophists which often ignore the complexity of physics. Hence it is better to pursue a doctrine which nature itself suggests, by any kind of experience to the judgement of human senses on the basis of irrevocable reasoning. This is as follows. It is a fact to be read in Lucan's epic that the ancient Egyptians tried with greatest effort to find and witness by human sight the source of the Nile and then, by knowing the place of the spring, fully understand the reason of the flooding.[23] But after many frustrated failures they gave up; those who were sent out for exploration, exhausted by the long way through the huge desert, were finally prevented from following the river's origin any further due to the heat of the burnt zone. Thus unable to continue the journey, they scarcely made it back home.

[23] *Phars.* X,268ff.

Credendum itaque est Nili alveum mediam dividere perustam et in australi temperata fluxus habere principium. Ex hoc namque fit ipsum in septemtrionali estate exuberantium Egiptum et sibi vicina larga copia irrigaret aquarum, quoniam his in locis quibus in superas prius funditur auras pluviosam constat esse 5 eiusdem hiemis gelidi infecunditatem. Hoc autem nusquam fit in septemtrionali habitatione, estas enim est. Sed nec in perusta fieri hoc quivis sanus imaginabitur. Restat igitur, ut eius ortum in australi media fore credamus. In ea enim, cum nos estatis sentimus estus, hiemps est, quoniam et nos hiemem sub eius estate patimur. Ergo hec est exundationis eius causa certissima, quoniam cum sol in septemtrio- 10 nalibus signis estatis tempora metitur, temperata australis hiemali torpet frigore, in qua quoniam Nilus oritur, pluviales in suo suscipiens torrentes alveo transcursa perusta Egipto solito copiosior invehitur, et australis hiemis larga maximus ubertate circumfusa terrarum occupat. At vero cum nobis hiemps est atque australis temperata estatis desiccatur et calet estu, remissior Nilus amisso configurantium 15 fastu aquarum maris decolorat littoream cerulitatem.

Patet igitur in australi eum ortum temperata in septemtrionalem per mediam perustam decurrere. Quod si ita est, nequaquam occeanus totam ambit perustam. Neque enim ipsum aquarum dulcedo ingressa egredi, sed et si medium divide- re mare dicatur, hoc quia nec puerilem equat simplicitatem, refellere argumentis 20 respuo, quod quivis vel indoctus facile poterit confundere. Quamobrem his omis- sis reliqua videamus. *Et de duabus mundi partibus que gravitatis pondere deprimuntur,* quantum ad suscepti pertinebat | honus operis, *satis explicatum est;* de reliquis deinceps prosequamur. (f. 6r)

[29] ¶*Leves mundi partes due sunt, aer et ignis.* [27] *Horum locus, ut diximus, pro- 25 prius est inter gravia et media.* [28] *Si qua enim levium particula in gravis inferatur locum, dimissa naturaliter insita levitate alienis dedignatur retineri atque in sue nature sedem avolat, usque dum in locum veniens qui duarum partium medius est, ibi tandem in sui elementi positione quiescit. Levium autem alterum altero levius est.* Minus leve aerem, levius vero elementum ignem accepimus, *que duo secundum levitatis portionem 30 disposita sunt.* Aer enim inter gravium et ignis sedem, ignis inter aerem et celum proprie levitate nature constitutus est. [30] *Et aerium corpus quidem super speram aque circumpositum est, cuius forma speralis est planam habens superficiem exteriorem,*

32 circumpositum] circumpositus (L)

We therefore must believe that the Nile's course crosses the burnt middle and has its origin in the southern temperate zone. And this makes it irrigate Egypt and its neighbour regions with great masses of excessive water during the summer in the north. For there must be the same rainy infertility of the winter in those places where it is flowing before with its level raised high into the air. But this occurs nowhere in the northern inhabitable region, because it is summer there. However, no one with a sober mind could think that this would happen in the burnt zone. Therefore, there remains for us to believe that its source is in the temperate part of the south. For when we feel the heat of summer, it is winter there; just as we suffer from winter's cold when it is summer in that region. This is the most certain explanation for its flooding, because when the sun passes through the season of summer in the northern signs, the temperate south will be struck with winterly cold. As the Nile has its source there, it collects massive rainfalls in its riverbed, then crosses the burnt zone and reaches Egypt more exuberant than usual. Enhanced to its maximum by the ample richness of the southern winter, it takes possession of the flooded land. But when it is winter in our part of the world while the southern temperate zone gets hot and torrid by summer heat, the Nile, with its conceited donors now dismissed, discolours the coastal blue of the sea more gently.

Hence it is clear that, having its source in the southern temperate zone, the Nile runs through the burnt central zone to the north. For this reason the ocean cannot flow around the burnt zone completely. For the Nile loses not the sweetness that has gone into its water. And if someone says that it divides the sea, this would reveal a not even childlike simple-mindedness and is not worth being refuted by arguments, because any however untaught person can easily refute it by himself. Thus passing this by, we shall consider the rest. But for the purpose of this treatise *we have said enough about the two elements of the world that are pressed downwards by the weight of heaviness;* we may now proceed to the remaining ones.

[29] *There are two light elements of the world, air and fire.* [27] *Their proper place, as we have said, is between the heavy and the medium.* [28] *For if some particle of the light ones is taken to the place of the heavy, once released, it naturally refuses by its innate lightness to be held back by particles of another kind but floats away to its natural place, until it reaches the middle between the two parts, where it will finally be in its proper place and come to rest. But of the two light elements, one is lighter than the other.* We have learned that air is a less light element and fire a lighter one, *and they are arranged according to the amount of their lightness.* For air is arranged between the heavy and fire, whereas fire, by the lightness of its nature, is arranged between the air and

interioris autem, que speralis profunditas dicitur, pars planitiem, pars asperitatem habet. Que enim superficies aeris superficiei aque iungitur, plana est. Superficiem enim aque liquidi elementi planam esse superius positum est. *Ea vero pars que terreo corpori iungitur, propter asperitatem diversi elementi violatur;* ibi nempe superficiei terre similis est. *Id autem minimum ad sue magnitudinem quantitatis habetur. At exterior aeris superficies, que speralis est* nec habet partem altera humiliorem, *iungitur interiori superficiei ignei elementi.*

[31] Sed cum IIII[or] sint mundi elementa, quorum ignis semper calidus sit, aer, terra et aqua susceptibilia sunt contrariorum. *Solum tamen aerem discordem mundi partem aiunt.* Terra enim et aqua frigiditatis et caloris, siccitatis et humoris susceptibiles cum sint, *maxime tamen aer diversorum et repugnantium accidentium commutabile subiectum invenitur. Qui cum sit naturaliter subtilis, quedam tamen eius pars spissa, quedam subtilis dicitur.* [32] *Spissitudo tamen eius terre et aque affinium fit accidentibus. Fit enim ex spisso fumo, qui de humido corpore ascendit solis calore imitato, quo ex fumo creantur nubes, ventus, nebula, nix, grando, pluvia.* [33] *Alteratur etiam eadem pars terre vicinio in caloris et frigoris varietate, atque is quidem qui terre proximus est calidior, medius temperatus, exterior calidissimus invenitur. Cuius rei causa est solis radii calor, qui a solido reverberatus terre corpore* in proximum aerem refertur, quique prius simplex et radius et calor fuerat geminatur. *Quanto igitur longius terram despicit, tanto minus est calidus, usque dum ignis ardori propiquans ab eodem calefit, cui quanto propinquior, tanto calidior.*

[34] ¶*Levissimum elementum ignis aeris speram circumtenet, cuius forma speralis est, interiorque eius superficies, que speralis eiusdem profunditas dicitur, exteriori iungitur superficiei aeris. Exterior autem superficies speralis est atque admodum rotunda, ut nichil in ea asperum, nichil ab equalitatis mensura discrepet. Hec corpori iungitur qui levitatem ignis coherceat,* ne pulcherrimam mundi formam dissolvat, de quo attendendum quoniam quanto a centro remotiorem elementariam vendicat sedem, tanto ceteris purius elementum nichil habet alieni ammixtum. Semper enim calidus est, neque unquam frigiditatis torporem sentit aut patitur. *Atque hec sunt que*

4 superficiei] superfitiei aque **11–12** commutabile] commutabilem **14** imitato] immitato **16** vicinio] vicino (L) is] his (L) **20** calefit] calo fit (L) **25** discrepet] *corr. ex* discepet (L)

the heaven. [30] *And the body of air is placed around the sphere of water, and its shape is spherical with an even outer surface, whereas that of its inner surface, which is called the 'spherical concavity', is partly even, partly rough. For that surface of the air that is tangent to the surface of water is even.* For we have said above that the surface of the liquid element water is even. *But that part which is tangent to the body of the earth is violated by the roughness of the different element,* because there it resembles the surface of the earth. *Compared to its size, however, this can be regarded as very small. But the outer surface of the air, which is spherical* and has no part lower than another one, *is tangent to the inner surface of the fiery element.*

[31] There exist four elements of the world, among which fire is always hot, whereas air, earth and water can take opposite qualities. Yet, *only air is said to be an ambiguous element.* For even though earth and water can adopt coldness and heat, dryness and humidity, *air is found to be the most changeable by different and adversary accidents. Although it is naturally fine, nonetheless some part of it is said to be dense and some other thin.* [32] *But this density comes from impacts of the adjoining earth and water. For it is caused by dense fog, which rises from the humid body when the latter has adopted the heat of the sun and from which clouds, winds, mist, snow, hail and rain are made.*[24] [33] *Further, in the proximity of the earth that same element is changed by the alternating heat and cold, wherein the air that is closest to the earth is rather warm, the air in mean distance temperate, and the outermost air the hottest. The reason for this is the heat of the sun's radiation, which is reflected from the solid body of the earth and thrown back to the lower air,* and radiation and heat that were single-acting before are now doubled.[25] *Hence, the further away it is from the earth, the less warm it will be, until it approaches the heat of the fire and is warmed by the latter, the hotter the closer it comes.*

[34] *Fire, being the lightest element, encloses the sphere of the air; its shape is spherical and its inner surface, called its 'spherical concavity', is tangent to the outer surface of the air. Meanwhile, its outer surface is spherical and of such roundness that there is nothing rough and nothing deviates from perfect equality. It is tangent to the body that holds the lightness of fire in its place,* to prevent it from dissolving the world's beautiful structure. Concerning this, one has to keep in mind that the further away from the centre an element has its place, the purer it is from other elements, without intermixture of any other substance. For it is always hot and never feels or suf-

---

[24] Arab. distinguishes between the rising of fog and clouds, as an effect of warm air, and the condensation into rain and snow in cold air.

[25] The insertion anticipates an aspect of Stephen's later discussion of the warming effect of the sun, starting on p. 142:1.

de levium natura ad presens dicenda videbantur, ad aliud enim nostra festinat oratio. Quamobrem hec summatim tetigimus, ut ad eam de qua proposueramus disputare mundi partem congruo tandem ordine ascenderemus.

[35] ¶*Illa igitur mundi pars quam nec levem nec gravem,* hoc est mediam, *dicimus,* IIII^{or} *ut dictum est elementa cohibet et claudit, atque circa ea* | *eterno infatigabilique motu discurrit. Hec autem pars ea est, quam celum vulgato nomine apellari audimus, in quo luna, sol et omnia astra micant.* [36] *Celi formam speralem due circundant sperales paralelleque superficies, quarum centrum mundi est. Harum interior speralis profunditas est equaliter rotunda, que colligatur et iungitur exteriori ignis superficiei. Exterior et omnium altissima ipsa quoque speralis et recte rotunda omnia includit corpora. Huius et interioris centrum idem est quod et mundi, quemadmodum proxime positum est.* Sed cum IIII^{or} elementorum, quorum duo gravia, duo levia, motus gravium in centrum, levium a centro sit, huius partis que gravis et levis media est motus qui sit, ambigi non oportet. Quoniam neque in centrum, non enim gravis est, neque a centro, quia nequaquam levis, circa centrum rationis invenitur certissimis argumentis. [37] *Hoc igitur corpus de loco, quem orientem, ad locum, quem occidentem dicimus, citissime movetur, suoque impetu movet omnia celestia corpora rotundo motu,* eaque ipsa contra eius cursum nitencia in occidentem refert secumque oriri cogit et occumbere.

Hactenus generaliter de omnibus mundi partibus gratia sperarum, quo facilior ad earum noticiam fieret accessus, breviter pauca perstringendo disputavimus. Restat nunc de proposita questione singularis disputatio. Cui quoniam occulta et subtilia sunt de quibus agitur, ingenii medullas adhibere oportet, ut que longe sunt assensu sola collegisse ratione, quisquis es qui nostra non fastidis, gratuleris promtiorque ad alia, si qua sunt, occultiora animi capacitatem provoces.

---

2 proposueramus] *corr. ex* proposueamus (L) **21–22** occulta] oculta

fers the rigour of the cold. *This is what for the moment seemed necessary to say about the nature of the light elements,* because we strive to discuss other things. We therefore touched on this only very briefly so that finally, and in proper order, we can proceed to that part of the world which we had announced to talk about.

[35] *That part of the world which we call neither light nor heavy,* that is, the medium, *encloses the four elements, as we have said above, and keeps them in place and revolves about them in an eternal and tireless motion. This is the part which is commonly called the 'heaven',*[26] *where the sun, the moon and all the stars are shining.* [36] *The spherical shape of the heaven is enclosed between two spherical and parallel surfaces, whose centre is that of the world. The inner one is the spherical concavity, which is equally round to match and enclose the outer surface of fire. The outer surface, which is the highest of all and also spherical and perfectly round, encloses all bodies. Its centre and that of the inner one is again the centre of the world, as we have just stated.* But whereas for the four elements, two of which are heavy and two light, the motion of the heavy ones is towards the centre and that of the light ones away from it, one should not wonder about the motion of the medium between heavy and light. Since this motion can neither be directed towards the centre, as it is not heavy, nor away from it, because it is not light, it is by most certain arguments found to be around the centre. [37] *This body is thus moved very fast from what is called 'east' to what is called 'west' and, by its momentum, moves all celestial bodies in a circular motion;* it even moves those bodies westward that tend in the opposite direction and forces them to rise and to set in its own direction.

So far we have spoken generally and briefly about all parts of the world, in order to introduce the spheres and to facilitate an understanding of them. Now there remains to give a detailed discussion of the announced subject. It will be necessary to bring up much intellectual effort for this, as we will deal with abstract and complex topics, so that whoever you may be, not weary of us, can pride yourself on having grasped by pure reason what is difficult to conceive and that you may bring up your mental strength more readily for other, more demanding subjects, if there are any.

[26] 'heaven': Arab., 'orb' (*falak*). Stephen's translation, 'celum,' corresponds with his systematic avoidance of the term *orb* in connection with the celestial spheres.

[38/40] <***C***>[(L)]*elum in multas divisum esse partes* antiquorum experimentis comprobata indagatio tradidit, *cuius principalis divisio* IX *habet partes, speralia scilicet corpora, quorum aliud ab alio clauditur. Iunguntur enim superficies interiorum exteriori spere et exteriorum interiori sibi invicem. Harum omnium centrum sperarum centrum mundi,* [41] *quarum queque celum a maioribus dicitur.*

[42] Novem igitur celi sunt, *primum quorum nobisque proximum et celum lune vocamus,* in quo luna posita mutuate mundo radios lucis refundit. *In secundo vero Mercurium, in tertio Venerem, in quarto solem,* qui planetarum VII medius est, a quovis extremo quartus intuetur. *Quintum deinde* solis spere exterius proximum *Mars rutilus, sextum* blandi luminis *Iuppiter,* VII *autem ceteris altior Saturnus vendicant.* VIII *vero celum est in quo fixarum multitudo stellarum posita est. Horum altissimum et ultimum nonum est, quod omnes alios tenet et circumcingit,* ipsum nullo corpore inclusum. [43] *Habent autem omnes he spere discordem motum. Quedam namque in occidentem ab oriente, quedam contra nituntur.* Et de motu VIII sperarum suis in locis dicemus. Interim ad nonam totum stilum convertimus.

[44] *Nonum, quod ultimum positum est celum quodque alios continet, movetur celeritate quamplurima ab oriente in occidentem, cuius motus super duos fit immobiles fixosque polos. Huius tanta est celeritas, ut .ld. horarum spatio plenum perfitiat cursum.* Quecumque enim oriente sole eius partes fiunt in eundem, antequam surgat orientis circulum, revertuntur. Si exempli gratia in linea recti circuli sol in primo puncto sit Arietis, eo oriente idem surgit in quo esse | dicitur punctus signi. Sol autem a sua spera in orientem pleno sub die .oi. sexagenariis et .h. secundis infertur. Fit igitur, ut prior eodem die primus occumbat punctus Arietis sole qui consurrexerat. Prior ergo necesse est ad orientis minimum discrimen festinet. *Plenum igitur circuitum uno die faciens suo motu ab oriente in occidentem omnes que inter ipsum celorum speras continentur refert.*

Hos in quibus movetur polos, quos fixos esse dictum est, non rei iteratione

1–2 comprobata] comprobatam (L) **3–4** exteriori] interioris (L) **18** .ld.] id (L)

1.3 <The Heaven>

Config. 3: The Orb

[38/40][27] The investigation by the ancients, proved by experience, teaches us that *the heaven is divided into many parts; its main division is into nine parts, which are spherical bodies that enclose one another. The surfaces of inner spheres are tangent to an outer sphere and those of outer spheres, to an inner one, mutually. The centre of all these spheres is the centre of the world,* [41] *and each sphere has been called a 'heaven' by our ancestors.*

[42] Hence there are nine heavens, *the first of which, being closest to us, is called the 'heaven of the moon',* in which the moon is placed and reflects borrowed rays of light to the world. *In the second one Mercury is found, in the third Venus and in the fourth the sun,* which is the central one of the seven planets and the fourth from either end.[28] *Then the fifth heaven,* the next one outside the solar sphere, *is claimed by red Mars, the sixth by Jupiter* with its charming light, *and the seventh by Saturn, higher than the others. The eighth heaven is where the multitude of the fixed stars is placed. The ninth is the highest and outermost of the heavens, which keeps and encloses all others,* but itself is enclosed by no other body. [43] *But all these spheres have different motions. For some of them move from east to west and some in the opposite direction.* We will speak about the motion of the eight spheres at their proper places; first, however, we will discuss only the ninth.

[44] *The ninth heaven, which is placed outermost and which encloses the others, moves at an utmost speed from east to west about two unmoved and fixed poles. Its speed is so fast that it performs a complete revolution in the span of 24 hours.* For whatever parts of it lie on the eastern horizon when the sun is rising, they will return to the same position before the sun rises again. For example, if the sun is on the equator in the first point of Aries, the point of the sign in which it is said to be rises together with the sun. But during one entire day the sun is moved by its own sphere by 0;59,8 degrees eastwards. As a consequence, the first point of Aries, which on the same day had risen together with the sun, will set earlier than the sun. Hence it necessarily rushes back to the east earlier by a small difference. *Therefore, while making a full revolution in one day, it takes along by its motion from east to west all the heavenly spheres that are enclosed in it.*

It[29] is not repeating the same thing when we say that the poles about which it

---

[27] Stephen omits passage 39, where Ibn al-Haytham discusses the various meanings of *falak*. The passage has become obsolete by Stephen's more distinct terminology. The content of passage 38, originally belonging to the previous chapter, has been combined by Stephen with passage 40.

[28] Cf. Cic. *Somn.* 4 and Macr. *Somn.* I,19,14f.

[29] Stephen elaborates again on the relation between the poles of a sphere and its rotating motion, which will form the basis of his later criticism of Ibn al-Haytham's description of the planetary spheres, starting on p. 340:6.

immobiles testamur, sed ut id certum sit eos et fixos esse et non alterari. Diversa quippe sunt fixum et immobile, nec eadem significantia inveniuntur. Omne namque immobile fixum est, quod non convertitur. Quedam enim fixa sunt, que motum habent. Nam etsi non a se, non enim iam fixa haberentur, ab eisdem tamen in quibus fixa sunt moventur. Que quidem ratio in omnium sperarum preter nonam polos cadit. Moventur enim omnes spere certo circuitu in proprium motum. Sed quod certum habet circularem circuitum, in duobus fixis punctis moveri supra ostensum est. Habet ergo queque spera II^{os} proprios polos, in quibus spera currente ipsi suo non moventur motu. Nam si motu sue spere moverentur, recte non poli; quia si poli, verissime ab illius motu moveri dicentur. Fixi igitur sunt, non tamen immobiles. Moventur namque quarundam poli sperarum uno, nonnullarum duobus accidentalibus motibus. Qui uno tantum, VIII^{ve} sunt et IIII^{te} poli, ceteri omnes duobus, cuius rei causa est, quod octave spere poli inclinantur a polis none, qui circumfixi in sue spere motu perstent, semperque eidem parti none suppositi sint. Eius tandem celeritate ab oriente in occidentem moventur. In huius spere polorum radio solaris spere poli, que IIII^{ta} est, positi sunt, ideoque nullam aliam nisi quam hii quoque motionem norunt. Sed de hoc et de aliarum polis sperarum convenientior et uberior suis in locis demonstratio fit.

None autem spere poli, que ceteras omnes claudens a nulla clauditur, fixi sunt a motu sue spere sicut aliarum quoque poli. Quia vero eorum spera suo scilicet et non alterius motu movetur, quippe que ultima est corporum et omnes alias movet, immobiles sunt. Nam quod non moveantur, idem semper status eademque eorum in quolibet loco altitudo immutabilis manifestat. Recte igitur fixi immobilesque dicuntur.

[45/46] *Horum alterum septemtrionalem, alterum australem aiunt. Septemtrionalis ille est, qui terre, que septemtrio dicitur, superpositus est, queque sola in terre orbe, etsi non omnis, habitatur. Hic itaque septemtrionalis polus super septemtrionalem habitationem videtur, ab australi nunquam visus mundi parte.* Iam vero quis australis dicatur, manifestum credimus, quoniam septemtrionali oppositus australibus locis supereminet. Quod quemadmodum septemtrionalis nunquam videt in austro semperque in septemtrione, *ita nunquam australis in septemtrionali habitatione* semperque ab au-

4 haberentur] habentur (L) **14–15** tandem] tamen

moves and which we said to be fixed are also unmoved; we say this to make clear that they are both fixed and also unchanging. For there is a difference between 'fixed' and 'unmoved', which terms do not denote the same. Everything unmoved is fixed, but the inversion is not true, as some things that are fixed have a motion. They do not move by themselves, in which case they would not be considered as fixed, but they move by those things to which they are fixed. And this is the case for the poles of all spheres except the ninth, because all spheres move by their own motion in a certain revolution. But we have shown above that everything that has a certain circular motion is moved about two fixed poles. Hence each sphere has two poles which, while the sphere is running about them, are not moved by its own motion. For if they were moved by the motion of their own sphere, they would not be the poles really, whereas even if they are the poles, they can still rightly be said to be moved by the motion of that other sphere. Thus, they are fixed but not unmoved. The poles of some spheres are moved by one, others by two occurring motions. Those that are moved only by one motion are the poles of the fourth and the eighth spheres, whereas all others are moved by two motions, because the poles of the eighth sphere are inclined from the poles of the ninth and, stable in the motion of their own sphere, stand firm and always at the same place of the ninth sphere, by whose fast rotation they are then moved from east to west. On the axis of this sphere are placed the poles of the sphere of the sun, which is the fourth sphere, for which reason these poles experience no other motion than the former ones. But this and the poles of the other spheres can be discussed better and in more detail at their proper places.

The poles of the ninth sphere, which encloses all the others but itself is not enclosed, are fixed with respect to the motion of their own sphere, in the same way as the poles of the other spheres. But since their sphere—being the outermost body, which moves all other objects—is moved by its own but not another sphere's motion, they are also unmoved. For it is clear that they are not moved, as they always have the same position and the same altitude as seen from any place. Hence they are rightly said to be fixed and unmoved.

[45/46] *One of them is called the 'north pole' and the other one the 'south pole'. The north pole is that which is located above the northern part of the earth, which is the only, though not completely, inhabited region on the globe of the earth. Therefore, the north pole is always seen above the northern inhabited region but it is invisible from the southern part of the world.* It should be clear by now that the south pole is called in that manner because it stands opposite the north pole, above the southern part. And in the same way as the northern one never faces southward but always to

stro *conspicitur.* Fixus enim uterque et immobilis in eadem perseverant constantia. Et nona quidem spera ac poli eius huiusmodi sunt.

[47] *Aliarum vero, que* VIII *sunt, sperarum quedam moventur in orientem, quedam contra. Earum queque, sive sit in orientem seu in occidentem motus eius, super duos proprios fixosque movetur polos.* [48] *Sperarum autem* quibus VII president planete *pars plures, pars pauciores habet partes sperales, de quarum propriis accidentibus motibus disseremus, cum de earum principalibus speris singillatim disputabimus.* Nunc autem de circulorum origine eorumque situ et necessaria cognitione videamus.

¶In rerum subsistentes natura nulli habentur in celo circuli, de quibus hoc in libro dicturi sumus, sed imaginarii. Intellectus enim humani facultas mirabilis que non sunt imaginatione constituit, quo ad eorum que sunt intelligentiam veri similibus paret viam. Ad indagandam | igitur occultam subtilemque celestium sperarum positionem, ad cognoscendum solis aliarumque stellarum locum et magnitudinem in superficiebus celestium sperarum, vetustiores circulos imaginati sunt, quorum opera que perspicax animi vigor desiderabat agnovit. Eorum autem imaginatio in hunc fit modum. [49] *Celi spere omnes in duobus moventur, ut positum est, fixis polis* in certum circularemque motum, *que dum moventur, earum quisque punctus qui in superficie est facit in discursus loco atque in superficie spere quemdam imaginarium circulum, per quem et in quo idem semper punctus nullatenus declinans volvitur sperali motu. Id ita esse, ut aiunt, verissimis ostenditur in libro geometrie rationibus. Unaqueque etenim superficies que speras invicem iunctas dividit et terminat, facit in singulis earum circulum.* [50] *Quoniam igitur cum spere moventur super fixos polos quisque punctus in earum superficie circulum circinat, faciet unaqueque stellarum que in eis sunt in spera circulum per suum centrum. Huius circuli si imaginaveris augmentari superficiem usque dum totum secet mundum et dividat, faciet in superficie alti celi imaginarium circulum. Fiunt etiam aliter circuli. Si enim imaginemur superficiem linee que transit per centrum cuiusvis stelle et centrum mundi crescentem totum mundum dividere, faciet hec divisio in superficie alti celi circulum magnum, cuius centrum et mundi unum est.* Magnos autem circulos dicimus, qui quamlibet speram in duo media dividentes idem habent quod et spera centrum. *Erunt igitur in hunc modum in superficie alti celi multi imaginarii circuli, quorum origo a qualibet horum que dicte sunt sumitur imaginatione. Eorum quodque habet proprias et figuras et proprietates,* de quibus inferius dicetur.

f. 7v

---

2 eius] *add.* (L) **7** singillatim] sigillatim **24** imaginaveris] *corr. ex* imagineaveris (L) **25** secet] seccet

the north, *the south pole is never seen from the northern inhabited region,* whereas it is always visible in the south. For either of them remains fixed and unmoved, always in the same position. Such are the ninth sphere and its poles.

[47] *Of the other eight spheres, however, some move to the east and some to the west. Each of them, moving eastward or westward, is moved around its own two fixed poles.* [48] *But of the spheres* which are governed by the seven planets, *some have more and others fewer spherical parts, whose particular occurring motions we will discuss when we deal with their elementary spheres one by one.* Now let us look at the origin of the circles, their positions, and how to determine them as required.

The circles which we will speak about in this book do not exist as natural objects in the heaven, but they are imaginary. For the human mind has the astonishing faculty to conceive by imagination things that do not exist, in order to facilitate through analogy an understanding of the things that do exist. Thus, as a means to investigate the inscrutable and complex arrangement of the heavenly spheres and to determine the position and size of the sun and the other stars on the surfaces of the heavenly spheres, the ancients imagined circles through which the curious intellect could recognise what it sought. They are to be imagined in the following way. [49] *All spheres in the heaven move, as we have said, about two fixed poles* in a certain circular motion, *and by their motion every point on their surface describes an imaginary circle in its observed position and on the surface of the sphere. On that circle the same point revolves steadily without any deviation. The validity of this common statement is demonstrated with undeniable proofs in the book of geometry.*[30] *For every plane that divides adjoining spheres and delimits them produces a circle on each one of them.* [50] *Thus, as the spheres move around fixed poles and, therefore, every point on their surface draws a circle, every star that is located on them will produce a circle by means of its centre. If you imagine the plane of that circle magnified until it cuts through the entire world, it will produce an imaginary circle on the surface of the high heaven. Circles are also produced in other ways. For if we imagine a plane on a line that passes through the centre of a star and the centre of the world to be extended such that it divides the entire world, this section will produce a great circle on the surface of the high heaven whose centre coincides with that of the world.* We call those circles 'great circles' which divide any given sphere into two halves while having the same centre as the sphere. *In that way there will be many imaginary circles on the surface of the high heaven, whose origin is from any imagination of what we have said. And each of them has its own configuration and properties,* as will be discussed below.

[30] Arab.: 'the books of the geometers.' Stephen seems to think of a particular text.

[51] *Magnum itaque celum quando movetur suo proprio celerique motu, omnes qui super sunt ipsi puncti facient in superficie eiusdem multos imaginarios circulos, quorum omnium superficies paralelle sunt,* hoc est, ab invicem equaliter distant per omnes sui partes. *Eorum poli sunt duo poli mundi, super quos eorum in cuius superficie fiunt spera volvitur.* At centrum cuiusvis scire cum volueris, in radio mundi invenies, qui ab altero usque ad alterum mundi polum rectus extenditur. *Horum circulorum inequalis et multiplex est magnitudo. Nam quanto polis propinquiores sunt, tanto minores, quanto autem longius ab utroque recedunt, tanto latiores. Maximus vero omnium ille est circulus, cuius superficies transit super centrum mundi. Centrum enim mundi ipsius hic. Ille est, qui ab utroque mundi polo equalibus undique distat spatiis. Quamobrem eius superficies totum in duas equales partes mundum partitur. Per centrum quippe mundi transit,* ut dictum est. [52] Ceteris igitur cum parallellis omnibus maior est, quapropter magnus circulus dicitur, *sed et circulus recti diei. Hoc ideo, quoniam cum centrum solis in eo volvitur, fit ubique terrarum equa diei et noctis super easdem horas divisio,* quod nos equinoctium dicimus, *duobus punctis terre exceptis, qui in recta sub mundi polis linea positi sunt, id est in radio, quorum alter septemtrionali, alter australi polo subiacent.* Illis etenim in locis sex mensibus nox totidemque dies continuus protelatur. Quamdiu nempe sol in septemtrionalibus exaltatur signis, a capite scilicet Arietis usque finem Virginis, perpetuus dies septemtrionem continuaque nox austrum tenet. In australibus signis converso fit modo. Qua de re in sequentibus nobis latius explicandum erit. Nunc ad circulos revertimur. *Magni igitur circuli sive recti diei duo poli sunt poli mundi, quorum alterum septemtrionalem, alterum australem diximus.* [53] *Polum autem circuli* non id intelligimus, quod polum spere, sed punctum superficiei tantummodo | quod est in polo mundi. *Ab eo enim si quotvis imaginarias eduxeris lineas ad circulum usque tendentes, omnes equales erunt longitudine.* Hoc a quolibet polo faciens unius omnes invenies quantitatis lineas a polis usque circulum. Medius igitur est utriusque poli recti diei circulus. *Alios autem circulos, qui inter magnum circulum et polos utrimque continentur, temporis circulos vocamus.*

f. 8r

10 spatiis] patiis **20** re] *corr. ex* te (L)

[51] *When the great heaven moves by its own fast motion, all the points on it will produce many imaginary circles on its surface with all their planes being parallel,* that is, they are equally distant from one another in all their parts. *Their poles are the two poles of the world, around which their sphere revolves on whose surface they are produced.* If you want to know the centre of any of them, you will find it on the axis of the world, which extends straight from one pole of the world to the other. *The size of these circles varies considerably, because they are smaller, the closer they are to the poles, whereas the further from both poles they lie, the larger they are. And the largest of all is the circle whose plane passes through the centre of the world. For in this case the centre of the world is also the centre of the circle. This is the circle that is equally distant in all its parts from either pole of the world. Hence its plane divides the world into two halves, because it passes through the centre of the world,* as has been said. [52] Since it is larger than all other parallel circles, it is called the 'great circle'; *but also 'equator'. This is because, when the centre of the sun revolves in it, everywhere on earth there will be an equal division of night and day into equal hours,* which we call an 'equinox', *except for those two points on the earth that lie on the straight line under the poles of the world, i.e. on the axis, one of which lies under the north pole and the other under the south pole.*[31] For in those points night and day are extended to last six months each. Because as long as the sun stands in the northern signs, that is, from the head of Aries until the end of Virgo, the north is held by permanent day and the south by continuous night. The opposite occurs [when the sun stands] in the southern signs. We will need to discuss this in more detail later; now we return to the circles. *The two poles of the great circle, or: equator, are the poles of the world, one of which we have called the north pole and the other one the south pole.* [53] We do not consider *the pole of the circle* as being identical with the pole of the sphere but merely as a point of the surface which is located at the pole of the world.[32] *For if you draw an arbitrary number of imaginary lines from that point until the circle, they will all be of equal lengths.* And no matter from which pole you do it, you will find all the lines from the poles to the circle to be of equal lengths. The equator thus lies in the middle between the two poles. *The other circles, however, which are contained between the great circle and the poles on both sides are called 'time circles'.*

---

[31] 'except... south pole': the passage is not contained in the Arabic MS Y, whereas in K it has been marked as a quote from another version. It appears also in L1 and H6, H8 (tr. Ibn Paṭer).

[32] Ibn al-Haytham describes the pole of a circle as a point that is equally distant from every point on the circle's circumference, which would be the case for any point on the axis through the centre of the circle. Stephen, like the translators of H1 and L1, gives a more specific definition, which further demands that the pole lies on the sphere to which the circle is the equator. This agrees with Ibn al-Haytham's own later use of the term in passage 64.

[54] *Omnes vero circuli quicumque in spera positi sunt in .xp. partes imaginarias equaliter dividuntur. Harum queque pars gradus dicitur,* qui per .p. partes, quibus sexagenarie nomen est, distribuitur, easque in totidem secundas, ipsa in eius numeri tercias, eoque pacto quantum progredi partiendo volueris, nomen ab ordine parti-
5 bus impones. *Partes autem quibus gradus nomen est impositum partes quoque temporis dici possunt.* [55] *Dividitur etiam cuiusvis quantitatis circulus omnis in .ld. partes, quarum queque* XV *gradus prime divisionis complectitur, cui nomen hora est, .ld. quippe horarum totus est dies, plenus scilicet circulus.*

[56] Apparet igitur ex his que premissa sunt, quoniam *quando magnum celum*
10 *volvitur, refert secum omnium stellarum globos, quarum queque super unum eorum qui positi sunt in alta spera paralellorum movetur. In quo cum quilibet celi punctus* XV *gradus, plenam scilicet horam, transierit, .ld. pars celi surrexisse atque sic unius spatium hore complevisse dicitur. Cum vero totum transierit circulum, completas aiunt .ld. horas.* [57] *Transit autem idem punctus totum circulum unoquoque die; a magno enim circuitu*
15 *volvitur. Per unum autem plenum equalemque ipsius celi circuitum* vel paulo plus *fiunt, ut dictum est, dies et nox, ideoque .ld. horis distenditur.*

[58] His imaginando completis *alium quoque veteres magnum imaginati sunt circulum per centrum terre et mundi polos transeuntem.* Huius imaginatio eo loco facta est, qui sub recto positus circulo equaliter habitationis terminis circularibus undi-
20 que distat. *Dividit autem hic circulus recti diei circulum in duo media super duos diametricos punctos,* alter punctus orientis, alter est occidentis. *Totum itaque mundum in duas dividens equales partes eius centrum sibi vendicat.* [59] *Omnem autem mundum in duo alia media dividi per recti diei circulum dictum est. Dividitur ergo terre globus ab his duobus circulis in* IIII^{or} *equales partes, quarum due in septemtrione, due in austro.*
25 [60] *Septemtrionalium altera ab animalibus incolitur* ceteris tribus perpetuo dampnatis heremo. *Circulus autem terre qui fit per crescentem superficiem recti diei circuli, cui et semper subiacet, dicitur recta linea. Medietas eiusdem que dividit habitationem ab austro apellatur longitudo habitationis, que est .tr. graduum.* [61] *In medio huius longitudinis super lineam rectam locus est quem medium terre ponentes, eidem in celo oppositum*
30 *punctum tholum vocant.*

---

12 scilicet] solis **27** que] quia **30** tholum] *corr. ex* tolum (L)

[54] *Now, all circles on the sphere can be divided into 360 equal imaginary parts. Each of these parts is called a 'degree',* which itself is divided into 60 parts called 'minutes', each of which is again divided into the same number of 'seconds', and these into the same number of 'thirds'; analogously, depending on how much you may want to subdivide further, you will name the fractions according to their order. *The parts that have been named 'degrees' can also be called 'parts of time'.* [55] *A circle of any size can also be divided into 24 parts, each comprising 15 degrees of the first division, which are called an 'hour', whereas an entire day corresponds to 24 hours, that is, a complete circle.*

[56] It becomes clear from the above that *when the great heaven revolves, it takes along with itself the bodies of all stars, each being moved on one of the parallel circles that are placed on the high sphere. Whenever a point of the heaven has passed 15 degrees on such a parallel circle, that is, one hour, one says that the 24th part of the heaven has risen and has thereby completed the span of one hour. And when it has passed the entire circle, one says that 24 hours have been completed.* [57] *The same point passes through the entire circle once every day, as it is moved by the great revolution. But by one complete revolution,* or a bit more,[33] *of that heaven, night and day, as is said, are made. Hence it spans 24 hours.*

[58] Once these circles had been arrived at by imagination, *the ancients imagined another great circle, which passes through the centre of the earth and through the poles of the world.* It is imagined from that point on the equator which is equally distant in any direction from the circular borders of the inhabited region. *This circle divides the equator into two halves at two diametrically opposite points,* one being the eastern point and the other the western point. *Thus dividing the world into two equal parts, it takes the centre of the world as its own.* [59] *But it was said that the entire world was divided into other two halves by the equator. These two circles thus divide the globe of the earth into four equal parts, two of which are in the north and two in the south.* [60] *One of the northern parts is inhabited by living creatures,* whereas the remaining three are condemned to eternal solitude. *Now, the circle on the earth which is produced by the extended plane of the celestial equator and which always lies beneath it is called the 'terrestrial equator'.*[34] *That half of it which separates the inhabited region from the south is called the 'longitude of the inhabited region', which measures 180 degrees.* [61] *In the middle of that longitude, on the terrestrial equator, is the position which is defined to*

[33] A similar insertion, 'approximately,' which considers the difference between the sidereal and the solar days, is also found in all Hebrew manuscripts.

[34] 'terrestrial equator': lit., 'straight line'.

[62] *Sed et circulus quem et per centrum et polos mundi transeuntem posuimus, est orizon loci illius quem tholum aiunt, vocaturque circulus orizontis recte linee. Hic autem mundum, ut dictum est, in duas equales distribuit medietates. Altera igitur earum eo in loco semper videtur in quo medium terre est, altera semper occultatur.* [63] *Ad huius circuli similitudinem quicumque mundum in duo secabunt media dividentes quod videtur ab eo quod occultum est, orizontes eorumdem in quibus faciunt locorum dicentur. Circuli autem orizontis medie terre polus est is punctus, quem tholum mundi posuimus.* [64] *A centro enim mundi egressa linea atque per medium terre punctum transiens, si ad medium usque celum extendatur, quem tangit in superficie punctus mundi tholus est, qui etiam capitis punctus dicitur, propterea quod super caput illis est qui in medio terre habitant.* f. 8v | *Atque hic punctus magni illius circuli superior polus est, quem orizontem medie terre constituimus,* inferior pedum punctus. [65] *In hunc autem modum et orizontes et eorum polos ubique inveniri licebit. Singuli etenim puncti, qui in superficie terre sunt, singulos habent punctos oppositos in superficie magni celi, qui eorundem locorum capitis puncti non absurde dicuntur. Hii igitur quorumcumque circulorum poli sunt, eos ipsos orizontem loci illius faciunt, in quo punctus superficiei terre puncto capitis subiacet.*

[66] Que cum ita sint, manifestum credimus *unumquemque punctum qui in superficie terre est suum habere orizontem, qui in eodem loco mundum in duo media dividit.* Omnis enim orizontalis circulus mundum in duas dividit medietates. Orizon namque est magnus circulus mundum in duo media dividens, quorum alterum semper videtur, alterum nunquam, sed occultum latet. Illud autem plurimum adtendendum est, que sit recti diei circuli positio super orientalis circuli superficiem. Nam aliquando rectis angulis, sepissime altero acuto, altero lato super ipsum invenitur. Que res dierum noctiumque plurimum facit discordiam. [67] *Et orizon puncti recte linee quicumque super duos mundi polos transit, rectis super se angulis circulum recti diei semper videt.* Discursus enim eius est per punctos capitis et pedum, qui orizontis poli sunt. [68] *Reliqui vero orizontes omnes alterum vident polum, alterum occultant;* neque enim per polos transeunt. *Quamobrem circulus recti diei super illos*

---

6 orizontes] *corr. ex* oriontes (L) dicentur] diceretur (L) **9** tangit] tangii superficie] superfitiem **24** noctiumque] noctuumque **25** recte] terre

be the 'middle of the earth', whereas the point in the heaven opposite to it is called the 'dome'.[35]

[62] *Meanwhile, the circle which we have assumed to pass through the centre and the poles of the earth is the horizon to that place which is called 'tholus', and it is called the 'horizon of the equator.' This circle, as has been said, divides the world into two equal halves; one of them is always seen from the middle of the earth, whereas the other one is always concealed.* [63] *Analogous to this circle, any other circles that cut the world into two halves and thereby separate a visible half from a concealed one are called 'horizons' of those places where they cause this; and the pole of the horizon for the middle of the earth is that point which we have defined as the tholus.* [64] *For a line which originates from the centre of the world, through the middle of the earth and prolonged until the heaven, will touch the surface at the tholus of the world, which is also known as the 'zenith',*[36] *as it stands above the heads of those who live at the middle of the earth. This point is the upper pole of that great circle which we have established to be the horizon of the middle of the earth,* whereas the lower pole is the nadir.[37] [65] *In this manner, it will be possible to find the horizons and their poles for every place. For every single point on the earth has a particular point opposite itself on the surface of the great heaven, which is very reasonably called the 'zenith' of that place. To whatever circles these points are the poles, they produce those circles as the horizon for that place where the point on the surface of the earth lies right under the zenith.*

[66] As this is the case, it should be clear that *every point on the surface of the earth has its own horizon, which at that place divides the world into two halves.* For every horizon divides the world into two halves; because the horizon is a great circle which divides the world into two halves, one of which is always visible and the other never, but hidden and concealed. But most of all we need to consider the position of the celestial equator above the horizon. For occasionally it is found at right angles to it, but mostly in either an acute or an obtuse angle. This causes much disparity between days and nights. [67] *Only the horizon of any point on the terrestrial equator, as it passes through both poles of the world, sees the celestial equator at right angles above itself.* For the latter's course is through the zenith and the foot point, which are the poles of the horizon. [68] *All other horizons see only one pole while concealing the other.* For they do not pass through the poles. *That is why the*

[35]Stephen makes correct use of the term 'tholus', whose corrupted form 'torus' in L1 irritated Millàs Vallicrosa; cf. *idem*: *Los traducciones orientales*, p. 288, note 1.

[36]'zenith': lit., 'head point'.

[37]'nadir': lit., 'foot point', corresponding to the Arabic term 'samt al-qadam', which term, however, does not appear in *On the Configuration*.

inclinatur. Hec inclinatura alia est in septemtrionalibus, alia in australibus orizontibus. *In septemtrionalibus quippe, quando septemtrionalis polus videtur,* australis occultatur, *inclinatur circulus recti diei in austrum super orizontem,* et est latus eorum versus septemtrionem angulus, acutus ad austrum. *Contra fit in australibus orizontibus. Australes etenim orizontes suum vident polum* septemtrionali occulto. *Quamobrem ab ipsis magnus circulus inclinatur in septemtrionem.*

[69] ¶*Est et magnus circulus quidam, quem imaginamur transiens super duos mundi polos et punctos capitis* et pedum, polos scilicet orizontis. *Hic dicitur circulus meridiei in eodem, per cuius polos transit, orizonte. Qui ea causa meridionalis dictus est, quoniam cum centrum solis in eo super terram cito motu magne spere figitur, media diei pars eo loco transiit, media superest, sicque cum sol medietatis illius eiusdemque circuli, qui sub terra est, quempiam punctum tetigerit, medium preterierit noctis.* [70] *Hunc autem circulum si imaginatione crescentem usque centrum mundi duxeris, secabit superficies eius superficiem sui orizontis in duo media* ab eodem pari iure divisa. *Dividunt igitur sese hii duo circuli in duobus diametricis punctis, alter quorum in septemtrione est, unde et septemtrionalis dicitur, alter in austro australis.* [71] *Circulus quoque recti diei orizontem in duas equaliter seiungere partes dictum est in duobus diametricis punctis, quorum orientalis alter est, alter occidentalis.*

[72] *Habet igitur quisque orizon* IIII^{or} *punctos, per quos in* IIII^{or} *equales quadrantes distribuitur. Quorum primus est orientalis, secundus* huic oppositus *occidentalis, tercius septemtrionalis,* cui diametrice quartus opponitur australis. Quadrantum autem duo sunt septemtrionales, quorum fines terminantur in puncto septemtrionali, duo australes, quos dividens australis discriminat. Eorum duo sunt orientales, qui puncto orientis iunguntur, duo occidentales, qui a puncto occidentali nuntiantur. *Recta vero linea, que extenditur a puncto* | *occidentis usque orientalem punctum, dicitur linea orientalis et occidentalis* et commune dividens circuli recti diei et orizontis. [73] *Sed que a puncto septemtrionis usque australem, dicitur linea septemtrionis et austri et linea meridionalis, et est communis divisio circulorum meridiei et orizontis.*

[74] ¶*Quartum quoque magnum imaginamur circulum,* cuius maxima est in astronomia opera, per cuius facultatem altitudinem solis et quarumlibet stellarum super orizontem cognoscitur. *Hic ergo transit per punctum capitis, qui est polus orizontis et centrum mundi* ipsum in duo media dividens. *Eius positio super orizontis circulum rec-*

---

13 crescentem] crescentrem (L) **15** diametricis] diametris (L) **16–17** orizontem] *corr. ex* oriontem (L) **17** diametricis] *corr. ex* diametris (L) **20** oppositus] positus (L)

celestial equator is inclined above to them. This inclination is different in northern horizons than in the southern ones. *In the northern horizons, since the north pole is visible* and the south pole is concealed, *the equator is inclined southwards above the horizon,* and its angle towards the north is obtuse while that towards the south is acute. *The contrary occurs in the southern horizons. For the southern horizons see their own pole,* whereas the north pole is concealed. *For this reason the great circle is inclined northwards.*

[69] *There exists another great circle which we imagine to pass through both poles of the world and the zenith* and the nadir, that is, the poles of the horizon. *It is called the 'meridian' to that horizon whose poles it passes through. It is called meridian because, when the fast motion of the great sphere places the sun on this circle above the earth, half of the day has passed for that place while the other half remains to come; similarly, when the sun reaches any point of that circle's half which lies below the earth, half of the night has passed.* [70] *If you imagine this circle extended until the centre of the world, its plane will cut the plane of its horizon into two halves,* while being in the same way divided by the latter. *These two circles thus divide each other at two opposite points, one of which is in the north, hence called the 'northern point', and the other one in the south and called the 'southern point'.* [71] *Also the equator was said to divide the horizon into two equal parts at two opposite points, one of which is the 'eastern point', the other one the 'western point'.*

[72] *Therefore, every horizon has four points by which it is divided into four equal quadrants. The first is the eastern point, the second*—opposite to the first—*the western point, the third is the northern point,* diametrically opposite from which lies *the fourth, the southern point.* Two of the quadrants are in the north and terminate at the northern point, and two in the south separated and defined by the southern point; two are towards the east, meeting at the eastern point, and two in the west, named after the western point. *And the straight line that extends from the western to the eastern point is called the 'east-west line'* and the common intersection of the equator and the horizon. [73] *The line from the northern to the southern point is called 'north-south line' and also 'meridional line',*[38] *and it is the common intersection of the meridian and the horizon.*

[74] *Furthermore, we imagine a fourth great circle,* which is used more than any other in astronomy, by means of which the altitude of the sun and any other star above the horizon can be determined. *It passes through the zenith, which is the pole of the horizon, and the centre of the world,* dividing the latter into two halves. *Its position*

[38] Arab. add.: 'It is that line which is drawn on the surfaces of sundials.'

tos habet angulos ab eodem in duo divisus media super duos diametrice oppositos punctos, amborumque sese medias dividunt superficies, quorum communis est divisio recta illa linea, que est inter duos punctos, in quibus ab invicem ligantur circuli, cuius medium centrum obtinet. *In hunc itaque modum imaginatus magnus circulus* 5 *altitudinis circulus nuncupatur,* [75] *omnesque quos imaginatio duxerit similes, circuli alti dicentur. Duos vero punctos, in quibus orizontis dividit circulum, partium punctos vocamus,* hoc est, in quam partem idem sit circulus; *lineam rectam, que interest duobus his punctis, partis lineam;* [76] *arcum autem circuli orizontis, qui est inter punctum australem vel septemtrionalem et inter punctum partis, arcum partis. Quandoque etiam* 10 *arcum circuli orizontis, qui est inter punctum orientis aut occidentis et inter punctum partis, dicimus arcum partis,* sed et arcum extractum ab orizonte. Volvitur autem hic circulus in puncto capitis et puncto pedum quemadmodum et stellarum centra, per que eum discurrere videt imaginatio. Et quoniam de imaginatione circulorum none spere et eorum in invicem proportionibus satis dictum est, signorum circulus 15 eiusque partitiones quomodo fiant, videamus.

13 que eum] quem eos videt] vestra (L)

above the horizon is at right angles and it is divided by the horizon into two halves at diametrically opposite points. And also the planes of both circles divide each other in half along that straight line, whose middle is at the centre, between the two points where the circles are bound to each other. *The great circle that is imagined in this way is called the 'altitude circle',* [75] *and all circles that are imagined analogously will be called 'altitude circles'. The two points where it divides the horizon are called the 'points of the directions',* that is, the direction towards which that circle lies, *whereas the straight line between these points is called the 'line of the direction'.* [76] *And the arc on the horizon from the southern or northern point to the point of the direction is named the 'arc of the direction'. The arc on the horizon from the western or eastern point to the point of the side is occasionally called 'arc of the direction', too,* but also the 'arc taken from the horizon'. Now, this circle revolves about the zenith and the nadir in the same manner as the centres of the stars through which it is imagined to pass. But as we have said enough about the imagination of the circles on the ninth sphere and their mutual relations, we will now consider how the ecliptic and its partitions are defined.

[107] <***U***>[(L)]*numquemque* VII *planetarum suam habere qua defertur speram superius ostensum est, singulasque speras proprio moveri motu propositum est. Sed et ex superioribus manifestissime intellectum putamus, quoniam cuiusque centrum stelle motu spere, in qua est circumvolvens, facit imaginarium circulum in sue spere corpore.* Sol vero suam habet speram, que proprio circunfertur motu. *Quando igitur spera solis movetur ab occasu in ortum*—huiusmodi enim motus ipsius est—*centrum solis motu spere circumlatum facit in eadem spera magnum circulum.*

[108] *Hunc imaginati sunt* astronomie peritissimi *in eam crescere magnitudinem, ut totum in duo media mundum dividat. Facit igitur in superficie magni celi magnum circulum, quod idem cum mundo centrum sortitur. Hic est, qui signorum circulus dicitur,* cui similem in spera fixarum imaginamur. Omnes enim hi tres magni circuli in eadem superficie positi sunt, et qui in spera solis, que quarta sit, cuius spere solis poli sunt, et qui per eiusdem fiunt augmentationem, alter quorum est in superficie magni celi, alter autem superficiei octave spere insidet, que multitudinem stellarum gerit, quorum sunt idem poli qui et ipsius octave. [109] Tercius horum, qui altior et in superficie magni celi videtur, *secat circulum recti diei in duo media super duos diametricos punctos. Alter dicitur punctus equalis veris, alter equalis autunni,* et qui veris est, caput Arietis, qui autunni, Libre. In his enim eisdem temporibus solis centrum figitur.

[110] *Imaginamur etiam magnum circulum, qui transeat super duos polos circuli recti diei, qui sunt poli mundi tangentem quoque duos polos circuli signorum. Hic secat circulum signorum super duos* | *diametrice oppositos punctos,* qui inclinantur a recto circulo, *alter in septemtrionem, alter in austrum. Atque ille, cuius est in septemtrionem inclinatio, dicitur punctus mutationis estatis,* qui et ipse punctus est capitis Cancri, *cuius autem videt auster inclinationem, punctus est mutationis hiemis* et capitis Capricorni. Hii itaque a duobus punctis equalitatis distant equaliter, utrimque scilicet .s. gradibus, IIII^{ta} unius pleni circuli parte.

f. 9v

10 qui] quod (L) **12** qui] que in] *add.* (L) quarta] quanta **13** qui] quod (L) **17** Alter] *corr. ex* Aliter (L) **21** tangentem] tangente (L)

1.4 <The Ecliptic>

Config. 5: The Orb of the Ecliptic

[107][39] *It has been shown above that each of the seven planets has its own sphere by which it is moved, and we have stated that every single sphere moves with a particular motion. We also believe that it has become totally clear from the above that the centre of every star, when revolving by the motion of the sphere to which it is fixed, produces an imaginary circle inside the body of its sphere.* Yet, the sun has its sphere which revolves by its own motion. *Therefore, when the sphere of the sun moves from west to east*—for such is its motion—*the centre of the sun is revolved by the motion of the sphere and produces a great circle in that sphere.*

[108] Most experienced astronomers *imagined this circle extended to such a size that it divides the entire world into two halves. It thus produces a great circle on the surface of the great heaven which happens to have the same centre as the world. This is the one that is called the 'ecliptic',* to which we imagine a corresponding circle on the sphere of the fixed stars. All these three great circles are placed in the same plane; the one in the solar sphere, which is the fourth sphere, whose poles are those of the sphere of the sun, as well as the circles which are produced by extending that first one, one of which lies on the surface of the great heaven and the other one, on the surface of the eighth sphere, which sphere bears the multitude of the stars, which have the same poles as the eighth sphere itself.[40] [109] The third one of them, which is the highest and which is seen on the surface of the great heaven, *cuts the equator into two halves at two diametrically opposite points. One is called the 'point of the spring equinox' and the other, the 'point of the autumnal equinox';* that of spring being the head of Aries and that of autumn being the head of Libra. For the centre of the sun is stuck to these points at those times.

[110] *We further imagine a great circle that passes through the two poles of the equator, which are the poles of the world, and that touches also the two poles of the ecliptic. This circle intersects with the ecliptic at two diametrically opposite points* which are inclined from the equator, *one to the north and one to the south. The one that is inclined northwards is called the 'point of the summer solstice',* which is the point at the head of Cancer, *whereas the one whose inclination is seen by the south is the 'point of the winter solstice'* and the head of Capricorn. Hence these points are equally distant from both equinoctial points, that is, by 90 degrees, the fourth part of a

[39]The content of *Config.* 4 (passages 77–106), is postponed in the *Liber Mamonis* to follow the content of *Config.* 6 (passages 140–159).

[40]Arab. add.: 'The two poles of this circle are called the poles of the ecliptic orb.'

[111] *Est et alius magnus circulus, quem imaginamur transire per polos circuli signorum et duos punctos equalitatis. Dividunt igitur hi duo magni circuli circulum signorum in* IIII*or* *equales partes,* quorum alter per IIIIor polos, duos mundi et duos circuli signorum, et punctos mutationum transit, alter a duobus polis circuli signorum habens initium circulum signorum medium dividit in principiis Arietis et Libre, scilicet punctis equalitatis. *Harum partium quem primum ponimus quadrantem,* is est, qui a capite Arietis usque caput Cancri, *a puncto scilicet equalitatis veris usque ad punctum mutationis estatis, continetur. Hic nempe quadrans tempus veris a reliquis facit differre temporibus. Cum etenim sol proprie spere motu in huius initium quadrantis intraverit, ver initiare dicitur.* Huius temporis eo usque protenditur spatium, usque dum sol huius quadrantis transcurso arcu Geminorum et Cancri tenet confinia. [112] *Secundus autem quadrans,* quem ab initio Cancri usque finem Virginis ponimus sive initium Libre, *inter punctum mutationis estatis et punctum equalitatis autunni, estivi temporis quadrans est.* Fit enim intrante sole in huius quadrantis principium veris finis estatisque initium. *Cuius tam longum tempus esse contenditur, quamdiu sol in eo quadrante movetur.* Solis nempe centro ultima eiusdem tenente estatis calor terminatur. [113] *Quadrans vero tercius,* cui est initium caput Libre finisque ultima Sagittarii, *is est, qui inter punctum equalitatis autumpni et punctum mutationis hiemis conspicitur. Hunc itaque autumpnalem quadrantem aiunt ea de causa, quoniam sole in ipsius quavis particula morante, autumpni tempora perlabuntur. Quorum tanta est longitudo, quantum in huius quadrantis arcu transcurrendo sol temporis spatium consumit.* [114] *Quartus deinde quadrans,* qui et ultimus, a primo capitis Capricorni puncto usque ad extremitatem Piscium surgit, et *est inter punctum mutationis hiemis et punctum equalitatis veris,* quem nos quadrantem hiemalem non absurde possumus dicere, eo quod per ipsum a ceteris tribus hiemis differat temporibus. *Cum etenim quadrantis eius quasvis partes solis centrum sua illustrat presentia, hiemis torpent omnia frigoribus. Totum igitur hiemale tempus in* IIII*ti* *quadrantis transigendo arcu solis centrum conficit. Atque in hunc modum quos posuimus circuli signorum* IIII*or* *quadrantes anni solaris tempora discernunt.*

Punctos autem mutationis estatis et hiemis dicimus, non quod in ipsis estatis aut hiemis alterentur tempora, hoc enim esset eorundem mutatio, sed quod veris et autumpni finita temperie, tempus in estatem mutatur aut hiemem. Idem super

9 motu] motum

complete circle.

[111] *There is still another great circle, which we imagine to pass through the poles of the ecliptic and the two equinoctial points. These two great circles thus divide the ecliptic into four equal parts;* one of them passes through the four poles, that is, the two of the world and the two of the ecliptic, and through the points of the solstices, whereas the other one starts from the two poles of the ecliptic and divides the latter at the beginnings of Aries and Libra, that is, at the equinoctial points. *Of these parts, the quadrant which we put first* is the one from the head of Aries until the head of Cancer, that is, *from the spring equinox until the summer solstice. This quadrant distinguishes the season of spring from the other seasons. For when the sun, by the motion of its own sphere, enters into the beginning of this quadrant, it is said to initiate spring.* The time span of this season extends until the moment when the sun has passed through that quadrant and stands at the border of Gemini and Cancer. [112] *The second quadrant,* which we assume to be from the beginning of Cancer until the end of Virgo or the beginning of Libra, *between the point of the summer solstice and the point of the autumnal equinox, is the quadrant of summertime.* For when the sun enters into the beginning of this quadrant, spring ends and summer begins. *The latter lasts as long as the sun is moving in this quadrant.* For when the centre of the sun has reached the final limit of that quadrant, the heat of the summer ends. [113] *The third quadrant,* which begins at the head of Libra and terminates at the end of Sagittarius, *is that which is seen between the point of the autumnal equinox and that of the winter solstice. It is called the autumnal quadrant, because when the sun stands in any part of it, the times of autumn are lapsing. They last just as long as it takes the sun to pass through the course on the arc of this quadrant.* [114] *Finally, the fourth quadrant,* which is also the last, reaches from the first point of the head of Capricorn until the end of Pisces, and *it lies between the point of the winter solstice and the point of the spring equinox;* we can very reasonably call it the winter quadrant, because it distinguishes the winter from the other three seasons. *For when the centre of the sun illuminates by its presence any parts of that quadrant, everything is struck by the cold of winter. The entire season of winter is thus determined by the centre of the sun passing through the arc of the fourth quadrant. In this way the four quadrants of the ecliptic which we have just presented define the seasons of the solar year.*

However, we do not call the points of the summer and winter solstices[41] such because the climate of summer or winter would change in them, which would mean a transformation of the seasons themselves; it is rather because the climate

[41] 'solstices': lit., 'changing points'.

punctis equalitatis veris et autumpni intelligendum est; neque enim tunc magis sunt equalia veris et autumpni tempora, sed terminatis et finem habentibus inequalitatibus sive caloris estatis sive hiemis frigoris, in quibus temperies temporum initiatur, ver dicitur et autumpnus.

f. 10r 5 | [115] *Quando itaque sol movetur per ferentis spere motum,* de qua posterius dicendum est, *incipit moveri a quolibet puncto sui circuli circumferturque a motu sue spere, usque dum in eundem a quo incepit moveri revertatur punctum. In quem cum redierit, solaris nobis anni completum esse tempus intimatur.* Omnium enim IIII[or] pervolavit arcus quadrantum. [116] *Initium autem anni solaris ponunt in capite Arietis, ut* 10 *tunc scilicet annus eius incipiat, cum centrum ipsius in puncto equalitatis veris movetur.* Nec hoc quidem ratione sed nature insectatione dictum est. Sicut enim animalium individua a sue creationis in matrum officinis articulo habent principium, a quo usque certum tempus crescunt, peractoque statu incipiunt minui et dissolvuntur, ita quoque secundum creaturarum insensibilium de matre terra productionem an- 15 nus solis dispositus est. Intrante namque sole caput Arietis plante oriuntur, estate maturantur, autumpno decidunt, hieme frigoris torpore stringuntur. Nam etsi quas estate oriri videas aut hieme, hoc amplissimo eorundem fit frigore aut calore, que temperantur, rerum frigiditas estatis calido, aliarum calor hiemis frigido. Sed etsi qua sunt plurimum calida et humida, hec temperantur frigido autunni et sicco, 20 tumque deinde ad ortum festinant. Nichil enim quod maximam habet humorum intemperantiam oritur, nisi prius temperetur. Ea vero temperies plurima fit in veris temperato tempore. Recte igitur anni solaris initium in puncto equalitatis veris positum est. Et hec quidem super septemtrionalem habitationem sic fiunt. Nam in australibus locis contra fit. Septemtrionalis etenim estas austro hiemps est, no- 25 straque hiemps illic estatis ardor maximus. Sed de his hactenus. Nunc ad nostra revertamur. *Postquam igitur sol ingressus punctum capitis Arietis, ascendit in septemtrionem; veris est tempus, usque dum ascensu in septemtrionem completo Cancri primi puncti tangit confinia, cui quanto fit propinquior, tanto maiorem nobis evaporat in septemtrionali habitatione fervorem.* Cuius rei que causa sit, inter philosophos non modica 30 fuit contentio. Aliud enim Aristotelicis quibusdam, aliud autem aliis visum est, quorum controversias inserere non videtur inutile, verum brevi, ne a proposito longius deviemus.

---

18 frigiditas] frigidas (L) **23** hec] hoc **26–27** septemtrionem] septemtrione (L) **27** ascensu] ascensum (L) septemtrionem] septemtrione

changes into summer or winter with the end of the moderation by spring and autumn. In the same way one must interpret the points of the spring and autumnal equinoxes.[42] For spring and autumn are not most equal then; it is rather that when the inequalities have come to an end, be it those of summer's heat or winter's cold, and temperate climate begins, this is called 'spring' and 'autumn'.

[115] *When the sun is moved by the motion of the deferent sphere,* which we need to speak about later, *it starts moving from any point of its circle and is moved around by the motion of its sphere, until it returns to the same point from where it started. When it has come back to that point, this indicates to us that one complete solar year has passed.* For it has passed through the arcs of all four quadrants. [116] *The beginning of the solar year is assumed to be at the head of Aries, that is, the year of the sun begins when its centre is moving at the point of the spring equinox.* This is not said for any logical reason but in analogy of nature. For just like individual animals, from the moment of their creation in their mothers' wombs,[43] have a beginning, from which they grow until a certain moment and, having passed their life-size, begin to shrink and fade away, the year of the sun is equally arranged also in accordance with the production of insentient creatures from mother earth. For when the sun enters into the head of Aries, plants begin to pullulate, in summer they ripen, in autumn they fall to the ground, and in winter they are constricted by cold rigor. Even if you see some plants pullulating in summer or winter, this is due to their surfeit of cold or heat which is becoming balanced; coldness of things by the warmth of summer, and heat of other things by the cold of winter. Even if some are extremely hot and wet, they are moderated by the cold of autumn and by dryness and will then urge to pullulate. For nothing that has a great excess of moisture will pullulate, unless it becomes moderated before. But this moderation happens mostly in the temperate climate of spring. The beginning of the solar year, therefore, is rightly placed at the spring equinox. But this occurs in this manner in the northern inhabited region, whereas in the southern places the contrary occurs. For the summer of the north corresponds to winter in the south, and our winter is the time of greatest heat there. But this is enough on that subject. Let us now return to our actual concern. *After the sun has reached the head of Aries, it rises to the north. Spring lasts until that rising to the north has finished and the sun has reached the point where Cancer begins, during which period it eradiates already more heat to us the closer it comes to that point.*[44] The

---

[42] 'equinox': lit., 'equality'.

[43] 'wombs': lit., 'workshops'. For a similar metaphor, cf. Macr. *Somn.* I,6,63: 'verum semine semel intra formandi hominis monetam locato hoc primum artifex natura molitur.'

[44] Arab. add.: 'The heat intensifies because of the persistence of the sun's heating of the air.' Stephen replaced the sentence with the following, extended discussion of different theories on the subject.

Horum primi antiquiores et ante Aristotilem, multique post ipsum sed non ex ipsius sequacibus, estatis calorem fieri propter moram solis super circulum orizontis dixerunt, huiusmodi habentes ad sue comprobationem sententie argumentum.

¶Solis centro in puncto Arietis aut Libre posito equalis fit eius super terram 5 mora occultationi eius sub terra. His in diebus anni tempus maiorem habet a calore et frigore temperiem. Eo autem in septemtrionalia ascendente signa maiores fiunt dies, quoniam maior est mora solis super terram, minor sub terra. Quibus in temporibus calorem aeris crescere nulli dubium est. Cum vero sol in ultimo Geminorum a primo Cancri gradu volvitur, qui locus eius maior est in septemtrio- 10 nalem exaltatio, maiorem quam in quibuslibet aliis diebus super terram consumit moram, in quibus maximum esse aeris calorem nulli dubium existimant. A capite vero Cancri usque finem Virginis quanto plus sol descendit, tanto minor fit eius super orizontem mora, tantoque calor aeris minuitur. In capite Libre temperatus est, quia equales dies et nox, quod equalitas more solis super terram et subtus facit. f. 10v 15 Sed a capite Libre usque finem Sagittarii descendente sole | in austrum minorantur dies ab equali. Minus enim sol super terram moratur, que res facit naturale frigus terre augmentari et vires, quas perdiderat, resumere. Parum enim sol super terram moratur, unde et ipsam parum calefacit, tantumque ab eo, ut habitari terra ab incolentibus possit, efficitur. Hoc autem frigus plurimum est sole in vicinitate capitis 20 Capricorni commorante. Minor enim super terram eius est mora. Inde ascendens ad altiora frigiditatem minuit aeris sua longiori presentia. Qui postquam venerit in punctum verne equalitatis, temperatus est nichil habens aer immodicum. Hii calidum et siccum solem confitebantur, nec aliter tantas temporum varietates facere, nisi innatum haberet naturalemque calorem.

25 Aliter autem Aristotelici complures senserunt, neque solis hec calore fieri arbitrabantur, sed per ipsius motum radii quasdam adhibentes rerum similitudines. Lapis, aiebant, ab aliquo proiectus si in terram oblique ceciderit, repulsus ab eadem consurgens latum angulum in aere facit. Si vero recte tamquam a puncto capitis decidat, rectus in aerem eadem via qua descenderat refertur, neque angulum

---

10 exaltatio] ex altero **11** calorem] *corr. ex* calorim (L) **20** Capricorni] *ex* Caprico *corr. in* Capricorno (L) **21** longiori] *corr. ex* longioris (L)

reason for this has been discussed very controversially among the philosophers. For some Aristotelians have believed in one theory and others in another. It seems very useful to include their conflicting views here; but in short, in order not to digress too far from our actual concern.

The first group are the older ones, before Aristotle, but also many after him who were not among his followers, who said that the heat of the summer was due to the duration of the sun above the horizon, giving the following argument as support for their theory.

When the centre of the sun stands at the point of Aries or Libra, its time span above the earth becomes equal to its concealment under the earth. This is the time of the year when the climate is most temperate between heat and cold. But when it rises to the northern signs, the days become longer, because the sun's period above the earth is longer and its period beneath the earth is shorter. And nobody will deny that in these times the air gets warmer. But when the sun stands at the final point of Gemini towards the first degree of Cancer, which is its highest elevation to the north, it stays above the earth for a longer time than on any other days, while they consider it unquestionable that the air reaches its greatest warmth in that time. But the further the sun descends from the head of Cancer until the end of Virgo, the shorter is its time above the horizon and the more decreases the heat of the air. At the head of Libra it is temperate, because day and night are equal as a result of the equal times of the sun above and below the earth. But when the sun descends southwards from the head of Libra until the end of Sagittarius, the days fall shorter from equality. For the sun stays shorter above the earth, which lets the natural cold of the earth increase and recover the strength which it had lost. For the sun stands only for a very short time above the earth and, therefore, warms it only a little and just enough to make it inhabitable. This cold is greatest when the sun stands near the head of Capricorn, because then is its shortest time above the earth. Then, rising upwards, it reduces the cold of the air by its longer presence, and when the sun has reached the spring equinox, the air is temperate and without any imbalance. These people said that the sun was hot and dry and that it would in no other way cause such a difference between the seasons, if it had not an innate and natural heat.

Yet, many Aristotelians thought differently. Using certain analogies, they did not believe that this was caused by the heat of the sun but by the motion of its rays. They said that a stone, if thrown by someone and falling on the earth in an oblique angle, bounces from the earth and, while rising again, produces a wide angle in the air. But if it falls vertically, like from the zenith, it will be thrown back on the same

facit. Maiori autem impulsu que recte cadunt in aerem revertuntur quam que oblique. Hoc enim angulus oblique et opposita recte cadentium regressio manifestat. Maiorem igitur motum faciunt que recte, quam que oblique cadunt.

Huic comparantes solis radium similitudini, cum in capite Capricorni eius
5 centrum volvitur, quod a puncto capitis longe nobis est, ipsius radium septemtrionalem percutere terram oblique ponunt, a qua repulsus latum angulum facit. Cum vero capitis nostri puncto propinquior est, rectior in terram cadit, eiusque resurgentis angulus acutus est; quanto longius recedit a puncto capitis, tanto latior. Omne autem quod movetur, motum facit. Neque enim moveri potest, nisi
10 id, in quo positum est, moveat. Motum autem rei recte in terram cadentis maiorem esse quam obliquorum manifestum est. Maiori enim vi fertur in terram quod, cum ceciderit, recte in aerem redit quam quod oblique. Magis ergo movet aerem solis radius, cum sol nostri capitis puncto propinquior est, minus cum remotior. Per motum autem calor fit, quantoque minor motus fuerit, tanto minorem, quanto
15 maior, tanto maiorem excitat calorem. Motus ergo radii solis cum sine angulo septemtrionalem percutit habitationem, quod fit in fine Geminorum a capite Cancri, plurimum habens impulsum ex plurimo motu maximum facit aeris calorem.

Atque hee sunt super hac re illorum dissentientes sententie, quorum neutros veritatis secutos arbitramur vestigia. Nam et hos, qui more solis super terram cau-
20 sam imponunt, plurimum errasse et Aristotelicos, qui motui radii tantum, a veritate deviasse videmus. Ceterum primi quidem recte, quod solem calidum dixerunt, et illud peroptime Aristotelici, quod de similitudine dictum est, in hoc uno maxime peccantes, quod nec solem calidum et motu radii calorem aeris fieri arbitrabantur.

Nobis autem alia occurrit utrisque partim oppugnans ratio. Nam et ab his
25 magnopere dissentimus, qui caloris augmentum mora solis super orizontem fieri dicunt, et ab illis, qui motu radii non etiam calore solis. Adversus quos nobis firmissima et patenti subnixe veritate rationes habentur. Non enim sine causa eorum refellimus sententias, et primo quidem, quare prima falsa sit ratiocinatio, post etiam adversus secundam contendemus.

30 Mora itaque solis super orizontis partes nequaquam fieri aeris calorem hac ra-

---

11 quod] qui (L)

way on which it had fallen and will not produce an angle. But those objects that fall vertically are thrown back into the air by a stronger impact than those that bounce in an oblique angle. This is obvious from the angle of obliquely falling objects and the direct reversal of vertically falling ones. Therefore, vertically falling objects produce a stronger motion than those that fall at an angle.

Comparing the radiation of the sun to this analogy, they say that when the centre of the sun stands at the head of Capricorn, which for us is far from the zenith, its radiation hits the northern part of the earth in an oblique angle and, when reflected from it, produces a wide angle. But when it stands closer to our zenith, it falls more vertically to the earth and its angle of reflection is acute, whereas it gets more obtuse, the further it removes from the zenith. But all that is moved produces a motion, for nothing can move unless that in which it is placed moves. But it has been shown that the motion of an object that is falling vertically is stronger than that of objects which are falling at an angle, because what returns into the air vertically after its impact on the earth has hit the earth with more force than what returns obliquely. The sun's radiation therefore moves the air more when it stands closer to the zenith, and less when further remote. Motion, however, produces heat, and the smaller the motion, the less heat is caused, whereas the greater the motion, the greater heat it causes. Hence, when the motion of the sun's radiation hits the northern inhabited part of the earth vertically, which occurs at that end of Gemini towards the head of Cancer, it has the greatest impact from greatest motion and, therefore, produces the greatest heat in the air.

These are their conflicting theories about this subject, which we think are both missing the line of truth. For we can see that those who attribute the cause to the duration of the sun above the earth have erred, just as those Aristotelians went astray who ascribe it only to the motion of the radiation. Now, the former correctly stated that the sun is hot, while also the Aristotelians' analogy is excellent. But they were fatally wrong in that single point that they did not believe that the sun is hot *and* that the heat of the air is produced by the motion of the radiation.

We, however, came to another explanation, which partly contradicts both of them. For we disagree fundamentally with those who say that the increase of heat is caused by the duration of the sun above the horizon and also with those who say that it is by the motion of the radiation and not also by the heat of the sun. Against both positions we have arguments based on incontrovertible and obvious truth. For we do not reject their doctrines without reason. We will first discuss why the first explanation is wrong; afterwards we will argue against the second one.

The following reason makes it evident to us that the heat of the air is by no

tione manifestum habebimus. Duas esse mundi glaciales partes omnis doctorum successio tradidit, quarum alteram septemtrionalem, alteram | dicunt australem, in mundi ultimis constitutas, pari numero temperatas, quarum mediam perustam aiunt zonam. Due igitur glaciales polis mundi supposite, perusta sub recti diei circulo in recta linea solis maximum patitur ardorem. Iuxta hanc a sinistro iacet septemtrionalis habitatio temperata ipsius incendio et septemtrionalis glacialis frigore. Pari ratione temperatur australis media peruste <et>[L] australis glacialis, sed tamen inhabitata. In perusta zona dies omnes equalitate gaudent, quod in sequentibus manifestissimum fiet. Equalis enim eademque mora solis super terram, nec unquam tardius aut tempestivius sol surgit aut occumbit. In temperatis enim dierum discordia est. Inclinatis enim in alteram partem orizontibus inclinatur recti diei circulus cum paralellis circulis. In his igitur mora solis super terram nunc maior, nunc minor videtur. In suppositis vero mundi polis, qui glatiarum zonarum mediis, VI continuis mensibus dies totidemque nox protenditur. Diem autem nihil aliud esse nisi solis lucentis super terram claritatem phisicum est. Quamdiu igitur dies est, sol super terram est. Dies autem, ut dictum est, sub polis VI mensium spatia continet. Id si non esse ita contenditur, paulisper concedi argumentandi gratia postulamus. In sequentibus enim sic esse nec aliter posse fieri demonstrabimus.

Sex igitur mensibus sol cum in alterum inclinatur polum, super illam terram est, cui idem supereminet polus. Si igitur calor aeris mora solis super terram fit, cumque equalis super et subtus moratur, temperatum sentimus aerem, qui sub recto circulo aer est semper temperatus, qui ab illo in utramque partem recedit aliquando calidior, aliquando frigidior, qui polis suppositus est VI mensibus calidissimus, totidemque frigidissimus esse inevitabili argumentatione conclude.

Hoc autem quanta involvatur erroris et falsitatis caligine, neminem qui aliquantulum in philosophia profecerit latere arbitror. Nam perusta quidem zona ma-

5 a sinistro] *corr. ex* sinistra (L)

means a result of the sun's duration above the parts of the horizon. Every tradition of learning teaches that there are two icy parts of the world, one called the northern one and the other the southern one, which are located at the extremities of the world, and that there is the same number of temperate parts, between which lies the so-called 'burnt zone'. The two icy regions thus lie beneath the poles of the world, whereas the burnt zone, lying on the terrestrial equator beneath the celestial equator, suffers the greatest heat of the sun. Next to it on the left[45] is the northern inhabited zone, which is temperate between the former's heat and the cold of the icy north. Analogously, also the southern region between the burnt zone and the southern ice is temperate, but, nonetheless, uninhabited. In the burnt zone all days enjoy equality, which will become clear in the following. For the time of the sun above the earth is equal and the same, and the sun never rises or sets earlier or later. But in the temperate regions there is inequality among the days. For when the horizons are inclined to one side, the horizon and the parallel circles will also be inclined. In these circles, therefore, the time of the sun above the earth is found sometimes longer and sometimes shorter. But in the places right under the poles of the world, being in the middle of the icy zones, the day extends over six consecutive months and the night over just as many. But it is a physical fact that a day is nothing else than the period during which the sun is shining above the earth. Hence, as long as it is daytime, the sun stands above the earth. But it has been said that under the poles a day comprises the period of six months. If anyone wanted to object that this is not the case, we ask for a little patience for the sake of the argument. Because we will show in what follows that it is true and cannot be any different.

During the six months in which the sun is inclined towards one of the poles, it stands above that part of the earth above which that pole is looming. Thus, if the heat of the air was due to the sun's duration above the earth and if the air was temperate when the time of the sun above the earth equals its time beneath, one would inevitably have to conclude that the air under the equator is always temperate, the air on either side of it sometimes hotter and sometimes colder, and the air under the poles for six months extremely hot and for the same time span extremely cold.

I think that no one who has made even little progress in philosophy will overlook how riddled this is with error and falsity. For the burnt zone receives period-

[45]Stephen associates 'north' with 'left', in agreement with Arabic usage. See also the south-up orientation of Stephen's diagrams in Fig. 2.4 (p. 244) and Fig. 3.7 (p. 278).

gis et minus circa caloris qualitatem suscipit. Magis quidem, cum sol in circulo magno ipsique proximis paralellis volvitur, minus autem, cum plurimum inclinatur aut in septemtrionem aut in austrum. Temperiem autem caloris nunquam aut novit aut suscipit. Hoc enim omnium testatur philosophorum doctrina. Sed et illud quoque, mundi polis suppositas terras perpetua dampnatas hieme nichil vivum nichilque animatum habere. De quibus, quoniam sola comprobantur ratione, cui tamen auctoritas accedit, satis dictum est.

De illa nunc que septemtrionalis est et temperata dicitur habitatio, videamus. Facilius enim sensui corporis ratio animi assensum accomodat. Hec autem in VII divisa est climata, quorum prius calidius, quod peruste zone iungitur, secundum minus, III^{um} adhuc calidum, IIII^{tum} autem temperatum, V^{tum} frigidum, VI^{tum} paulo plus, VII^{mum} frigidius. Quibus in omnibus si consideremus que sit inequalitas dierum, illud quoque peroptime intelligemus, in quo sit maior solis mora super terram. Et primum quidem quod vicinius est peruste viciniores habet equalitati dies et noctes. VII^{mum} autem, quod remotissimum est, ceteris omnibus maiorem habet discordiam. Maiorem itaque sol in eo super terram moram consumit quam in VI^{to}. Eodem modo de reliquis intelligendum est. Amplius autem secundum a primo, terciumque a secundo, quartumque a tercio et duobus primis maiores habent dierum discordias. Maiorem enim super suos orizontes moram solis patiuntur. Sed quartum temperatum, tercium calidum, secundumque a primo minus calidum | et ratione colligitur et vulgus sentit. Non igitur mora solis super terram quemadmodum dierum ita etiam frigorum et calorum facit discordiam.

Sed quoniam in hos satis dictum est, nunc etiam Aristotelicorum utrum vera sit, placet explanare sententiam. Hii, ut diximus, non solis calore sed motu radii calorem fieri argumentantur similitudinibus etiam sui erroris velantes fallatiam. Si enim ita quemadmodum ipsi aiunt verum est, illud iam impossibile illis occurret omnium stellarum globos aeris augmentare calorem. Habent enim omnes radios, quibus ubicumque noctem illuminant. Sed de ceterarum stellarum radiis iterum differentes de luna, cuius et maior radius est et maiorem efficit frigiditatem, videamus. Si solis radius non ex eo quod calidus est, sed quia motum facit, calefacit aerem, et lune radius. Habet enim ipsa radium, quo mundum illuminat, et in terram recte cadentem eademque iuxta illorum similitudinem via in aera resurgentem, qui tametsi non tanti sit impetus, quantum habet solis radius, habet tamen et impulsionem et motum. Ex motu autem, non ex calore radii solis aerem cale-

---

2 proximis] proximus (L) **10** climata] *corr. ex* clinata (L) **14** viciniores] *ex* vitiores *corr. in* vitiniores **23** nunc] non

ically more and less of the quality of heat; more when the sun stands in the great circle and in the parallel circles nearest to it, and less when it is most inclined to the north or to the south. But it never experiences or receives moderation, which is attested by the teaching of all philosophers; just as the fact that the lands under the poles of the world are condemned to eternal winter and bear nothing living or animate. We have said enough about these things, as they can be proved only by reasoning, which nonetheless enjoys authority.

Let us now take a look at the northern, so-called 'temperate', inhabited region. For the intellect adapts its consent most readily to the sense of the body. This region is divided into seven climates, the first of which is the hottest, because it is adjacent to the burnt zone, the second is less hot, the third still warm, the fourth temperate, the fifth cold, the sixth a bit more, and the seventh very cold. If we consider the variation of the daylengths in all these climates, we will thereby also clearly recognise in which of them the duration of the sun above the earth is longest. The first one, close to the burnt zone, has days and nights closest to equality, whereas the seventh and most remote one has a greater difference than all the others. The sun therefore spends a longer time above the earth in that climate than in the sixth. The same is to be understood analogously for the other climates. Moreover, the second climate has a greater difference in the days than the first, the third a greater than the second, and the fourth a greater than the third and the former two, because they experience a longer time of the sun above their horizons. But that the fourth climate is temperate, the third warm and the second less hot than the first is perspicuous by reasoning and commonly felt. Therefore, the sun's duration above the earth does not produce a difference of heat and cold in the same way as it does for the daylength.

But as we have said enough against these people, we now want to demonstrate also whether the Aristotelians' theory is true. They argue, as we have said, that heat is not produced by the heat of the sun but by the motion of the radiation, and they conceal their error under analogies. For if it were true as they say, they would also encounter that impossible circumstance that the bodies of all stars would increase the heat of the air. For they all have rays by which they illuminate the night everywhere. Yet, distinguishing further among the rays of the remaining stars, let us take a look at the moon, which has the strongest radiation and which causes the greatest coldness. If the radiation of the sun did not heat the air by being hot itself but because it makes motion, the same would apply also to the radiation of the moon. For the moon also has radiation by which it illuminates the world, and which falls to the earth vertically and which similar to those others is reflected

fieri positum est. Calefacit igitur motus radii lune aerem. Eodem modo de reliquis omnibus tam planetis quam fixis colligendum est.

Quod verum non esse huiusmodi ratio manifestat. Estatis tempus calidum esse et siccum nullus ambigit, cuius tamen noctes frigide sunt et humide. Omni etenim 5 nocti frigidum adiacent et humidum accidentia. Ea cuius fiant benefitio, sic ad- tendendum videtur. Luna in eodem signo soleque morantibus noctium frigus et humor parva sunt. Ea cum a sole ante et retro .d. signis removetur, per rectum sci- licet quadrangulum, medium habet sui luminis in terra defixum, quibus noctibus maior est aeris frigus et humor quam cum sole luna sinodante. At vero cum et 10 plenum sui orbis in terra figit lumen et tota lucet nocte, maiorem nocti ministrat et frigiditatem et humiditatem. Super quo qui nobis non adquiescit, disciplinalibus virgis magis videtur quam philosophanti rationum examine, qui tantam secum ad philosophiam attulit infantiam, ut quod rusticorum turba fatetur, sibi ipsi dubi- tabile atque aliter esse videatur. Quapropter non talibus sed paulo plus provectis 15 loquimur.

Per lunam igitur noctium fit frigus et humor, non quemadmodum luna quove tempore incrementa patiuntur aut detrimenta. Patet ergo non motu radiorum calo- rem fieri. Nam si motu fit calor, et lune. Radium etenim habet. Sed radius eius, ut dictum est, frigiditatem facit et humidum, non igitur calorem. Ergo neque motu 20 radii solis calorem fieri aeris concedi necesse est. Alia igitur querenda est verior- que sententia, que nobis verissimam alterationis aeris promat causam. Utrosque etenim falsa sectasse demonstratum est.

At nobis quidem videtur ex utrorumque rationibus temperanda veritas, et eo- rum scilicet, qui calidum aiunt solem et siccum, et illorum, qui ex eius propin- 25 quitate puncto capitis fieri aeris calorem dixerunt. Radius enim solis calidus est et siccus, cuius fit calore aer calidior. Eius in terram percussio iuxta superiorem Ari- stotelicorum de lapide similitudinem consideranda est. Quanto enim sol a puncto nostri capitis longius recedit, tanto minor eius calor est, quanto autem propinquior, tanto calidior. Cum etenim longe a nobis est, eius calidus radius septemtrionalem 30 habitationem lenius percutiens in surgendo latum facit angulum. Unde et debiliter f. 12r cadens debilius surgit. Cum vero nostri capitis puncto proximus, maiori | in terram

14 plus] post **31** cadens] surgens

into the air on the same way as it has come. Although it does not have as much impetus as the sunlight, it still has impact and motion. But it was assumed that the air was heated by the motion, not from the heat, of the sun's radiation. Therefore, the motion of the moonlight would heat the air. The same is to be understood for all other planets and fixed stars.

But reason shows that such is not the truth. Nobody questions that the summer is hot and dry, whose nights are nonetheless cold and moist. For every night is affected by cold and wet as accidents. By whose gift these accidents come about is to be considered in the following way. When the moon and the sun stand in the same sign, coldness and humidity of the nights are little. When the moon moves to the fourth sign ahead or behind the sun, that is, at a right angle, it has half the impact of its light on the earth, in which nights coldness and humidity of the air are stronger than when the moon was in synode with the sun. But when the moon shoots to the earth light from its full disc and shines during the entire night, it gives the strongest coldness and humidity to the night. Whoever dissents from us in this point seems suited to a school teacher's birch rather than to a critical philosopher, as he approaches philosophy with such childishness that he puts into question and denies what even the crowd of farmers confirm. We therefore do not speak to people of that kind, but to the more advanced.

The coldness and the humidity of the nights thus result from the moon, whereas they experience increase and decrease not in the same manner or at the same time as the moon. Hence it is obvious that heat is not caused by the motion of the rays. For if heat was caused by motion, then also by that of the moon, which also has radiation. But its radiation, as we have said, causes coldness and humidity, hence no heat. One thus has to admit that the heat of the air is caused not by the motion of the sun's radiation either. We therefore need to seek another and truer theory, which provides us the truest cause of the changing of the air. For it has been shown that either group pursued false beliefs.

It seems to me that the truth lies in a mixture of the arguments of either group, of those who say that the sun is hot and dry and of those who say that the heat of the air was caused by the sun's proximity to the zenith. For the radiation of the sun is hot and dry, and its heat makes the air warmer; whereas, its impact on the earth must be considered according to the Aristotelians' aforementioned analogy of a stone. For the further the sun removes itself from our zenith, the weaker is its heat, whereas the closer, the hotter. For when it stands remote from us, its hot radiation hits the northern inhabited region more gently and, when rising, produces a wide angle. Thus impinging weakly, it is thrown upwards even weaker. But when it

radium cum calore vi impellit. At terra solidum et durum corpus fortiter cadentem fortius impellit quam obliquum. Sicque fit, ut in estate geminati maiori impetu et calor et radius maiorem aeri conferant claritatem et calorem quam hieme. Unde fit, ut in calidis climatibus nulla sit in estate pluviarum effusio, quoniam quanto sol puncto capitis propinquior fit, radius eius in terram cadens acutiori angulo surgit et altius a terra repellitur, totusque aer ille calefit, in quo nubes et pluvia non habent fieri. In his vero climatibus, que maius habent, scilicet V°, VI° et VII°, solis radius oblique terram percutiens minus acutum facit angulum surgens, nec in altum a solido terre corpore repellitur. Unde non multum calefit aer superior, in quo terre fumus in nubium conversus naturam imbrium effundit sepe largitatem. Nam si aliquando et in his estas fiat sicca et in calidioribus humida, reliquorum fiet planetarum impulsu, qui solis calorem minuunt et augmentant.

Amplius sole in Ariete posito aut Libre principio, in quibus facit equinoctium, calor radii solis non idem apud omnes sentitur, maior enim his, qui in primo climate eique proximis, minor quinta et duobus reliquis. Mora tamen eius omnibus super terram equalis videtur. Cum vero in Cancri capite, calidissimus his, qui a recti circuli linea .lc. gradibus et .me. absunt sexagenariis, videtur. Terram enim recte radius percutit, et absque angulo in aera refertur, quia sol in puncto capitis volvitur, VII^{mo} autem climati paulisper temperiei extendens in calore mensuram. Qui cum in Capricorni capite commoratur, VII clima glaciali fere zone relinquit, quoniam parum super orizontes surgens tamquam ab obliquo illud respicit, atque ideo latum angulum radius eius faciens parum aerem a frigoris nimietate temperat. Et de reliquis planetis eadem ratio est. Saturnus enim frigidus est et siccus, Mars calidus et siccus, Iuppiter et Venus atque Mercurius temperati, sed Iuppiter paulo calidior, Venus paulo frigidior, luna autem frigida et humida. Quos omnes temporum facere varietates ipsorum cum sole coniunctiones manifestant. Saturnus etenim si cum sole sinodet, estate calorem temperat, hieme frigiditatem augmentat. Idem de reliquis intelligendum est secundum suarum qualitates proprietatum. Maximas tamen luna facit cum sole temporum diversitates. Ipsa enim ceteris frigidior et humidior quanto terre propinquior maiorem fundit claritatem, tanto copiosius aliis in frigidum mutat aerem et humidum. Estate quidem, quia sol in septemtrionalibus habitat signis, eius fit plenilunium in australibus, longeque a puncto capitis remota minorem nocti confert frigidum. Descendente vero sole

---

2 geminati] geminatis **10** nubium] nubes **27** estate calorem] estatem calore

stands closest to our zenith, it sends its hot radiation to the earth with greatest force. The earth, however, being a solid and hard body, reflects the forcibly falling radiation stronger than the oblique one. Thus doubled in summer by a stronger impact, heat and radiation give more brightness and heat to the air than in winter. As a result, there is no rainfall in the hot climates during summer, because the closer to the zenith the sun is standing, the more acute is the angle by which its radiation that falls on the earth is reflected and the higher it will be thrown back, and all the air becomes so warm that clouds and rain cannot form. But in the zones with a greater distance between the sun and the zenith, that is, in the fifth, the sixth, and the seventh climates, the radiation of the sun hits the earth obliquely and is reflected in a less acute angle and is not thrown back to a high altitude by the solid body of the earth. Hence the higher air is not heated much, where thus fog from the earth is turned into clouds and often pours out masses of rain. And if summer is sometimes dry in these climates while being humid in the hotter zones, this has been effected by the other planets, which reduce and increase the heat of the sun.

Moreover, when the sun stands in Aries or at the beginning or Libra, at which positions it causes an equinox, the heat of the radiation of the sun is not felt equally everywhere, but hotter by those living in the first and the adjacent climates whereas less so in the fifth and the remaining two. Nevertheless, its time above the earth is the same everywhere. Conversely, when it stands in the head of Cancer, it is hottest for those living 23;35 degrees away from the equator, where the radiation hits the earth vertically and is thrown back into the air without an angle, because the sun stands in the zenith, whereas for the seventh climate it raises the temperature for a short while. But when the sun stands in the head of Capricorn, it almost cedes the seventh climate to the icy zone, as it hardly rises above the horizon but looks at that climate like from the side; thus producing a wide angle, its radiation can moderate the air only a little from the excess of coldness. The same reasoning applies to the remaining planets. Saturn is cold and dry, Mars hot and dry, Jupiter, Venus and Mercury are temperate, with Jupiter being a bit hotter and Venus a bit colder, whereas the moon is cold and wet. Their conjunctions with the sun make it evident that they all cause changes to the weather. For when Saturn stands in conjunction with the sun, it moderates the heat in summer and increases the cold in winter. The same is to be understood for the other planets according to their characteristic qualities. The greatest differences in the climate, however, are caused by the moon in relation to the sun. Being colder and wetter than the others, the moon makes the air colder and wetter than the others, the closer to the earth it

de septemtrionalibus ad australia, quia eius plenilunium fit semper in opposito, ipsa ascendit septemtrionalia, fitque puncto capitis propinquior et sol remotior. Quamobrem diei minuitur calor et siccitas, noctiumque augmentatur frigus et humidum. Cum autem sol in Sagittarii fine vel Capricorni capite diem illuminat, tunc 5 in Cancri capite aut Geminorum ultimo plenum habens luna lumen nostri puncto capitis valde proxima plurimum facit frigus et humidum, dieique parvus est calor, quia solis centrum a puncto capitis remotissimum.

Illud quoque mihi dictante rationis iuditio rectissime dici posse videtur, quoniam sub yberno tempore si fiat in nocte plenilunium, noctem humidiorem illam 10 esse frigidioremque proximis noctibus. Eademque mihi videtur sinodi lune et solis f. 12v ratio. | Idem etiam, si fiat in die, intelligi oportet. Nam in estate, quod ego id compertum habeo, plenilunialem noctem humidiorem esse et frigidiorem, sinodalem vero diem minus calidum et siccum. Permutantur tamen hec aliquando aliorum planetarum concursu.

15 Sed nos longius gratia eorum, que nostratum auribus ignota erant, pervagati sumus. Puto autem facile me veniam habiturum, si et utilitas eorum, que inserta sunt, et nostri intentio animi pensabitur. Nam honestare Latinitatem totius, si posset fieri, subtilitate philosophie cum desiderem, si propter tempus aliquid de eius archanis inscius non videor reprehendendus, quamobrem veniam siquidem tamen 20 in hoc peccatum postulo, ut volens utilitatis aliorum causa minus fortassis iusto digressus sum, ex hoc saltem commissis abluar, quoniam desiderantissimo nostratum utilitatibus inserviam animo. Nunc, quoniam omnem premissam questionem dissolutam arbitror, ad propositum redeo.

17 honestare] honestate

emits its greater brightness. In summer, when the sun stands in the northern signs, full moon occurs in the southern signs, far remote from the zenith, and thus brings little cold to the night. But when the sun descends from the northern signs to the southern ones, and as the full moon always occurs opposite the sun, the moon climbs the northern signs and comes closer to the zenith while the sun gets further away. For that reason, heat and dryness of the day get less whereas cold and wetness of the nights increase. But when the sun illuminates the day from the end of Sagittarius or the head of Capricorn, the moon is full when standing at the head of Cancer or the end of Gemini, closest to our zenith, and thus produces a maximum of cold and wetness, wereas the heat of the day is little, because the centre of the sun stands furthest from the zenith.

This seems to me to be said most truly and dictated by rational judgement, also because a full moon in a night in winter causes that night to be wetter and colder than the proximate nights. The same reasoning seems to me to apply also to conjunctions of the moon and the sun. The same should also be understood if it happens at daytime. For I have experienced by myself that in summer a night with full moon is wetter and colder, whereas a day with a synod is less hot or dry. However, this changes occasionally by the positions of the other planets.

We have digressed very lengthily for the sake of matters that were yet unheard by our people. However, I believe that I will easily be granted clemency, once the usefulness of the insertions and the intention of our mind will be considered. For I would wish to ennoble Latinity with all the subtleties of philosophy, if only this was possible. If I thus ignored some of its secrets because of the time, one should not blame me; for the same reason I ask for clemency if I have done wrong when for the usefulness of other things I have deliberately digressed less than possibly adequate. I should be pardoned for my errors, at least because I am serving the benefits of our people with keenest mind. Now, as I believe that every outlined problem has been solved, I return to the proposed subject.

Sole igitur in primo commorante quadrante, quod est a puncto equalitatis veris usque punctum mutationis estatis, veris tempus est, cuius, ut positum est, initium in aliis locis calidius, quibusdam autem minus sentitur. Hoc autem tempus eo usque perdurat, donec solis centrum caput Cancri ingrediatur. Cui quanto propin-
5 quior est, tanto calidior in septemtrionali habitatione. Propinquior enim fit nostri puncto capitis. [117] *Ingresso autem sole secundum quadrantem fit aer calidior. Quod ideo fieri iure dicimus, quoniam aer cotidianis frequentibusque calefactus caloribus semper caloris patitur augmentationem propter calefatiendi maximam frequentiam, usque dum longius a puncto capitis sole recedente temperari calor aeris incipit.*

10 [118] *Pretergresso vero equalitatis autumpni puncto sol in austrum inclinans a capitis puncto elongatur, aerisque temperies frigido mutatur accidenti. Quanto autem sol amplius elongatur in austrum, tanto magis infrigidatur aer, usque dum tangat initium mutationis hiemis, in quo magnum fit frigus in septemtrionali habitatione propter solis remotionem a puncto capitis.* [119] *In medio vero ultimi quadrantis morante sole maxi-*
15 *ma fit aeris frigiditas propter parvissimam calefactionis frequentiam. Cuius quadrantis transcurso arcu ad punctum equalitatis veris, in quo annum inceperat, revertitur. Hocque modo cum suum transierit circulum, annus solaris completus dicitur.*

Sed de zodiaco apud Macrobium non eque omnia. Dicit enim solum hunc circulum ceteris linea circumductis latitudinem spatii adipisci potuisse, tantam
20 scilicet quantam utrimque porrecta .kb. signorum formantia imagines sidera occupant, nec ultra hanc .kb. signorum latitudinem distendi. In quo si Macrobianam sequimur traditionem in astronomia, multum nobis adversantur inconvenientia. Ponamus igitur eorum partem, quemadmodum apud illum est. Nulla hoc posse sine periculo teneri ratione ostendamus, ut inveteratus diutissime error sane cedat
25 veritati.

Zodiacum eam non habere latitudinem, cuius est testis Macrobius, per porrecta sidera hinc videri licet, quoniam cum ab eo positum sit IIIes esse in zodiaco circulos, II^{os} qui latitudinem, III^{um} qui medietatem terminet, et tercium maximum et eclipticum, reliquos autem altrinsecus positos alterum a septemtrione, alterum
30 ab austro, medio minores, intelligendum erit fortasse hos extremos aut per amfractus aut recte circumductos. Si recte, paralelli medii et sui utrique sunt. Si per
f. 13r amfractus, nunc propius, nunc vero longius accedunt. | Horum utrumque non esse demonstrabimus, atque sic, cum nichil tercium sit, significacio non concedi

12 infrigidatur] infridatur (L) **14** medio] medium (L)

While the sun is staying in the first quadrant, which extends from the point of the spring equinox until the point of the summer solstice, it is spring time, whose beginning, we have said, is felt hotter in some places and less so in others. This season lasts until the centre of the sun enters into the head of Cancer. The closer it comes to this point, the hotter it gets in the northern inhabited region, because it is getting closer to the zenith for us. [117] *But when the sun has entered into the second quadrant, the air gets hotter. We rightly say that this happens, because the air becomes heated up by the frequent daily hot periods and thereby experiences a permanent increase of heat by the very high frequency of heating, until the sun has receded substantially from the zenith and the heat of the air therefore begins to become temperate.*

[118] *But when the sun has passed the point of the autumnal equinox and inclines southwards, it removes itself from the zenith and the moderation of the air is changed by the upcoming cold. And the further the sun removes itself southwards, the more will the air become cold, until the sun reaches the beginning of the winter solstice, which causes great cold in the northern inhabited region due to the sun's distance from the zenith.* [119] *But the greatest cold of the air occurs while the sun resides in the middle of the last quadrant, because of the very low frequency of heating. Having passed through the arc of this quadrant, it returns to the spring equinox, where it had begun the year. Once it has thus passed through its circle, the solar year is said to be complete.*

But what is read in Macrobius about the zodiac is not consistent. For he says that whereas all other circles are described by a line, only this circle has a certain width, namely as much as the stars which form the images of the twelve signs spread to either side, whereas it does not exceed this width of the twelve signs.[46] If we follow the Macrobian tradition in this respect, in astronomy, we will face much inconvenience. Let us assume the arrangement of the signs according to his theory, and we will show that this cannot be held by any consideration without danger, hence this long-settled error may duly give way to the truth.

That the zodiac does not have that width according to the extended stars, as attested by Macrobius, can be seen from the fact that, since he says that the zodiac comprises three circles, two of which defined the width and the third one the middle, with the third one being the greatest and identical to the ecliptic and the other two lying on both sides of it, one to the north and the other to the south and both smaller than the central one, one has to conclude that these outer circles are drawn either straight or crooked. If drawn straight, they are parallel to the central circle and also to one another. If they are drawn crooked, they sometimes come closer

[46] *Somn.* I,15,8–10 (L).

latitudinem nostra legentibus equo animo patebit. Paralellos non esse hos medii neque recte circumductos ex ipsius verbis Macrobii facile datur intelligi. Ait enim: «quantum lata dimensio porrectis sideribus occupabat spatii, duabus lineis limitatum est, et III ducta per medium ecliptica vocatur.» Quantum, inquit, porrecta occupabant sidera. Illud autem inequale est, neque tantum occupant spatii in austrum aut septemtrionem Cancri sidera quantum Leonis aut Geminorum vel Tauri, ipsaque horum inter se diversa spatii mensura. Minus enim inclinantur ab ecliptico in austrum aut in septemtrionem sidera signi unius sideribus alius; quapropter minus occupant spatii, quorumdam vero signorum sidera amplius, unde et plus occupabunt. Sed duabus utrimque lineis tantum limitatum est, quantum porrectis sideribus occupant. Minus igitur pars zodiaci, quam Cancer micat, spatii quibusdam aliis habebit, minus enim sidus eius occupat. Quamobrem non recte circumducuntur linee extremorum circulorum, sed per amfractus. Quod autem signorum sidera alia aliis minus ampliusve porrigantur aut in septemtrionem ab ecliptico aut in austrum, in presentia docere non attinet. Alterius enim est negotii, quod suo tempore et loco, si sic quidem gratia divina annuat, verissime demonstrabimus.

Per amfractus autem hos nequaquam circumductos esse circulos ex ipsius alias inserta sententia docemur in .a. commentariorum libro eo in loco, quo de lune sub septenario numero per zodiacum discursu disputat in hec verba: «Hunc itaque numerum, qui in quater septenos equa sorte digeritur, ad totam zodiaci latitudinem emeciendam remetiendamque luna consumit. Nam .g. diebus ab extremitate septemtrionalis hore oblique per latum meando ad medietatem latitudinis pervenit, qui locus vocatur eclipticus, .g. aliis ad imum australem a medio dilabitur.» Que si diligenter attenduntur, tantam esse latitudinem zodiaci ex Macrobii verbis deprehendemus, quantum utrimque ab ecliptico, quem nos signorum circulum dicimus, luna inclinatur. Draconis etenim circulus, quem de celestibus solum circulis supreme spere cursu ferri agnovimus, ab oriente scilicet in

11 Cancer] Cancri **19** Hunc] Hinc **20** quater] quantum

and sometimes diverge. We will show that neither of these is the case and, since there is no third alternative, the consequence that a width cannot be admitted will be evident to our readers equanimously. From the words of Macrobius himself we can easily conclude that those circles are neither parallel to the central one nor drawn straight. For he says: 'Two lines define the lateral extension of the stellar signs, whereas a third one, drawn centred between them, is called the ecliptic.'[47] He speaks of the 'lateral extension of the stellar signs'. But this extension is uneven, and the stars of Cancer do not take up as much space to the north or to the south as those of Leo, Gemini or Taurus, which themselves differ again from one another in their extensions. For the stars of one sign are less inclined from the ecliptic to the south or the north than those of another sign and, therefore, take up less space, whereas the stars of other signs are spread further and thus take up more space. But two lines define on either side just as much as the signs occupy with their spread stars. Then the part of the zodiac where Cancer twinkles takes up less space than some others, as its sign occupies a smaller area. Therefore, the lines of the outer circles cannot be drawn straight but crooked. However, that some stars of the signs reach less or further than others from the ecliptic to the north or to the south is not to be discussed at present. It belongs to another task, which we will demonstrate most accurately at its time and place, if the grace of God permits so.

Yet, that the circles are by no means drawn crooked is something that we can say from a statement of his which is inserted elsewhere in the first book of his commentaries, where he discusses the course of the moon through the zodiac in relation to the number seven, using the following words: 'It takes the moon this number, which falls equally into four times seven, to cross and return through the entire width of the zodiac. For during seven days it moves obliquely from the northern extremity through the latitude to the middle of the width, which place is called the ecliptic, and during further seven days it slides down from the middle to the southern lowness.'[48] If we consider this attentively, we will realise from Macrobius' words that the width of the zodiac equals the amount by which the moon inclines to both sides from the ecliptic, which we call the 'circle of the signs'.[49]

---

[47] *Somn.* I,15,10 (L): 'quantum igitur spatii lata...'

[48] *Somn.* I,6,53 (L): 'hunc etiam numerum, qui in quater septenos aequa sorte digeritur, ad totam zodiaci latitudinem emetiendam remetiendamque consumit. nam septem diebus ab extremitate septemtrionalis orae oblique per latum meando ad medietatem latitudinis pervenit, qui locus appellatur eclipticus, septem sequentibus a medio ad imum australe delabitur.'

[49] Stephen's preferred term for the ecliptic is 'circulus signorum'. As synonyms he also uses 'signifer' and '(circulus) zodiacus'.

occidentem, centrum lune semper gestat, nunquam alias inclinatur. Hic autem cursum suum perficit in .kh. annis plus fortassis minusve, inclinaturque ab ecliptico medietas eius in septemtrionem, altera in austrum, maiorque eius inclinatio in utramque partem est .d. graduum .nf. sexagenariarum. Totum igitur eclipticum eius 5 utraque maior inclinatio in .kh. annis commeat. Sed quemadmodum in verbis Macrobii habetur, luna ab extremitate septemtrionali zodiaci cursum aliquotiens incipit, ipsa autem semper in circulo draconis est.

Draconis igitur circulus, eo quo amplius inclinatur in septemtrionem, latitudinem zodiaci terminat semper. Idem facit altera in parte ecliptici altera ligantis 10 medietas. Erit igitur latitudo zodiaci inclinatio draconis in utramque partem a circulo signorum .d. graduum .nf. sexagenariarum, totaque simul .i. graduum .mb. sexagenariarum, eruntque paralelle que ipsum terminant altrinsecus linee medio et in invicem. Quamobrem non per amfractus sed paralellice circunducitur.

Duplex ergo est Macrobii sibique invicem repugnans sententia, quare neutram f. 13v 15 tenendam censeo; neque eam, que | per amfractus, quia illi secundo posita obviat, neque eam, que paralellice circunducit circulos, primam enim adversantem sentit; terciamque utrique, quare nullam habere latitudinem circulus signifer comprobatur. Cuius non modica pars ex ipso elicitur Macrobii libro. Nam cum sint .g. vage, quarum .e. sunt planete, .b. sol et luna, que adversus none spere motum nituntur, 20 omnes in zodiaco esse semper perhibetur. Quod qua ratione possit subsistere, cum earum nulla significis inmisceatur sideribus, quippe que longe inferius ferantur, Macrobii est sententia quamvis earum esse in Ariete cum Arietem, in Tauro cum Taurum desuper habuerit.

Cui rei nequaquam resistimus, sed esse quasdam eorum, que secundum su-25 perius positam latitudinem nullo in signo quibusdam temporibus vehantur. Hee sunt Mars et Venus, quarum altera inclinatur modo in austrum, modo in septemtrionem .i. gradibus, altera autem in austrum aliquando .g.. Inclinatur igitur .d. gradibus .kd. sexagenariis a zodiaci latitudine utrimque, Mars vero .b. gradibus et .kd. sexagenariis in austrum. Si ergo zodiaci certa est latitudo, eaque ipsa est quam 30 superius ex Macrobii verbis per cursum lune invenimus, erit aliquod tempus, quo

---

26–27 septemtrionem] orientem **27** autem] aut (L) **28** .b.] *corr. ex* .db. (L)

For the circle of the dragon, which is the only celestial circle that we recognise as being moved by the course of the highest sphere, that is, from east to west, always bears the centre of the moon and never inclines differently. It completes its revolution in 18 years, more or less, and one half of it is inclined northwards from the ecliptic, the other half southwards, whereas its greatest inclination to either side is 4;46 degrees. Thus, within 18 years either of its greatest inclinations passes along the entire ecliptic. But as we read from Macrobius' words, the moon begins its course several times from the northern extremity of the zodiac, while it is always on the circle of the dragon.

The circle of the dragon, therefore, where it inclines furthest to the north, always defines the width of the zodiac. The same does the other half of the ligant on the other side of the ecliptic. The width of the zodiac thus equals the inclination of the dragon towards either side from the ecliptic, which is 4;46 degrees, and altogether 9;32 degrees, and the lines which define it on both sides are parallel to the middle and to one another.[50] Hence it is not drawn crooked but parallel.

Macrobius' doctrine is therefore ambiguous and contradictory in itself, for which reason I think that neither statement is to be maintained; neither that which draws the circles crooked, because the second statement refutes it, nor that which draws them parallel, because it has the former statement standing against it. A third alternative would be in contradiction to both, hence it is proved that the circle of the signs has no width. A great part of this can be gained from Macrobius' book itself. For as there are seven wandering stars, five of which are the planets and two are the sun and the moon, which all move against the motion of the ninth sphere, it is said that all of them are always within the zodiac. How this can be possible, although none of them belongs to the stars which constitute the signs but each moves on a sphere far beneath, Macrobius explains by saying that any of them is in Aries when Aries stands above it, or in Taurus when Taurus is above.[51]

We fully agree with this, but we maintain that there are some wandering stars which, according to the above given width, stand in no sign at all at certain times. These are Mars and Venus, one of which sometimes inclines southwards and sometimes northwards by 9 degrees, and the other one southwards by up to 7 degrees. It thus inclines by 4;14 degrees to either side beyond the width of the zodiac, and Mars by 2;14 degrees to the south. Therefore, if the zodiac has a particular width and if this is the same which according to Macrobius' words we

[50]Cf. *Somn.* I,21,10.
[51]*Somn.* I,21,2 (L).

neque Mars neque Venus in zodiaco ferentur. Sed neque Mars neque Venus extra signa quandoque vehuntur. Quid ergo? Maiorem zodiaci latitudinem ponemus Macrobiana? Non. Sed alio quodam modo celum sperasque omnes dividere oportet, quo quecumque in .h. spera sunt stelle et .g. in signo aliquo invenire possimus, zodiacum autem, quem nos signorum omnium circulum dicimus, nullam habere latitudinem, sed linea quemadmodum alios circunductum.

In quo adhuc quorundam plurimum errat opinio. Ex quadam enim, que in somnio Scipionis a Tullio, quem ego gravissimum confiteor philosophum, ponitur, solis cursus subtilitate, circulum hunc in quo solis semper motus est per amfractus ductum tergiversantes autumant. Quod si securum et nullatenus a veritate devium obstinaciter tenent, videant, ne fiat non eclipticus. Si eclipticum confitentur, nequaquam per amfractus. Illud quoque hoc falsum esse comprobat, quod a capite Capricorni usque Cancrum sub meridionali circulo in dies fit altior, quemadmodum ab eodem Cancro usque Capricornum cotidie humilior. Quod si per amfractus curreret inter hec II° tropica, variatim nunc altior, nunc vero humilior per meridionalem transiret circulum. Absit autem a nobis credere signiferi lineam, que circumducitur, nunc propius, nunc vero longius centro abesse. Non enim iam ferentis spere motu sed sui ipsius moveretur. Quod frivolum esse nullus ambigit. Sed solis amfractus Tullianus sermo significare voluit eius nunc in septemtrionem, nunc in austrum a recti diei circulo inclinationem, sive quod nunc terre propinquior, modo vero feratur remotior vel utrumque simul. Fiunt enim hec nec vacant pondere. Sed de his hactenus.

Nunc divisionem zodiaci qua ratione fieri conveniat ad vitanda superius incommoda posita, videamus. Illud igitur oportet meminisse nos signiferum in .d. quadrantes divississe, quorum primus est ab Ariete usque Cancrum, secundus a Cancri capite usque Libram, abinde tercius usque Capricornum, a quo IIII^{tus} usque Arietem. Horum quemque in III^{a} dividimus signa hoc modo. [120] *Duos magnos imaginamur circulos transeuntes per utrumque polum circuli signorum. Hii* II^{os} *quadrantes, alterum veris alterum autumpni, oppositos* | *in .c. equales partes dividunt. Imaginamur etiam duos alios eiusdem magnitudinis circulos a polis circuli signorum incipientes, qui duos reliquos quadrantes estatis solis et hiemis quemque in* IIIes *partiuntur partes.*

f. 14r

4 quo quecumque] *corr. ex* quocumque (L) **15** tropica] topica (L) **16** que] qui (L) **24** nos signiferum] vos significatum (L)

have found from the course of the moon, there will be some moment when neither Mars nor Venus move within the zodiac. But neither Mars nor Venus ever move out of the signs. What then? Shall we define a greater width of the zodiac than the Macrobian one? Not at all! One rather needs to subdivide the heaven and all the spheres in some other way, such that all the stars on the eighth sphere and the seven planets can be found standing in a sign, whereas the zodiac, which we call the cirle of all the signs, has no width but is drawn as a line like the other circles.

In this respect, some people are still fatally wrong. From a detail about the course of the sun which is given in *Scipio's Dream* by Tullius Cicero, whom I esteem to be a most respectable philosopher, they perversely state that this circle on which the sun permanently moves was crooked.[52] If they stubbornly maintain this as the certain and absolute truth, they should take care not to refer to it as the ecliptic. For if they say that it is the ecliptic, it cannot be crooked. The falsity of this is also proven by the fact that, if measured at the meridian, it rises higher day by day from the head of Capricorn until Cancer, just as it gets lower every day from Cancer to Capricorn. Now, if it ran crooked between these two tropics, it would cross the meridian changeably, sometimes higher and sometimes lower. And forbid that we believe that the line of the ecliptic is sometimes closer and sometimes further from the centre! For it is no longer moved by the motion of the deferent sphere but by that of its own. The falsity of this cannot be questioned by anyone. Instead, when speaking about the crook of the sun, Tullius Cicero intended to refer to its changing inclination, sometimes to the north and sometimes to the south from the equator, or that it sometimes comes closer to the earth and sometimes removes itself, or to both effects together. For such phenomena occur and they are of considerable importance. But we will speak no more about these things.

Now let us consider how a division of the zodiac can be made such that the above mentioned problems are avoided. We should remember that we have divided the ecliptic into four quadrants, the first of which is from Aries until Cancer, the second from the head of Cancer until Libra, from where the third extends until Capricorn, from where the fourth extends until Aries. We divide each of them into three signs in the following manner. [120] *We imagine two great circles passing through either pole of the ecliptic. These circles divide two opposite quadrants, that of spring and that of autumn, into three equal parts. We also imagine two further circles, of the same size and starting from the poles of the ecliptic, which divide the remaining*

[52] Cic. *Somn.* 12,4: '... solis anfractus reditusque.'

[121] *Sunt igitur omnes magni circuli, qui transeunt super duos magnos polos circuli signorum ipsum dividentes, sex, qui in eum modum sunt imaginati, ut signorum circulum in .kb. equalia membra dividant. Partiuntur etiam altissimam speram eiusque superficiem in .kb. equales partes, quarum quamque a circuniacentibus duorum circulorum discrimi-*
5 *nant. Dicuntur autem he* XII *partes alte spere .kb. signa. Sed et partes .kb. circuli signorum signa nominamus,* unde et idem circulus circulus signorum dictus est. [122] *Quisque vero arcus .kb. partium circuli signorum in* XXX^{ta} *distribuitur partes imaginarias, quas gradus appellamus. Duodecies vero* XXX^{ta} *in .xp. gradus ascendit.*

[123] *Imaginamur quoque horum sex circulorum superficies totum secare mundum*
10 *et dividere cuiusque planete sperarum quamque in .kb. equales partes, que eodem modo signa dicuntur.* [124] *Sed et circulum signorum imaginamur omnium speras dividere planetarum in* II° *media et facere in cuiusque superficie circulos, qui sunt paralelli magni signorum circuli, qui est in superficie magne spere, quorum quisque dicitur signorum circulo circulus similis,* [125] *et dividitur* VI *circulorum, de quibus dictum est, superficiebus*
15 *in .kb. equales partes, quarum queque signum dicitur. Dividitur igitur spera fixarum in .kb. equales partes, que octava est, et dicuntur signa.*

[126] *Hac in spera multe sunt stelle, per quarum positionem quedam formantur imagines.* XII^{cim} *igitur partium quedam pars eiusdemque mediana stellas habet, quarum positio Arietis imaginem exprimit. Eam igitur partem in magno celo signum dicimus Arietis,*
20 *arcumque circuli signorum, qui est in medio eius, signum Arietis vocamus.* [127] *Hoc signum primum est trium partium veris quadrantis, in cuius primo puncto initium ponimus circuli signorum ipsiusque punctum equalitatis veris. Est igitur punctus equalitatis veris initium circuli signorum et capitis Arietis.*

[128] *In secunda autem parte, que Arietem sequitur in spera, Tauri stelle posite sunt,*
25 *ideoque pars, que est super illis in magna spera, signum Tauri dicitur, arcusque circuli signorum, qui eiusdem subest imagini, signum Tauri vocamus.* [129] *Sic habet queque* XII *partium in spera fixarum positione stellarum formatas imagines, vocanturque signorum queque imaginis nomine, quam stelle figurant, eodemque nomine dicitur arcus circuli signorum, qui eidem subest imagini.*

30 [130] *Tercium igitur signum Gemini dicitur,* quia in eo Geminorum formatur imago, *quartum Cancer. Est autem hoc estatis primum signum, cuius est initium punctus mutationis estatis. Quintum deinde Leo signum sequitur, post quem Virgo* VI^{tum} *conspicitur. Virginem autem* VII^{mum} *Libra sequitur, que est primum signum in autumpnali qua-*

---

3 .kb.] .kh. (L) **14** superficiebus] superfities **23** initium] *corr. ex* initum (L) **24** spera] spere (L)

two quadrants, of summer and winter of the sun, into three parts. [121] *Hence the great circles that pass through the two great poles of the ecliptic and intersect with the latter are six in total, which are imagined such that they divide the ecliptic into twelve equal segments. They also divide the highest sphere and its surface into twelve equal parts and distinguish each part from the surrounding ones of two circles. These twelve parts of the high sphere are called the twelve 'signs'. But we also call the twelve parts of the ecliptic 'signs',* for which reason that circle is called the 'circle of the signs'. [122] *Every arc of the twelve segments of the ecliptic is divided into thirty imaginary parts which we call 'degrees'; whereas, twelve times thirty multiplies to 360 degrees.*

[123] *We further imagine the planes of these six circles to dissect the entire world, dividing each of the spheres of every planet into twelve equal parts, which are called 'signs' in the same way.* [124] *We also imagine the ecliptic to divide the spheres of all planets into two halves and to produce on the surface of each circles that are parallel to the great circle of the ecliptic, which lies on the surface of the great sphere, each of which is called a 'similar circle of the ecliptic',* [125] *and each is divided by the planes of the six aforementioned circles into twelve equal parts, each of which is called a sign. The sphere of the fixed stars, which is the eighth sphere, is thus divided into twelve equal parts, which are called 'signs' [cf. Fig. 1.2].*

[126] *On this sphere there are many stars by whose position certain images are formed. One of the twelve parts contains in its central region stars whose position describes the image of a ram. We therefore call that part of the great heaven the 'sign of Aries', and also the arc of the ecliptic which passes through the middle of that sign is called the 'sign of Aries'.* [127] *This is the first sign of the three parts of the spring quadrant, at whose first point we place the beginning of the ecliptic and its point of the spring equinox. Hence the point of the spring equinox is the beginning of the ecliptic and the head of Aries.*

[128] *In the second part, however, which follows Aries on the sphere, the stars of a bull are located; hence the part on the great sphere above them is called the 'sign of Taurus', and also the arc of the ecliptic that lies under its image is called the 'sign of Taurus'.* [129] *In this way, each of the twelve parts on the sphere of the fixed stars has images that are formed by the position of stars, and each sign is called by the name of the image which is formed by the stars, and the same name is given to the arc of the ecliptic which is under that image.*

[130] *The third sign is called 'Gemini',* because the image of twins is formed in it; *the fourth is called 'Cancer'. It is the first sign of summer, with its beginning being the point of the summer solstice. As the fifth sign follows 'Leo', after which 'Virgo' is seen as the sixth. Virgo is followed by the seventh sign, 'Libra', which is the first sign in the*

drante, ipsiusque initium punctus equalis autunni. VIII*vum* *post Libram surgit Scorpius, quem architenens, nonum scilicet signum, insequitur, quem Sagittarium aiunt, ipsumque* corniger *Capricornus* X*mus*. *Hoc est primum signum hiemalis quadrantis, in cuius capite primus est punctus mutationis hiemis. Post Capricornum autem* rectus *oritur Aquarius,*
5 *qui in ordine signorum* XI*mus* *est. Ipsum autem Pisces ultimi sequuntur,* XII*mum* *scilicet signum, finis signorum circuli, cuius signi finis est punctus equalitatis veris.* Et circulus
f. 14v quidem signorum et XII eius partes, signa scilicet, ita disposita sunt. | [131] *Quando igitur movetur centrum solis proprie spere motu, semper circunfertur signorum sub circulo nullam declinans in partem. Cum itaque imaginamur rectam lineam a centro terre*
10 *surgentem eamque per centrum solis transeuntem ad magnam usque extendi speram, tangit in ea quempiam punctum circuli signorum. In quo momento centrum solis esse dicimus atque in eo signo et gradu signi, in quo idem punctus fuerit.*

[132] Ad aliarum vero stellarum nequaquam persepe hec valet ratio. Per signa etenim diffuse, non sub circulo signorum, feruntur. *Quamobrem ad eorum signum*
15 *inveniendum in circulo signorum imaginamur lineam erigentem se a centro mundi et transeuntem per centrum stelle in eamque crescere longitudinem, ut superficiem magne spere tangat. Hic igitur punctus, quem huiusmodi linea in ultimo celi tetigerit, in aliquo est* XII *signorum, per que eiusdem superficies spere in* XII *equales distribuitur partes. In eodem ergo signo stella esse dicitur.*
20 [133] Cognito autem in qua sit partium XII, ut in quo gradu sit possimus advertere, *magnum imaginamur circulum, qui incipiens a polis circuli signorum transit per punctum, qua lineam* in superficie magni celi *tetigisse dictum est. Hic circulum signorum in duas dividit medietates per arcum signi* et oppositum transiens ipsius, *in quo stellam esse posueramus. Illo igitur in gradu signi stella esse perhibetur, cuius quamvis*
25 *partem isdem dividit et tangit circulus. Hunc aiunt esse locum longitudinis stelle* a capite Arietis. [134] *Sed etsi qua egreditur de centro mundi linea et transit per centrum stelle aliquem* VI, *de quibus diximus, circulorum tetigerit, qui circulum signorum in* XII *partes dividunt, ipse locus stelle in celo dicitur. Punctus vero quem idem circulus in circulo signorum tetigerit, stelle locus in circulo signorum ponitur, eoque modo in capite alicuius*
30 *signi esse perhibetur.*

[135] Et quoniam de longitudine stellarum dictum est, de lato dici tempus et ordo postulat. Latum igitur stellarum dicimus non earum a recti diei circulo inclinationem, sed qua a signorum circulo removetur longitudinem. *Est autem latum stelle arcus circuli, qui transit per polos signiferi circuli et punctum, quem linea a centro*

---

6 signorum] signum (L) **28** dicitur] *in margine* Nota

autumn quadrant, and its beginning is the point of the autumnal equinox. As the eighth sign 'Scorpio' rises after Libra, being followed by the archer as the ninth sign, which is called 'Sagittarius' and which itself is followed by the horned *'Capricorn', being the tenth. The latter is the first sign of the winter quadrant, at the head of which the first point is that of the winter solstice. After Capricorn rises* upright *'Aquarius', who is the eleventh in the order of the signs. He is followed by 'Pisces', as the last and twelfth sign and the end of the ecliptic, with the end of this sign being the point of the spring equinox.* This is how the ecliptic and its twelve parts, that is, the signs, are arranged. [131] *When the centre of the sun is thus moved by the motion of its own sphere, it is always taken around on the ecliptic without inclination to any side. Therefore, when we imagine a straight line rising from the centre of the earth and passing through the centre of the sun and extending to the great sphere, it touches there some point of the ecliptic. In that moment we say that the centre of the sun is in that sign, and in that degree of the sign, where this point lies.*

[132] But this method applies hardly ever to the other stars. For they move through the signs in an irregular manner and not under the ecliptic. *For that reason, in order to determine their sign on the ecliptic, we imagine a line rising up from the centre of the world and passing through the centre of the star and prolonged to such a length that it touches the surface of the great sphere. The point which is defined by such a line on the outermost heaven is thus in one of the twelve signs by which the surface of that sphere is partitioned into twelve equal parts. In that same sign, therefore, the star is said to be.*

[133] Once we know in which of the twelve parts the star is, in order to determine in which degree it is, *we imagine a great circle starting from the poles of the ecliptic and passing through the point where we said that the line touched* on the surface of the great heaven. *This circle divides the ecliptic into two halves, as it passes through the arc of the sign*—and its opposite—*in which we had assumed the star to be. One says that the star is in that degree of the sign whose any part this circle divides and touches. This is called the 'position in longitude' of the star* from the head of Aries. [134] *Also if a line that originates from the centre of the world and passes through the centre of the star touches one of the six above mentioned circles which divide the ecliptic into the twelve parts, that point is called the 'position in the heaven' of the star. The point where this circle touches the ecliptic is defined as the star's position on the ecliptic, and it is said in that way to be at the head of some sign.*

[135] As we have discussed the longitude of a star, consecutive order demands now to discuss the latitude. As the latitude of the stars we do not denote their inclination from the equator, but their distance from the ecliptic. *The latitude of a star is an arc of the circle that passes through the poles of the ecliptic and through*

mundi egressa et per centrum stelle transiens designat. Tantum enim a circulo signorum longe est, quantum est a puncto usque circulum. Arcus igitur hic tante longitudinis est, quantum est inter punctum et zodiacum circulum. Hic arcus latum dicitur stelle, quia a signifero circulo tot in austrum aut septemtrionem elongatur gradibus, quo-
5 tis idem protenditur. *Atque in hunc modum invenitur locus fixarum errantiumve in longo et lato.*

[136] *Imaginamur etiam alios* VI *circulos paralellos recti diei circuli dividentes circulum signiferum in capite primis punctis signorum, quos signorum dicimus circuitus ob eam rem, quod initia signorum volvuntur super illos. Horum primus a septemtrione circui-*
10 *tus est Cancri tangens zodiacum in uno puncto, qui caput est Cancri et punctus mutationis estatis* atque finis inclinationis signiferi in septemtrionem, de qua post docebimus. [137] *At vero secundus a septemtrione remotior ad circulum recti diei accedens circuitus est Geminorum et Leonis, quia in eo ipsorum primi puncti semper volvuntur. Tercius trium septemtrionalium circulorum spere proximus Tauri caput et Virginis incessanter*
15 *gerit. At ab austro primus remotiorque a recto in austrum circuitus est capitis Capricor-*
f. 15r *ni signiferum | diverberans uno in puncto, quem mutationem hiemis vel punctum capitis Capricorni* seu finem inclinati zodiaci in austrum dicimus. *Secundus autem Sagittarii atque Aquarii capita gestat. Tercius australium recto propinquior Piscium et Scorpii portas videt. At horum medius a quovis ultimo quartus circulus est recti diei, per quem*
20 *Arietis et Libre capita volvuntur semper.* Sunt igitur VII paralelli signiferum dividentes, quorum medius maior ipsum dividit medium, reliqui VI altrinsecus positi tres in septemtrionem, IIIes inclinantur in austrum.

[138] *Hii igitur* VII *circuli orizontis secant circulum iuxta sue positionis proportionem, et est inter quosvis eorum duos proximos arcus quidam orizontis, quem largitatem*
25 *orientis in oriente* et largitatem occidentis in occidente *aiunt eorum qui inter circulos eosdem sunt signorum atque in eisdem arcubus orizontis oriuntur aut occumbunt. Cuius rei causa est, quoniam magne spere motus movet omnes alias suo motu, facitque surgere quemlibet signiferi punctum aut occumbere in duobus punctis circuli orizontis. Movetur autem quisque signiferi punctus in paralello magni circuli.* [139] *Movetur enim caput*
30 *Arietis* et Libre *semper in circulo recti diei.* Moventur igitur semper quilibet alii signiferi puncti in paralello eiusdem. *Punctus igitur capitis Tauri movetur in circuitu Tauri,* in quo et capud Virginis, *et quisque punctus Arietis vel Virginis movetur in pa-*

---

8 quos] quo **9** initia] initio **12** remotior] remortior (L) **14** circulorum] circulum (L) **17–18** Sagittarii] Sagittarius (L) **18** Aquarii] Aquarius (L)

the point which is indicated by the line which originates from the centre of the world and passes through the centre of the star. Its distance from the ecliptic is as large as the distance from that point until the ecliptic. This arc therefore has the same length as the interval between the star and the zodiac circle. This arc is called the 'latitude' of the star, because the latter lies by the same amount of degrees to the north or the south of the ecliptic as what is spanned by that arc. *In this manner the position of the fixed and the wandering stars in longitude and latitude is determined.*

[136] *We now imagine further six circles, parallel to the equator and dividing the ecliptic at the head in the first points of the signs, which we call the 'circuits' of the signs, because the beginnings of the signs revolve on them. The first of them from north is the circuit of Cancer, touching the zodiac in a single point, which is the head of Cancer and the point of the summer solstice* and also the termination of the inclination of the ecliptic to the north, about which we will speak later. [137] *The second one, further from north and closer to the equator, is the circuit of Gemini and Leo, because their first points always revolve on it. The third of the three northern circles, lying closest to the equator,*[53] *bears unremittingly the heads of Taurus and Virgo. The first circuit from the south, and furthest south from the equator, is that of the head of Capricorn, hitting the ecliptic in a single point, which we call the winter solstice or the point of the head of Capricorn* or the termination of the inclined zodiac to the south. *The second one bears the heads of Sagittarius and Aquarius. The third of the southern circles, which is closest to the equator, sees the gates of Pisces and Scorpio. But the central one of them, being the fourth circle from either end, is the equator, on which the heads of Aries and Libra revolve permanently.* There are thus seven parallel circles which divide the ecliptic, the central and largest of which bisects it in half, whereas the remaining six lie to either side of it, three towards the north and three towards the south.

[138] *These seven circles thus intersect with the horizon according to their position, and between any two neighbouring circles there is an arc of the horizon which in the east is called the 'amplitude of rising'* and in the west, the 'amplitude of setting' *of those signs which lie between those circles and which rise or set on these horizontal arcs. The reason for this is that the motion of the great sphere moves by its motion everything to other places and makes every single point of the ecliptic rise or set at two points of the horizon. But every point of the ecliptic is moved on a circle that is parallel to the equator.* [139] *For the heads of Aries* and Libra *always revolve on the equator.* Accordingly, any other points of the ecliptic move parallel to that. *The point of the head of Taurus is moved on the circuit of Taurus,* on which also the head of Virgo is moved, *and*

[53] 'equator': lit., 'sphere'.

ralello quolibet eorum, qui sunt in circuitu Tauri et Virginis et rectum circulum. Surgens igitur horum quilibet punctorum in quolibet orizontis puncto in eo arcu, qui inter duos hosce continetur circulos, occumbit in alio, quantumque longe a puncto orientis circuli recti diei oritur, que largitas orientis dicta est, tantam invenimus inter eiusdem et recti circuli occasum occidentis largitatem. *Atque in hunc modum quilibet arcus orizontis eorum qui sunt inter duos proximos* VII *paralellorum, dicitur orientis largitas* aut occidentis *eorum que inter sunt signorum.* In his enim, quemadmodum de Ariete et Virgine dictum est, oriuntur aut occumbunt. Et de his quidem dictum est. Reliqua vero in sequentis libri disputatione disseremus.

Explicit liber primus; incipit prologus secundi.

every point of Aries or Virgo is moved on some parallel between the circuit of Taurus or Virgo and the equator. Hence every point of them that rises from any point on the horizon on that arc between the two circles which enclose it will set at another point, and the distance of its rising point from the rising of the equator, which has been said to be the amplitude of rising, will be found equal to the amplitude of setting, between the setting points of the star and the equator. *In the same manner, every arc of the horizon between two neighbouring circles of the seven parallel circles is called the amplitude of rising* or setting *of the signs between them.* For they rise and set in these arcs, as we have demonstrated for Aries and Virgo. This has been discussed now. The rest will be dealt with in the discussion of the following book.

Here ends Book I, and the prologue of Book II begins.

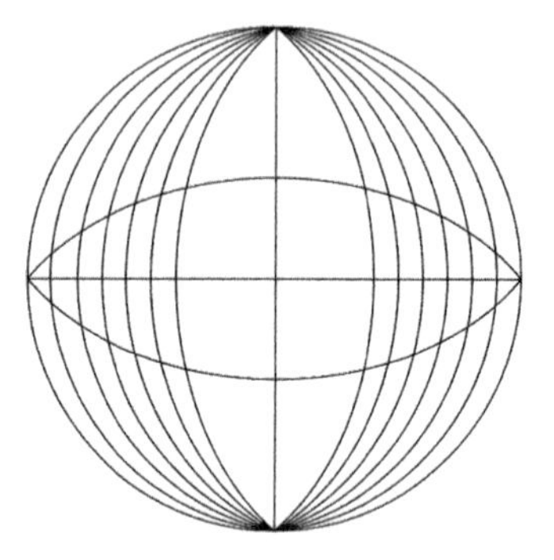

Fig. 1.2: From MS Cambrai 930, fol. 43v. The diagram shows the division of the heaven into the twelve zodiac signs. The diagram has been left incomplete; see Fig. B.5 on p. 422 for a more complete variant. It has also been wrongly titled 'figura (*expunged:* eg ec) retrogradacionis' and appears very late in the manuscript, on one page with Fig. 4.13 of Mercury's sphere

Book II

D. Grupe, *Stephen of Pisa and Antioch: Liber Mamonis*, Sources and Studies in the History of Mathematics and Physical Sciences,
https://doi.org/10.1007/978-3-030-19234-1

In astronomie mihi suscepta disputatione laboranti, de qua pauca certe habet Latinitas, eorumque pleraque erroris obfuscata caligine, obici fortassis animus doctis poterit arrogans in invidia, quod in Macrobium inter philosophantes non mediocrem totiens acrius invehar, eoque amplius, quod usque ad hec tempora omni caruerit obtrectationis livore. Quibus vellem satis esset mea cognita voluntas, intelligantque me Latine tradere facultati nostratum incognita auribus archana, que cum frequentibus vigiliis diuturnis cogitationum recessibus exquisita comparaverim, quorum Macrobium aut inscium fuisse video aut intellecta perversa depravasse expositione. Horum alterum, cum ad filium suum, quem sapientia sua sapientiorem fieri vellet, scriberet, fuisse dicendum non est. Nemo enim dilectum sciens perverse instruit. Non igitur intellecta veraciter depravasse, sed non intellexisse potius et ignorasse iudicandum est. Quamobrem non mihi in huius artis peritia philosopho sed cum inscio contencio est. Quod si et intellexisse et intellecta commode et desiderato nature ordine exposuisse dicetur, neque enim ipsos assumpturos membrum tercium existimo, intellecta depravasse, legant prius nostra quam distrahant, atque tum demum, cum nos | in sententie nostre subsidium attulimus rationum cum Macrobianis in unum tamquam certaminis discrimen collatarum, equa lance partiantur conflictum, agentis scilicet et defendentis seu testis. Omissa per sancta iuridicaria sedentes constantia omni vacuum passione parem confligentibus assensum prebeant, enodatisque meis Macrobiique sententiis, quodque in sui patrocinium intellectus uterque asserit, quem de statu deiectum aut forte inferiorem animadvertunt ipsum aut inscium horum aut intellectorum depravatorem iudicent, ac tamen demum cuius sit secte amicitior assensus sub veritatis indagine.

f. 15v

In Macrobium igitur nostra iccirco maior est animadversio, quoniam apud nostratum opinionem ceteris ipsum copiosiorem in astronomia et sentio et relatum per quamplurimos est. Quamobrem cum is artem teneret ceterisque amplius peccasse acutius intuentibus deprehendatur, eum iustius reprehensionis lima corripimus, quod cum alios precellere credatur, inveterato iamdiu errore Latinitas

2 eorumque] *corr. ex* eorum (L) **10** non] *add.* (L) **15** membrum] menbrum **19** iuridicaria] nundicaria **21** quem] quam **22** forte] fortem (L) animadvertunt] animadvertit intellectorum] *add. in marg.* (L) **23** amicitior] amiticior **29** inveterato] inveteratio (L)

2.1 <Preface to Book II>

As I am working on the commenced tract on astronomy, a subject about which the Latin world has only little certain knowledge and most of it contaminated with obscuring error, I may be accused of envious arrogance towards the learned, because I often criticise Macrobius so sharply, who is quite eminent among those who philosophise; especially, since up to our time he has never suffered jealous resentment. I wish those critics knew my intention well enough and recognised that for the benefit of our own world of letters I translate into Latin unknown findings that have not yet been heard of by our people and which I have obtained over many nights and days in thoughtful seclusion, and which I see were also unknown to Macrobius or, if he knew them, had been corrupted by his distorted account. We can exclude the second case, because Macrobius wrote for his son to make him wiser through his own wisdom,[1] and no one would deliberately teach a beloved person wrongly. One therefore must suppose that he did not understand or know the truth rather than falsified anything that he had understood correctly. For that reason I do not argue on this field with a philosopher but with an ignoramus. But if they say that Macrobius did understand astronomy and set it out appropriately and in the required natural order—for I do not think that they would approve the third alternative, namely that Macrobius falsified recognised facts—they should not refuse my arguments unread. Rather, after we have presented our view with argumentative support brought together like for a contest with Macrobius' arguments, they shall finally by fair weighing settle the conflict, that is, the one between plaintiff and defendant or witness. Bringing up firmness of character, they shall hold court by sacred legal ceremony and, without favouritism, grant equal and unbiased support to the conflicting parties. When my own and Macrobius' statements are presented, and also what either of us brings forth as support for the understanding of his view, they shall judge whom they recognise as refuted or less convincing, or as unknowing of the facts, or as falsifier of well-understood facts, and, finally, whose side deserves more willing approval, by exclusive consideration of its truth.

The reason why we pay so much attention to Macrobius is that he seems to me—and by many is said to be—the most comprehensive author on astronomy. If he thus had the skills but on closer inspection can be found to have failed more than any others, we denounce him with even better reason; for since he is believed to excel over the others, Latin scholarship has inflicted a long fostered error on

[1] *Somn.* I,1,1 and II,1,1 (L).

sequaces implicuit.

Atque hec quidem precipua causa in ipsum nominatim conquerendi mihi est; secunda autem, quod cum eius traditioni contraria sentirem, que diuturnam adepta fidem ea tamquam validissimis niteretur, propinquavit in eis vero et novitas, 5 et novitatis sepius comes nocet invidia, rationibus quamplurimis confirmari desiderabant, ut si quidem ipsis edax livor fideri derogaret, quandoque fida rationum remearetur custodia, eosdemque detractantes, si forte aurem accomodent, aut ad eorum que dicimus inclinent assensum aut erubescentes acrioribus urgeant stimulis, cum animadvertunt se ubertati nostri presidii nulla posse ratione resistere. Nam 10 quod et consuetudo earum sit virium, ut vel vitiis quoque ad optinendam quamvis falso iusticie sedem opitulentur apud hominum opinionem, et nova, licet virtutis opinione et honestatis pulcherrimo signata sint cirographo, tamquam iniquitatis plena veneno popularis absorbeat animus, cum multa philosophorum accurate copioseque dicta factaque, tum Solonis quoddam eximium precipuus nobis testis 15 esse poterit. Is enim cum Atheniensibus sepius interpellantibus, ut legum aliqua sanciret instituta, quibus eorum res publica populique mores et privatorum status regerentur, rennueret, tandem optimatum bonorumque flexus precibus se quod postulabant facturum pollicitus est, si iurarent ipsi omnisque Athenarum populus se decennio, quicquid ipse servaret, servaturos. Quibus postulata concedentibus, 20 quippe quos in Solonem nulla iniustitie argueret suspitio, quem et sectatorem veritatis et iusticie, non otiosum censorem persenserant, leges promulgat, scriptas tradit, ipse urbe egressus post decennium rediit. Cuius in adventu civitas plurimum letata causas more accuratissime inquirit, dat Solon: ne deleret mutatumve iri sineret quodpiam in his, que ipse exquisitissime sub equilibritatis ratione descripserat. 25 Nec eum hec fefellit ratio. Nam populus inconsueta primo graviter ferens detrahebat his, que non noverat. At ubi consuevit, cupidissimo amplexatus animo et illa summa fovit diligentia, et auctorem plurima honestavit gloria. Igitur cedente novitate, evanuit invidia, quia in amarissimis educatis, etsi mel, quia novum, amarum, consueta autem dulcia, licet amara sint, videantur, ubi quis desuescens f. 16r 30 malum paulo amplius bonum affectaverit, animus in | contraria vertitur.

Insurgat igitur quantumlibet inter cetera iustior vitio invidia, que cuius est ip-

---

3 sentirem, que] sentiremque **9** animadvertunt] animadvertit **10** earum sit] eorum fit **14** quoddam] quendam **15** cum] *add.* (L) sepius] *corr. ex* sepeius (L) **23** Solon] Salonem **24** quodpiam] quempiam **27** illa] illam

credulous disciples.

This is my main reason to argue against him in person. A second reason is that since I disagree with his tradition, which has gained long-lasting credit and relies on this credit like on incontrovertible arguments, whereas with my opposing view also comes Novelty, whose frequent companion Resentment often acts destructively, my theory needed support by as many arguments as possible, so that if biting envy should refuse to accept it but the reliable care of reason is still followed, those who denigrate our view, once they lend us their ear, will either change to our doctrines or continue the argument hotheaded with increasing aggressiveness, when they recognise that they cannot resist our strong position by any thinkable argument. For many comprehensive and detailed works and actions of the philosophers, as well as Solon's outstanding example, give us clear evidence that customary use has such a powerful effect that it lets even vices, however ill-founded, obtain the rank of justice in peoples' opinion, whereas everything new, even if its noble and honest character could be documentally certified, will be perceived by the mind of the public as if it were soaked with the poison of mischief. Solon was assailed by the Athenians with the request to enact a legal corpus that would enable a regeneration of their state, of the people's moral constitution and of the status of the citizens. Solon used to refuse, but finally persuaded by the pleas of the high-ranked and the good, he promised to fulfill their request if they and the whole Athenian people swore to observe for one decade any rules that he himself would observe. They agreed to this condition as they had no suspicion of injustice against Solon, whom they had experienced as a follower of truth and justice; thus Solon enacted laws and had them written down, and after a subsequent ten-year exile from the city he came back. The city cheered at his return and asked Solon to tell them precisely the reason for that time span, to which he answered: that no one would abolish or leave to corruption any of the statutes which he had written down wisely and under consideration of social balance. And he was absolutely right in that point. For the people first disliked the unaccustomed rules and criticised those things that they did not know. But when they became used to them, they willingly accepted them and observed them with scrupulous diligence and honoured the initiator with the greatest glory. Thus, as novelty faded, resentment vanished. Just as for those who were brought up with bitter food—to whom honey will seem bitter, because it is a novelty, whereas the familiar will appear sweet in spite of its bitterness—whenever someone is being weaned from something bad and desires the good a little more, the mind changes to the opposite.

Therefore, envy may arise—however more just it is than vice in other respects–

sum continuis excoquit doloribus nec patitur in quiete securum. Illius autem, in quem sit, aut vix aut numquam obfuscat bona. Nichil enim nobis ad perpetuitatem offundet caliginis, quoniam, ut speramus, cedente novitate cedent etiam invidentes, iustiorque censura sequentium, quibus nostrum neuter minus erit cognitus, 5 quicquid livor infuderit maledictorum, absterget.

Sed ne totum in prohemiis utilitatis parum conferentibus tempus consumamus, que dicta sunt sufficiant. Funem ab anchora solventes demus vela ventis, procedatque divina annuente gratia propositi operis desiderabilis et continua disputatio.

---

2 numquam] umquam

which tortures its object in permanent agony and does not allow it carefree placidity. But this envy hardly ever darkens the merits of its victim. For nothing can overwhelm us with darkness forever, because hopefully the enviers will disappear, once the novelty has gone, and a fairer judgement of posterity, to which neither of us will be less known, will wipe off whatever the envy of the defamers has inflicted on us.

But to keep us from wasting all our time on proems that do not contribute much useful, the above shall be sufficient. Untying the rope from the anchor, we turn the sails to the wind and with God's mercy the longed-for discussion of the proposed work may steadily proceed.

Incipit secundus liber.

In primo igitur libro ab universitatis mundane corpore incipiens a terris usque celum gradatim sermo ascendit, in quo circulorum qua ratione circunducantur diversitates ostenso ad zodiaci tandem imaginarios circuitus particionesque reversus in eodem primi libri prolixitatem terminavit. Consequens ergo videtur atque ordo precipiendi necessarius subsequenter de inclinatione eius a recti diei circulo disserere, ut non frustatim et conscisse sed iugabili competentia totum quod elaboraverimus innectatur. Dehinc autem de lato terrarum et longo diversitatisque dierum multiplici causa atque alto altitudinis et orientalium varietate dicemus.

¶Signifer igitur atque recti diei circulus, quem equinoctialem aiunt, magni sunt circuli non easdem habentes superficies aut sese recto secantes angulo, sed acuto. Horum ab invicem his in partibus, quibus magis separantur, tanta est distantia, quantum eorundem sepositi sunt poli. Dividunt autem sese medios in duobus diametrice oppositis punctis, atque ab invicem incipiunt separari. Hanc separationem a recti diei circulo inclinationem signiferi dicimus, medietas cuius in septemtrionem, altera in austrum inclinatur. [140] *Est igitur inclinatio arcus magni circuli, qui transiens super utrosque mundi polos secat signiferum rectique diei circulum in duo media ab eisdem divisus.* Huius circuli arcus inter duas proximas medietates circulorum zodiaci et recti positus inclinatio est signiferi a magno circulo, ut in his medietatibus, que sunt a puncto in Ariete primo usque caput Libre, arcus, qui intererunt, inclinatum monstrant signiferum in septemtrionem. [141] *Habent autem omnes zodiaci puncti inclinationem* II^{bus} *exceptis, quorum alter Arietis alter Libre caput est.* In his enim sese ligantes circuli unum sunt, in reliquis omnibus diversi. [142] *Quoscumque igitur imaginarios circulos circumduxeris orientes a polis mundi, tangent hii singulos punctos circuli signorum ipsumque secantes extendentes in circulum recti diei, eritque quod interest inter punctum, quo dividitur signifer ab hoc circulo, et punctum recti circuli, quem circumductus tangit circulus, arcus circuli, qui transit per polos mundi. Is igitur arcus inclinatio est puncti, quo zodiacum dividit circulus.*

---

7 competentia] competentie **24** Quoscumque] Quodcumque

Here begins Book II.

In Book I we started our discussion with the body of the world's entireness and ascended stepwise from the earth to the heaven, where we demonstrated how different circles are to be drawn there, before we finally came to the imaginary circuits and partitions of the zodiac, whose discussion brought the wide range of Book I to an end. As the next subject, and following the necessary didactic order, one should subsequently discuss the inclination of the zodiac from the equator, so that all the results of our effort do not stand frazzled as chunks, but neatly interwoven.[2] But first we will speak about the latitude and the longitude of places on the earth, the multiple causes of the different day lengths, the altitude, and the varying ascensions.

2.2 <The Inclination>

Config. 6: The Declination

The ecliptic and the equator, which is called the 'equinoctial' circle, are great circles that neither have the same plane nor intersect at a right angle but in an acute one. Their distance from one another at those points where they are furthest apart equals the distance between their poles. They divide one another into halves at two diametrically opposite points and then begin to diverge. This divergence from the equator is called the 'inclination' of the ecliptic, with one half of it inclining northwards, the other one southwards. [140] *The inclination thus is an arc of a great circle that passes through both poles of the world and cuts the ecliptic and the equator into two halves and which itself is divided by them.* The arc of this circle that lies between two nearest halves of the circles of the zodiac and the equator is the inclination of the ecliptic from the equator, and the arcs between those halves of the two circles that lie between the first point of Aries and the head of Libra show the inclination of the ecliptic towards the north. [141] *All points on the zodiac have an inclination except two, one of which is the head of Aries and the other one the head of Libra.* For in these points the circles are tied and coincide, whereas in all other points they are separate. [142] *Therefore, whatever imaginary circles you may draw from the poles of the world, they will all touch individual points on the ecliptic, where they intersect with it, and extend further to the equator; the interval between the point where this circle divides the ecliptic and the point of the equator that is touched by this drawn circle is an arc of the circle that passes through the poles of the world. This arc therefore defines the inclination*

[2] 'iugabilis competentia': Macrobian formulation for the linking force between mutually repelling elements, cf. *Somn.* I,6,24 et al.

[143] *Is autem arcus non ubique idem est,* sed in aliis maior, in aliis minor, que res inclinationum facit discordiam. *Hi igitur arcus, qui proximi sunt punctis equalitatis, in quibus circuli ligantur, ceteris, qui longius absunt, minorem habent longitudinem.* Recedentibus enim in sese invicem circulis minus fit inter utrumque spatium. Quanto f. 16v 5 igitur propius accedunt, tanto minores faciunt | arcus inclinationum; *quanto autem a punctis equalitatis longius seponuntur arcus puncto quolibet mutationum propinquiores, tanto maiorem efficiunt inclinationem.* [144] *Maior igitur inclinatio illa est, que est puncti, qui medius est medietatis circuli signorum, que est a puncto capitis Arietis usque Libram, inclinata in septemtrionem, oppositique eius in altera medietate, que in austrum* 10 *descendit. Sunt autem hi alter caput Cancri atque punctus mutationis estatis, alter oppositus caput Capricorni sive punctus mutationis hiemis.* Sicut enim equalitatum puncti inclinationem nullam habent fitque in ipsis diei et noctis equalis particio, sic hii maximam habent, in quibus progressionis solaris celebratur status, in altero in septemtrionem, in altero in austrum. *Magna ergo inclinatio est arcus magni circuli,* 15 *qui incipiens a polis mundi transit per polos signorum signiferum in duo media dividens in duobus oppositis punctis, alter quorum Cancri, alter est capud Capricorni. Estque is arcus inter punctum mutationis et circulum recti diei, et cuius est longitudo* .lc. graduum et semis et uncie.

[145] *Et tocius quidem quantitatis cuiusvis signi inclinatio per circulorum arcus col-* 20 *ligitur, qui a polis mundi orti transeunt per capita illorum.* Imaginamur igitur circulos transeuntes per polos mundi secantesque signiferum in capite cuiusvis signi et magnum circulum, eritque arcus inter duas sectiones positus inclinatio signi a magno, cuius finem caputve isdem tangens circulus ab sequentis precedentisve capite vel fine discriminat. Exempli causa, *transeuntem imaginamur circulum magnum per polos* 25 *mundi et Tauri caput. Tauri vero capiti opponitur diametrice Scorpionis initium. Transit igitur idem circulus per Scorpii caput. Arcus igitur huius circuli, qui est a Tauri capite usque punctum, quem isdem in magno tangit circulo, dicitur inclinatio totius Arietis,* non quod tanta cuiusque partis sit inclinatio, sed quoniam qui in eo sunt .m. graduum ibidem terminantur et longitudo et inclinatio. At vero huic recte oppositus 30 eiusdem circuli arcus ille est, cuius quantitas eadem initiumque huius initio et finis huius fini diametrice opponitur, aut postremo quevis illius pars huius eiusdem mensure parti. Is autem est, qui interest eiusdem circuli a Scorpionis primo puncto

---

1 que] quo (L)

of the point where this circle divides the zodiac.

[143] *But this arc is not the same everywhere,* but in some places it is larger and in others smaller, which fact causes unequal inclinations. *Those arcs which lie near to the equinoctial points, where the two circles are tied, have a shorter length than those which are more remote.* But when the circles converge, the distance between them becomes shorter. Hence the closer they come, the smaller inclination arcs they produce; *whereas the further the arcs lie from the equinoctial points and the closer to one of the solstitial points, the larger an inclination they produce.* [144] *The greatest inclination to the north therefore occurs at the mean point on the half of the ecliptic between the head of Aries and Libra, and to the south at the point opposite the first and on the other half. The first of these points is the head of Cancer and the point of the summer solstice, whereas the other and opposite one is the head of Capricorn, or the point of the winter solstice.* Just as the equinoctial points have no inclination and produce an equal partition of day and night, these other points correspondingly have the greatest inclination and in one of them the sun's highest ascent to the north is celebrated and in the other one, to the south. *Accordingly, the 'great inclination' is the arc of that great circle which starts from the poles of the world and passes through the poles of the ecliptic, dividing it into two halves at two opposite points, one of which is the head of Cancer and the other the head of Capricorn. This arc lies between the solstitial point and the equator and its length is* 23;35 degrees.[3]

[145] *The inclination of the entire length of any sign is obtained from the arcs of the circles that start from the poles of the world and pass through the heads of the signs.* We therefore imagine circles that pass through the poles of the world and cross the ecliptic at the head of every sign and also the equator. The arc between two intersections is the inclination of the respective sign from the equator, while the circle distinguishes the head or end of that sign from the head or end of the following or the preceding sign. For example, *we imagine a great circle passing through the poles of the world and the head of Taurus. But diametrically opposite the head of Taurus is the beginning of Scorpio. Hence the same circle passes through the head of Scorpio. Then the arc of that circle from the head of Taurus until the point on the equator that is touched by that circle is called the 'inclination of entire Aries';* not because this is the inclination of every point of Aries, but because the 30 degrees that Aries comprises end there with respect to both longitude and inclination. But the same circle's arc exactly opposite the former is the one with the same length but whose beginning

[3]'23;35 degrees': lit., '23 degrees plus a half plus a twelfth.' Arab., 'approximately 24 of the parts of which 360 constitute the whole circle.'

usque magnum circulum. *Is igitur arcus arcui inclinationis Arietis oppositus oppositi Arieti signi inclinatio est,* sicut de inclinatione Arietis dictum est. Eadem est in aliis signis ratio. [146] *Arcus enim circuli, qui secat in septemtrionali medietate signiferum per punctum, quem Tauri finem initiumque Geminorum ponunt, in australi autem per ca-*
5 *put Sagittarii, qui a puncto capitis Geminorum usque magnum est, circulum inclinatio Arietis simul et Tauri.* At vero ipsius oppositus a Sagittarii capite usque circulum recti diei inclinatio est Scorpionis simulque Libre. [148] *Circuli autem, qui transit per capita Virginis et Piscium, arcus, qui inter Virginis caput et rectum circulum clauditur, inclinatio est Virginis,* oppositus vero Piscium, [147] *sed qui per principia Leonis et*
10 *Aquarii, arcus, qui de ipso est inter caput Leonis et circulum spere, tota est inclinatio Leonis et Virginis,* cui oppositus Aquarii et Piscium. At vero eius, qui per capita Cancri transit et Capricorni, punctos scilicet mutationum, arcus a puncto mutationis estatis usque magnum circulum inclinatio est in ordine signorum Arietis, Tauri et
f. 17r Gemini; mutato autem ordine Virginis, Leonis et Cancri. Huic opponitur | a punc-
15 to mutationis hiemis, qui caput est Capricorni, usque rectum circulum ordinem servans signorum Libre, Scorpii Sagittariique inclinatio est. At per retrogradam positionem Piscium, Aquarii et Capricorni.

Manifestum autem credimus esse ex his, que superius posita sunt de inclinatione, quoniam duo sunt puncti signiferi, quibus nulla est inclinatio, quorum alter
20 is est, quem initium circuli signorum ponimus, punctus scilicet equalitatis veris et capitis Arietis, alter signiferi medius, quem punctum equalitatis autunni et capitis Libre esse dictum est. Ab his vero circulus signorum inclinari incipit a puncto equalitatis veris in septemtrionem, et a puncto equalitatis autunni in austrum, cuius completur inclinationis augmentum in punctis mutationum. Duo vero sunt
25 signiferi puncti, quibus eadem inclinatio est maiorque, in septemtrionem is, quem mutationis estatis dicimus, alter eius, quem in austrum punctus mutationis hiemis. Ab his autem signorum circulus ad circulum recti diei revertitur.

Hi vero, qui intersunt punctos equalitatum et mutationum, puncti omnes incli-

---

28 Hi] His (L) punctos] punctum

and end are diametrically opposite to those of the other or, finally, whose every part is opposite to the part of the same measure of the other arc. But it is the arc of the same circle from the first point of Scorpio until the equator. *It is therefore opposite to the inclination arc of Aries and thus equals the inclination of the opposite sign Aries,* just as we have said about the inclination of Aries. The same reasoning applies to the other signs. [146] *For the arc of the circle that crosses the ecliptic in its northern half at the point that is defined by the end of Taurus and the beginning of Gemini but in its southern half at the head of Sagittarius, and which thus extends from the head of Gemini until the equator, equals the inclination of Aries and Taurus together.* The arc opposite to this, from the head of Sagittarius until the equator, equals the inclination of Scorpio and Libra together. [148] *And the arc which is enclosed by the head of Virgo and the equator and which lies on the circle that passes through the heads of Virgo and Pisces defines the inclination of Virgo,* whereas the arc opposite to this is the inclination of Pisces; [147] *but on the circle that passes through the heads of Leo and Aquarius, the arc between Leo and the equator is the entire inclination of Leo and Virgo,* opposite to which is that of Aquarius and Pisces. But on the circle that passes through the heads of Cancer and Capricorn, that is, the solstices, the arc from the summer solstice until the equator is the inclination of Aries, Taurus and Gemini in the order of the signs, but in reverse order that of Virgo, Leo and Cancer. Opposite to this, from the winter solstice at the head of Capricorn, is the inclination of Libra, Scorpio and Sagittarius, if one follows the order of the signs, whereas in backward order it is the inclination of Pisces, Aquarius and Capricorn.

We think it is clear from what has been said above about the inclination that there are two points on the ecliptic that have no inclination, one of which is the one that we defined as the beginning of the ecliptic, namely the point of the spring equinox or the head of Aries, and the other is half-way through the ecliptic, which was said to be the point of the autumnal equinox or the head of Libra. From these points the ecliptic begins to incline, northwards from the spring equinox and southwards from the autumnal equinox, until the increase of inclination is complete at the solstitial points. But there are also two points on the ecliptic with the same greatest inclination; northwards the point which is called the summer solstice, and the other, southwards, which is called the point of the winter solstice. From these points onwards the ecliptic turns back towards the equator.

Now,[4] all the points between the equinoctial and solstitial points form sets of

[4] *Configuration*, passage 149, in which the symmetry of declinations on both sides of an equinoctial point is stated, has been replaced with the commentary to the preceding passages, 145–48, and with the following extended discussion of the subject.

nationis et largitatis orientis et occidentis quaterni sub quadranguli forma, sed non equalibus lateribus, respectum ad se invicem habent IIII^{or}, aut hi sunt, qui in participandis inclinatione et largitate quadranguli equalia IIII^{or} servant latera. Quorum enim punctorum equalis est inclinatio eadem et largitas orientis. Sunt autem hi, quorum primus medius est punctorum equalitatis veris et mutationis estatis; secundus puncti mutationis estatis et equalitatis autunni, III^{tius} equalitatis autunni et mutationis hiemis, IIII^{tus} mutationis et equalitatis veris. Horum tota est equalis in forma quadranguli inclinatio, ut equa etiam quadranguli servent latera. Quod hac videri licet ratione; punctus, qui equalitatis veris et mutationis estatis medius est, plenis .s. gradibus distat a quovis eorum, qui aut mutationis estatis aut equalitatis autumpni medius est aut mutationis hiemis et equalitatis veris. Cui oppositus est vero, qui medius est equalitatis autumpni et mutationis hiemis, .tr. gradibus scilicet ab eo longe positus. Cum etenim totus signifer in .xp. gradus equales divisus sit, eiusque quarta pars .s. sint gradus, .tr. vero medietas, atque IV^{or} puncti equalitatum et mutationum per quandranguli equilateris formam sese intueantur, ut .s. primus a secundo, secundus a tercio, tercius a quarto, IIII^{tusque} a primo gradibus distent, qui punctus equalitatis veris et mutationis estatis medius est, ab utroque .ne. gradibus distat. Idem de eo dicendum est, qui positus in Leonis medio medius est mutationis estatis et equalitatis autumpni. At .ne., quibus secundus a puncto mutationis estatis distat, iunctis his .ne., quibus primus removetur ab eodem, .s. fiunt gradus, IIII^{ta} scilicet signiferi pars. Idem quoque de reliquis duobus et ad hos et ad se invicem considerandum est.

At nunc quoquo modo quadrangulus fiat, eodem erit IIII^{or} inclinatio. Elimanda est ergo certa diligensque consideratio, qua IIII^{or} angulares punctos, qui eandem sortiuntur inclinationum quantitatem, possit inquirentis exercitium facile deprehendere. Oportet igitur horum IIII^{or} punctorum quadrangulum eiusmodi esse, ut cuiuscumque quantitatis II° erunt latera, que in uno terminabuntur puncto angu-

11 oppositus] appositus (L) **13** positus] positis **17** distent] *corr. ex* distant (L) **23** Elimanda] Eliminanda (L)

four points each, of inclination and amplitudes of rising and setting,[5] in the shape of a rectangle whose four sides are unequal, except if the points are those four which, by sharing the same inclination and amplitude, maintain four equal sides of the rectangle.[6] For these points have equal inclination and the same amplitude of rising. The first of them lies in the middle between the points of the spring equinox and the summer solstice, the second between the points of the summer solstice and the autumnal equinox, the third between the points of the autumnal equinox and the winter solstice, and the fourth between the points of the solstice and the spring equinox. Their inclination is entirely equal in the shape of a rectangle, while they also maintain equal side lengths of the rectangle. This can be understood from the following consideration. The point in the middle between the spring equinox and the summer solstice stands exactly 90 degrees from either of the two points in the middle between the summer solstice and the autumnal equinox or in the middle between the winter solstice and the spring equinox. But opposite to it lies the point in the middle between the autumnal equinox and the winter solstice, which means a distance of 180 degrees from the former point. For if we divide the entire ecliptic into 360 equal degrees, and a quarter of it equals 90 degrees and its half 180 degrees, and the four equinoctial and solstitial points stand relative to each other like on a square, such that the first stands separated from the second, the second from the third, the third from the fourth, and the fourth from the first, by 90 degrees each, then the point in the middle between the spring equinox and the summer solstice stands distant from either of the latter by 45 degrees. The same must be said about the point in the middle of Leo, centered between the summer solstice and the autumnal equinox. But when the 45 degrees by which that second point is distant from the summer solstice are added to the 45 degrees by which the first point is distant from this solstice, that yields 90 degrees, or a quarter of the ecliptic. The same consideration applies to the remaining two points, relative to the aforementioned ones as well as to one another.

Now, in whatever manner the quadrangle is formed, in the same way will be the inclination of the four points. We therefore need to develop an accurate and reliable method through which the effort of the enquirer can easily come up with the four corner points of equal inclination. The rectangle of these four points must be such that the lengths of two sides that meet at one corner are equal to the

[5]Cf. *Alm.* II,3.

[6]See Fig. 3.1, right, on p. 42 of the Introduction. A similar construct is described by al-Battānī, *Ṣābiʾ Zīj*, ch. 4.

lari, eiusdem sint reliqua II°, sed per opposita. Dico autem, si eorum fuerint duo, alterum breve, alterum longum, oppositum brevis breve, oppositum longi longum esse oportere. Unde fit, ut neque longa communem habeant sui finis terminum, ne- (f. 17v) que brevia, sed longorum alterum cum utrisque brevibus in duobus rectis | angulis 5 terminatur identidemque alterum in opposito. Eadem contingit ratio in brevibus. Oportet etiam huius quadranguli equales esse angulos nullumque esse in eo alterum altero acutiorem. Sed et illud considerandum, quoniam IIII^or^ laterum IIII communes terminos angulariter se intuentes oppositos esse necesse est, at vero qui in eodem brevi aut longo latere utrimque termini sunt, aut equaliter distantes 10 utrimque a recto circulo, alter in austrum, alter in septemtrionem, aut equaliter inclinatos in eandem partem.

Cum vero cuius propositum fuerit investigandum, qui sibi quaterni puncti in participanda inclinatione aut largitate orientis consentiant, primo quemlibet sibi punctum proponat in septemtrionem aut in austrum inclinatum duobus his excep- 15 tis, qui inclinationis finis sunt in utraque parte. Quo sumpto eius segreget oppositum intuitoque primo, quota quantitate circuli signorum longe sit a viciniori sibi puncto, quo ligantur signifer et magnus circulus in eadem zodiaci medietate, alterum sumat, qui eadem mensura distat ab opposito puncto ligantium se circulorum. Cuius sumens IIII^tum^ scilicet et punctum oppositum, IIII^or^ hos eandem habere in- 20 clinationis mensuram manifestissime predicabit. Exempli ergo causa, ut tota ratio clarius conliquescat, primum Tauri punctum sumo. Huius est oppositus primus Scorpii, quia et oppositum signum Tauro Scorpius. At vero primus Tauri punctus .m. gradibus distat a puncto equalitatis veris in ea positus medietate, que in septemtrionem inclinatur. Totidem autem gradibus distantem a puncto equalitatis 25 autumpni in eadem medietate si queramus, is est, qui Leonem terminat, cui opponitur is, qui in Aquario ultimus habetur. Horum igitur IIII^or^ eadem est inclinationis quantitas. Videamus tamen in his, si que de quadrangulo eorum posuimus, recte observari possint. Eorum itaque quadrangulum equa non habere latera videre licet, quoniam is, qui Tauri primus est, ab ultimo Leonis plus quam .s. gradibus 30 distat, .d. scilicet signis; totidemque primus Scorpii ab ultimo Aquarii, que duo in quadrangulo hoc sibi opponuntur latera. At vero ultimus Leonis a primo Scorpii minori graduum distat quantitate, duobus scilicet signis. Idem inveniri inter ultimum Aquarii Taurique primum licet. Horum quoque duorum laterum alterum alteri oppositum est. At hoc quod a capite Tauri est, quod etiam primum posuimus 35 et longum, iungitur utrimque in communi termino brevibus, secundo quidem in ultimo Leonis, quarto autem in primo Tauri, cui tercium latus longum oppositum

4 cum] tum **9** longo] *corr. ex* longuo (L) **12** investigandum] investigando **19** scilicet] solis (L) **21** primum] *corr. ex* primi (L) **33** laterum] latera **34** quod] qui

lengths of the other two lines, each to its opposite. I mean that if there are two sides, a short one and a long one, then the short one must be opposite the short one and the long one opposite the long one. As a result, neither the long sides nor the short sides can have a common terminating point, but a long side meets at each end with one of the two short sides at a right angle, and in the same way does the other long side opposite. The same reasoning applies regarding the short sides. Furthermore, the angles of this rectangle must be equal and none of them is to be more acute than any other. We also need to consider that the four points where the ends of the four sides produce the angles when they meet must be opposite each other, while those which are the ends at both sides of the same short or long side are lying equally distant to either side of the equator, one to the south and one to the north, or equally inclined on the same side.

Whenever somebody wants to find a set of four points that share the same inclination or amplitude of rising, he should first take an arbitrary point with an inclination to the north or to the south, except those two points that limit the inclination on both sides. Once that point has been chosen, he should pick out its opposite; next, having determined the former's distance on the ecliptic from the nearest intersection of the ecliptic and the equator, he should take the other point on the same half of the zodiac which is equally distant from the opposite intersection point of the circles. By taking the opposite of that previous one, as the fourth point, he can clearly say that these four points have the same amount of inclination. As an example, and to make the entire reasoning clearer, I take the first point of Taurus. Its opposite is the first point of Scorpio, because the sign of Scorpio is opposite to Taurus. But the first point of Taurus lies 30 degrees from the spring equinox on that half which is inclined northwards. But if we look for the point which on the same half is equally remote from the autumnal equinox, this will be the one that terminates Leo, opposite to which is the last point of Aquarius. These four points, therefore, have the same amount of inclination. But let us now see whether we can really observe from these points what we said about their rectangle. Obviously, their rectangle does not have equal sides, because the first point of Taurus stands further than 90 degrees from the last point of Leo, namely by four signs. By the same amount the first point of Scorpio is remote from the last one of Aquarius, as these two sides lie opposite to one another on the rectangle. But the last point of Leo is distant from the first one of Scorpio by a smaller amount of degrees, that is, by two signs. The same is found between the last point of Aquarius and the first one of Taurus; and again these two sides are opposite to one another. But the side that starts at the head of Taurus, which we defined

eisdem iungitur brevibus sed non eisdem terminis, secundo scilicet in primo Scorpii, IIII[to] vero in ultimo puncto Aquarii. Angulariter autem sese intuentes angulares poncti opponuntur, ut primus Tauri primo Scorpionis, ultimus Leonis ultimo Aquarii. At vero qui quodvis IIII[or] laterum puncti terminant, si latus quidem per 5 superficiem recti transeat circuli, alter eorum in austrum, alter in septemtrionem sub eadem inclinatur quantitate ut lateris secundi vel quarti termini. Si vero totum latus in septemtrionem sit aut in austro, in eandem equaliter partem utrique eidem termini inclinantur ut primi lateris, qui in septemtrione est, termini, primus punctus Tauri et ultimus Leonis, equaliter in septemtrionem inclinantur. Idem di- 10 cendum de tercii lateris terminis in austrum. His plene expositis illud adiciendum arbitror, quod si huius quidem, quem hic imaginamur, quadranguli IIII[or] angulares f. 18r puncti punctis equalitatum viciniores fuerint, laterum IIII[or] duo, quorum | alterum in septemtrione totum est oppositumque eius, quod totum in austro continetur, longiora, reliqua vero duo, que per superficiem transeunt recti circuli, breviora 15 sunt. At si longius a punctis equalitatis recedentes punctis immutationum propius accesserint, contra fiet. Breviora enim, que in altera tantum sunt parte, longiora vero que in utramque extenduntur inveniuntur. Et de inclinatione quidem signiferi ita iudicandum est, de qua etiam illud scitu dignum videtur, quod addimus.

[150] *Quisque enim punctorum zodiaci,* qui ab equalitatis punctis in alteram in- 20 clinantur partem motu supreme spere, que secum movet omnia, *movetur in aliquo paralellorum recti circuli,* qui continentur a circuitu Cancri vel circuitu Capricorni usque magnum circulum, quos paralellos eiusdem esse superius dictum est. *Paralellus autem is, super quem quivis zodiaci punctus movebitur, eadem movetur longitudinis mensura a magno circulo, quota est eiusdem, quod in ipso, puncti inclinatio.* Nam si ma- 25 iorem velit eam quis minoremve arbitrari, procedente rationis vestigio punctum eo in parallello non moveri manifestissime colliget, cum solem totum peragrare zodiacum XII signorum in unius anni reflexu constat. *Quando itaque sol in aliquem punctorum zodiaci motu sue ferentis spere, de qua suo docebitur in tempore, intraverit, movebitur in superficie circuli, qui paralellus magni est, quem isdem, in quo sol est, punctus circuli signorum motu alte spere facit, atque in eodem semper volvitur.* Supra vero 30 dictum reminiscimur solis centrum semper in zodiaci esse superficie, nunquam-

1 eisdem] *corr. ex* eis (L) **4** puncti] punctum **9** punctus] puncti inclinantur] inclinatur **19–20** inclinantur] *corr. ex* reclinantur (L) **21** recti] *corr. ex* cti (L) **23** zodiaci] zodiacus (L) **25** procedente] *corr. ex* proceden (L)

as the first and also as a long side, is connected at both ends to the ends of short sides, to the second side at the end of Leo and to the fourth one at the beginning of Taurus, and opposite to it lies the third side, as the other long one, connected to the same short sides but at other points; that is, to the second side at the beginning of Scorpio and to the fourth one at the end of Aquarius. And the corner points lie at opposite angles relative to each other, that is, the first one of Taurus to the first one of Scorpio and the last one of Leo to the last one of Aquarius. But the two points that terminate any of the four lines, if this line crosses the equatorial plane, have equal inclination, with one of them being inclined to the north and the other to the south, such as the ends of the second or the fourth side. Conversely, if the side lies completely to the north or to the south, both its ends are inclined to the same side, such as the ends of the first side, in the north, which are the first point of Taurus and the last point of Leo, which are equally inclined to the north. The same is to be said about the ends of the third side, towards the south. As this has now been explained comprehensively, I think it should be added that, if the four corner points of this imagined quadrangle lie closer to the equinoctial points, those two of the four sides of which one lies completely on the northern side and the other one, opposite the first, is completely located in the south, are the longer sides, whereas the remaining two, which cross the equatorial plane, are the shorter ones. But if the points move further from the equinoxes and approach the solstitial points, the contrary will occur. For the sides lying on only one side will be shorter and those extending to either side will be longer. This is how one should consider the inclination of the ecliptic, about which also the following details seem noteworthy.

[150] *Every point of the zodiac* which is inclined from an equinoctial point to one of the two sides *is moved* by the motion of the highest sphere, which takes everything with itself, *on a circle parallel to the equator* which is enclosed between the circuit of Cancer or the circuit of Capricorn and the equator, whereas the former two were said above to be parallel circles of the equator. *But that parallel circle on which any given point of the zodiac is moved rotates at the same distance from the equator to which the inclination of that point on it amounts.* For if somebody wants to assume that the inclination is greater or smaller, proceeding along the path of reasoning he will conclude that the point can by no means move on that parallel circle, because the sun undeniably travels through the entire zodiac of the twelve signs in a period of one year. *Hence, when the sun by the motion of its sphere, which we will talk about at its proper place, reaches a certain point of the zodiac, it will move on the plane of the parallel circle which that point of the ecliptic where the sun stands produces by the motion of*

que in alteram deflectitur partem. Erit igitur semper suppositum cuilibet zodiaci puncto. *Ex qua re illud manifestissime colligitur totam esse inclinationis solis a recti diei circuli superficie quantitatem, quota est eiusdem puncti circuli signorum inclinatio ab eodem adversus quem sol recta regione suppositus.* Unde etiam fit, quod <in>(L) inve-
5 stiganda solis inclinatione suppellectili utimur regula, quoniam in quolibet puncto signiferi volvitur, isque punctus equaliter longe positus est semper a magno <circulo>(L), solisque centrum zodiaci superficiem nunquam excedit, inclinatione puncti cognita ipsum quoque solem a magno circulo removeri totidem mensuris intelligimus. At non in quinque planetis eadem est ratio, lunaque paululum diver-
10 sa sectatur. De quibus in his dicemus locis, quibus quodque expediri desiderat. In primo autem libro dictum recolimus, quod quisque celi punctus alicui terre puncto recta superponatur linea eo modo, ut linea terre, que de centro mundi egreditur et extenditur usque in superficiem celi, quem in superficie punctus tetigerit, superpositus est illi, quem de eadem linea in superficie tangit terre. [151] *Latum terre*
15 *dicimus,* de quo paulo post docebimus, *quantitatem hanc, qua queque terre pars a transitu recti diei circuli in meridionali circulo distat.* Est vero latum alias quidem maius, alias vero minus. Unde fit, ut paralelli recti circuli super quemlibet latum terre transeant, non omnes quidem super illud, sed id super hoc semper, ille vero super illud. Idem de reliquis dicendum. *Quisque igitur zodiaci punctus habet inclina-*
20 *tionem equalem alicuius lato terre, isdemque omni die per punctum lati, quem punctum capitis dicimus, transit semel per moventem speram altissimam.* Omni etenim die semel pleno, ut dictum est, magnum celum volvitur circuitu, in cuius unaquaque circumductione omnes alie circumvolvuntur. Ex quo fit, ut is quoque punctus, cuius est inclinatio equalis lato eiusdem loci, in eandem, in qua et latum est, partem semel
25 per punctum capitis in die transeat. Ac nos quidem de hac tantum loquimur in presenti inclinatione, que in septemtrionem vergit. Ad cuius exemplum, que de
f. 18v australi intelligenda sunt, veraciter poterit quis | eliminare.

Transit ergo quisque punctus eius medietatis signiferi, que in septemtrionem

---

5 suppellectili] supplici **9** quinque] quemque (L)

the high sphere, and it will always revolve on it. But we remember what has been said above, namely that the centre of the sun always moves in the plane of the zodiac and never deflects to any side. Hence it will always stand beneath some point of the zodiac. *From this we can clearly conclude that the inclination of the sun from the plane of the equator will be as large as the inclination of the point on the ecliptic directly under which the sun is standing.* As a result, when determining the sun's inclination, we also make use of the helpful rule that in whatever point of the ecliptic the sun stands, since this point will always be equally distant from the equator and since the centre of the sun will never leave the plane of the zodiac, once we know the inclination of that point, we will know that the sun's distance from the equator has the same magnitude. But this does not apply equally to the five planets, and also the moon follows a slightly different order. We will talk about these things where each of them needs to be explained. But we repeat the statement from Book I, saying that every point of the heaven stands on a straight line above a certain point on the earth such that a line from the earth, drawn from the centre of the world and prolonged to the surface of the heaven, meets that point of the surface which stands above the point on the earth that is crossed by the same line on the surface of the earth. [151] *By the term 'terrestrial latitude',* which we will talk about soon,[7] *we denote that length by which every point on the earth is remote from the equator's passing on the meridian.*[8] This latitude is larger in some places and smaller in others. By consequence, the parallel circles of the equator all pass above a certain terrestrial latitude; not all above the same, but one always above one latitude and another one above another. The same is to be said about all others. *Thus every point of the zodiac has an inclination that is equal to some point's terrestrial latitude, and it moves through the point of that latitude which we call the zenith once every day by the moving highest sphere.* And once every day the great heaven revolves by one full circuit, as has been said, and by every revolution of it also all other spheres revolve. As a result, also the point whose inclination equals the latitude of that place, in the same direction as the latitude, revolves through the zenith once per day. So far, we speak only about that inclination which tends towards the north; analogous to this, everyone can reliably bring out what is to be understood for the south.

Thus once every day every point on the northern half of the ecliptic passes

[7] As Stephen postponed the content of *Config.* 4, here he needs to refer the reader to that later place.

[8] Arab. add.: 'it necessarily follows from that that the circles of declination correspond in the motion of the whole to the circles of latitude.' Stephen indicates this correspondence in his commentary at the end of the paragraph.

inclinatur semel in unoquoque die super punctum capitis terre illius, cuius lato eiusdem puncti inclinatio equalis esse dinoscitur. Solem vero per proprie spere motum in quemvis horum devenire et ostensum est et apertius dicetur. [152] *Quando itaque sol in quovis horum erit punctorum, qui in septemtrionem inclinantur, transit eodem die super punctis capitis loci illius, cuius latum eiusdem est longitudinis cum inclinatione puncti signiferi, in quo sol volvitur. Hec enim causa est, quare sol super punctum capitis eo in loco transeat, cui minus est latum magna inclinatione.* [153] *Et locus quidem ille, cui equale latum est magne inclinationi, que est* .lc. <gradus> .me. sexagenarie, *.a. semel in anno super punctum capitis solem videt. Quando enim sol in puncto mutationis estatis volvitur,* qui caput est Cancri finisque Geminorum, in quo in septemtrionem terminatur inclinatio, transit super punctum capitis eo in loco, qui paralello Cancri suppositus est. *Itaque fit, ut si qua res eo in loco ea ponitur coaptacionis diligentia, ut caput eius punctum capitis recte intueatur, nullam habeat umbram in meridie huius diei, sed undique solis circumferiatur radio. Aliis autem diebus, quibus in aliis signiferi punctis sol volvitur, recte posite rei umbram in septemtrionem defertur.*

[154] *At in his locis, qui latum minus magna habent inclinatione, alia est veritatis ratio. In illis enim bis in anno sol in punctum capitis transit. Id autem ea fit causa, quoniam paralelli recti circuli, qui signiferum tangunt, ipsum in duobus punctis dividunt,* duobus his exceptis quorum alterum Cancri, alterum Capricorni circuitum diximus. Ab his etenim duobus tantum in punctis tangitur, quos mutationis punctos aiunt, et estatis quidem punctus a circuitu capitis Cancri, mutationis vero hiemis a circuitu capitis Capricorni tangitur, que dicta sunt. *Reliqui vero omnes quotquot ipsum tangentes circuli, qui paralelli recti diei circuli duos eiusdem punctos tangunt, quorum eadem est inclinatio, quia eadem a recto circulo longitudo. In quemvis igitur horum, qui a quolibet paralello tangitur, quando sol punctum intraverit in septemtrionem descendens, in paralelli eiusdem volvitur superficie, in quo et punctus, qui centro eius superest, unde et eadem die super punctum capitis terre illius transit, qui equalem habet longitudinem a recto circulo inclinationi puncti, in quo solis centrum volvitur. Illo igitur, quo fit hoc, die eo in loco res recte posita nullam habet umbram in meridie,* quia sole posito in puncto capitis lumen eius corpus recte positum undique fixit.

[155] Ab eo quoniam solis inclinatio nondum completa est, *a puncto capitis sol*

---

4 inclinantur] inclinatur

through the zenith of that place on the earth whose latitude is found equal to the inclination of that point. But it has been shown and will be explained more extensively that the sun reaches each of these points by the motion of its own sphere. [152] *Thus, when the sun is in any of the points that are inclined to the north, it will on the same day pass through the zeniths of that region whose latitude is the same as the inclination of the point of the ecliptic on which the sun revolves. This is the reason why the sun reaches the zenith at such points whose inclination is smaller than the great inclination.* [153] *And the region which has a latitude equal to the great inclination, which is* 23;35 degrees,[9] *sees the sun once per year in the zenith. For when the sun moves through the summer*[10] *solstice,* which is the head of Cancer and the end of Gemini, where its northward inclination reaches its maximum, it passes through the zenith at that place which lies beneath the parallel of Cancer. *By consequence, if we align an object at that place so diligently that its head points upright to the zenith, it will not have a shadow at noon on that day, but it will be completely encircled by sunlight; on the other days, however, when the sun moves in other points of the ecliptic, the shadow of the upright placed object*[11] *is cast northwards.*

[154] *But at those places with a latitude less than the great inclination, a different explanation is true. For here the sun passes through the zenith twice per year. This happens, because the parallel circles of the equator which meet the ecliptic divide the latter at two points,* except those which we called the circuit of Cancer and the circuit of Capricorn. For these two are tangential to it in only two points, which are called the solstices, with the circuit of the so-called head of Cancer being tangential to the summer solstice and the circuit of the head of Capricorn, to the winter solstice. *All other parallel circles of the equator that meet the ecliptic meet two points of it which have the same inclination, because they have the same distance from the equator. Hence when the sun, on its way north, reaches any of these points that are met by a parallel circle, it moves on the plane of that parallel circle on which also the point above the centre of the sun is moving. For this reason, the sun passes on the same day also through the zenith of that place on the earth whose distance from the equator equals the inclination of the point where the centre of the sun is standing. Therefore, at that place and on that day when this occurs, an object placed upright has no shadow at noon,* because when the sun is standing in the zenith its light captures the upright body from all sides.

[155] Subsequently, as the inclination of the sun has not yet reached its full

[9] Arab.: 'approximately twenty-four parts.'

[10] The text follows the reading of Arab. K.

[11] Stephen seems to avoid intentionally the term 'gnomon' as used in the Arabic.

in septemtrionem incipit inclinari. Quamobrem eo in meridionali circulo super orizontis circulum posito umbra recte positorum non in septemtrionem, ut apud nos est, sed *in austrum defertur,* semperque maior fit, donec sol post estivi contactum solsticii in septemtrionem amplius prohibitus ferri ad recti diei circulum regreditur. *Regre-* 5 *diente itaque sole paralellumque, de quo diximus, ingresso, in quo punctus alter volvitur, qui eandem habet quam alter inclinationem, cuiusque idem paralellus circuitus est, sol iterum per punctum capitis transit,* et umbra recte positarum deficit. *Quo transgresso sol in austrum a puncto capitis inclinatur, umbraque in septemtrionali subiacet parte.* Et hec quidem de his dicta sunt locis, que minus habent latum magna inclinatione.

10 [156] *Verum in his, quorum lati magnitudo quantitatem magne vincit inclinationis, nunquam sol capitis attingit punctum, sed semper in austrum a puncto capitis inclinatur. Ob quam causam in his,* que novimus, *climatibus nunquam umbra in austrum sed sem-* f. 19r *per in septemtrionem defertur.* Hec autem, | que in septemtrionali inclinatione dicta sunt, diligens rerum examinator poterit in australi rationabili perscrutari interpre- 15 tatione, et *de inclinatione quidem solis que dicta sunt,* quantum ad susceptum videtur attinere tractatum, sufficiunt.

[157] ¶Planetarum inclinatio autem ceterarumque, que fixa feruntur, et lune alio quodam modo iudicanda est. *Et quidem earum inclinatio* vel longitudo lune *a recto circulo arcus est circuli, qui transiens per polos mundi et centrum* lune *secat si-* 20 *gniferum et rectum circulum, eiusque inclinatio a recto arcus is, qui inter eius centrum et rectum circulum clauditur.* At vero latum eiusdem et longitudo a circulo signorum per latum draconis ab eodem investigatur. Est autem draco circulus quidam, qui etiam ligans dicitur, de quo in sequentibus plenius tractabitur, qui none spere cursum sequitur. Unde fit, ut non semper caput eius in uno signorum sed in omnibus 25 possit inveniri. Huius inclinatio a zodiaco semper est eadem quemadmodum etiam signiferi a recto, sed mutatur, quemadmodum est cursus in signo, ut non semper eius magnum latum sit in Cancro aut Leone, sed nunc quidem in hoc, nunc vero in alio. Ex quo fit, ut et lune quoque latum, cum eius semper centrum in hoc sit circulo, nec inclinatur in quamvis partem, quemadmodum et solis centrum esse in 30 signifero dictum est. Aliquando quidem maius fit in Ariete, aliquando in Cancro eorumve oppositis, sepe vero in aliis, his scilicet, in quibus draconis latum maius esse deprehenditur. Est autem lune latum arcus circuli, qui transit per polos

---

8 septemtrionali] septemtriona (L) parte] parti (L) **11** nunquam] nusquam (L) **20** is] his (L) eius] *add.* (L)

extent, *the sun will start to incline northwards from the zenith. Therefore, when it is placed on the meridian above the horizon, the shadow of upright placed objects* will not be cast to the north, as in our regions, *but to the south;* and it will steadily become longer, until the sun has reached the summer solstice and thus is prevented from moving further north but turns back towards the equator. *On its way back, the sun will reach the parallel circle which we have just spoken about, where there revolves a second point which has the same inclination as the first and which has the same parallel circuit; in that moment the sun will pass through the zenith again,* and the shadow of upright placed objects will vanish. *Having passed through this point, the sun will incline southwards from the zenith and the shadow will fall to the north.* This has been said for places with an inclination smaller than the great inclination.

[156] *But in those places whose amount of latitude exceeds the amount of the great inclination, the sun never reaches the zenith, but is always inclined southwards from it. For that reason, in these climates,* which are known to us, *the shadow is never cast to the south but always to the north.* However, by reasonable interpretation, a diligent examiner of the things can study also for the southern inclination what we have said about the northern one. For the purpose of this treatise, therefore, *we have said enough about the inclination of the sun.*

[157] But the inclination of the planets, and of the other, so-called fixed, stars and of the moon must be considered in another way. *Naturally, their inclination* or the moon's distance *from the equator is again taken as the arc of the circle that passes through the poles of the world and the centre* of the moon[12] *and intersects with the ecliptic and the equator, with its inclination from the equator being the arc that is enclosed between its centre and the equator.* But its latitude and distance from the ecliptic is determined by the latitude of the dragon from the latter. The dragon, which is also called the 'ligant', is a circle which we will discuss more extensively in the following. It follows [in its direction] the course of the ninth sphere. Hence its head does not always stand in one and the same sign but can be found in any of them. Its inclination from the zodiac, just like that of the ecliptic from the equator, has always the same amount, but it changes according to its course in the sign, such that its great latitude does not always stand in Cancer or Leo, but sometimes in one sign and sometimes in another. In consequence, as the centre of the moon is always on this circle, the latitude of the moon cannot incline in any direction, just as we have said that the centre of the sun is bound to the ecliptic.

[12] A distinction between the moon and the planets is necessary for Stephen's subsequent discussion of the dragon and the different points of reference for the latitude and the inclination.

signiferi et centrum lune, draconem et zodiacum. Arcus is a quovis puncto draconis, in quo centrum lune volvitur, usque proximam medietatem circuli signorum continetur. *In planetis quoque et fixis et latum et inclinationem simili ratione investigare oportet. Inclinationem quidem a recto, latum autem a signifero, que etiam duorum opera* 5 *circulorum, quorum alter per polos mundi alter per polos signiferi transit, fient. Est igitur inclinatio cuiusvis planete vel fixe arcus circuli, qui est a puncto, in quo centrum stelle invenitur, usque rectum, si circulus per polos mundi transeat; si vero per polos circuli signorum, arcus eiusdem, qui inter centrum stelle et signiferum clauditur, latum eiusdem est a signorum circulo.*

10 [158] *Et planetarum quidem sive inclinatio* sive latum *habent discordiam sicut et solis inclinatio, que non eadem in omnibus, sed alia in aliis atque alia in aliis signis esse deprehenditur. Habet enim quemadmodum et sol suum quisque circulum, in quo semper sue rotunditatis centrum volvitur, qui a* signorum circulo *inclinatus* cum eodem in duobus diametrice oppositis punctis ligatur. Sic de signifero et recto dictum est. 15 *Intelligendum igitur est, quoniam quanto centrum* rotunditatis *cuiusque planete longius a puncto removetur, in quo circulus,* per quem fertur, *et signifer sese ligant, tanto latum eius atque inclinatio maiori protelantur quantitate,* quantoque propius cuivis sue ligationis accedit, minus est eius latum. Et circulorum quidem Saturni, Iovis et Martis a signifero latum non mutatur, idemque semper est. At Veneris et Mercurii circuli 20 nunc longius, nunc propius aliquando in superficie sunt signiferi. Considerandum illud quoque in planetis ad latum et inclinationem inveniendam plurimum videtur, quod nec soli contingit et luna ignorat, quoniam aliquando a suis circulis inclinantur nunc in septemtrionem, nunc in austrum, tantaque est eorum cursus varietas, ut vix possit his certum poni tempus, quo parem habeant sibi aliquo tempore aut lati f. 19v 25 aut inclinationis quantitatem. Et planetarum quidem | ac lune ita varia sunt tam latum quam inclinatio. [159] *Sed non idem in fixas repperitur. Hee enim discordem quidem habent inclinationem, quam a recto circulo longitudinem dicimus, eaque ipsa parva, enim vix in multa temporis longitudine cognoscatur. Cuius causa est eorundem*

1 draconem] *corr. ex* dragonem (L)

The latitude of the moon sometimes gets greatest in Aries, sometimes in Cancer or in their opposite signs, and often in others, always in those signs where the greatest latitude of the dragon is found. But the moon's latitude is the arc of the circle that passes through the poles of the ecliptic and the centre of the moon, through the dragon and through the zodiac. This arc is contained from that point of the dragon where the centre of the moon revolves until the nearest half of the ecliptic. *For the planets and the fixed stars a similar method is to be used to determine their latitude and inclination. The inclination is determined from the equator, and the latitude from the ecliptic, which is again done by means of two circles, one of which passes through the poles of the world, and the other through the poles of the ecliptic. Hence the inclination of any planet or fixed star is the arc on a circle from the point where the centre of the star is found until the equator, if this circle is drawn through the poles of the world; but if it is drawn through the poles of the ecliptic, the arc of this circle which is enclosed between the centre of the star and the ecliptic equals the star's latitude from the ecliptic.*[13]

[158] *Now, the planets' inclination* and latitude *vary, just as the inclination of the sun, which is not found equal in all signs but of one amount in some signs and of a different amount in others. For just like the sun, every planet has a circle on which the centre of its epicyclic sphere revolves and which is inclined from the* ecliptic[14] but bound to it at two diametrically opposite points. This is analogous to what has been said about the ecliptic and the equator. *One therefore must conclude that the further distant the centre of any planet's* epicyclic sphere *is from the point where its* deferent circle *is bound to the ecliptic, the higher will be its latitude and inclination;* and the closer it comes to one of its intersections, the less is its latitude. But the latitude from the ecliptic does not change for the circles of Saturn, Jupiter and Mars but is always the same; whereas, the circles of Venus and Mercury are sometimes closer and sometimes further from the plane of the ecliptic. It further seems most important when determining the planets' latitude and inclination to consider something that does not affect the sun or the moon, namely that they occasionally incline from their circles, sometimes to the north and sometimes to the south, and that their course is so irregular that one can hardly determine a certain moment at which they might have a latitude or inclination which agrees with their values for any moment. Such irregular is the latitude and also the inclination of the planets and also of the moon. [159] *But this does not occur in the fixed stars. Though having a*

---

[13]Stephen changes Arab. 'equator' (which appears also in all other manuscripts) to 'ecliptic'. This resolves an unclear use of terms in the Arabic; cf. Langermann's note to this passage.

[14]'ecliptic': Arab. 'equator'. Stephen replaced the term, as he addresses also the planets' varying latitude in this passage, which in the Arabic deals only with the declination.

tardus motus. At vero latum, a circulo signorum scilicet longitudinem, ita semper eandem habent, ut nulla possit in eo vel minima deprehendi discordia. Et de lato quidem lune et planetarum fixarumque hec in presenti sufficiant. In sequentibus uberius de eis disseretur. Que autem dicta sunt, inclinationis gratia tetigimus, ut 5 eas sicut et solem inclinatione non carere animadverterent hii, qui nostra non fastidiunt. Deinceps igitur de longo et lato terrarum dicendum videtur, ut sicut ad orientalium scientiam pertingamus.

slightly varying inclination—this is how we call the distance from the equator—this variation is very small, because it can hardly be recognised even over a long period. The reason for this is their slow motion. But their latitude, that is, their distance from the ecliptic, is always the same, such that not the smallest variation can be found in it. For the moment, this should be enough about the latitude of the moon, the planets and the fixed stars. We have touched on the above for the sake of the inclination to make those who are not weary of our discussion aware that these objects, just like the sun, are not without inclination. Now it seems necessary to speak about the longitude and the latitude of places on the earth, so that we can then proceed to knowing the ascensions.

Latum igitur terre dicimus longitudinem cuiusvis a recta linea, que sub circulo recti diei in superficie terre posita est. Solus enim ille locus circulo equinoctiali suppositus est, quem eiusdem circuli superficies, si totum dividit mundum, tangit in superficie terre faciens imaginarium circulum, qui eam in duo media terram dividit, quarum pars altera in septemtrionem, pars altera in austrum inclinatur. Et quoniam quivis terre magnus circulus in .xp. gradus dividitur, si circulum per polos mundi transeuntem secare mundum imaginaveris, facit eius circulum superficies in superficie terre, qui circulum recte linee in duo media dividens eadem per illum positione dividitur. Est igitur medietas tota in septemtrione, altera in austro. Quoniam vero circulus recti diei utrorumque polorum medius est, .s. scilicet gradibus ab utroque distans omnibus suis partibus, erit circulus terre, qui per eius fit superficiem, totidemque longe positis partibus distans a puncto terre, qui polo mundi suppositus est recta linea. Eorum quo fit, ut circulus terre, qui fit a superficie circuli per polos mundi transeuntis, in IIII^{or} equales dividatur quadrantes per IIII^{or} punctos, quorum duo sunt suppositi polis mundi, alter septemtrionali, alter australi, duo vero sunt communes dividentes circulorum, suntque oppositi diametrice, qui polis supponuntur, identidemque ligantes puncti circulos. Distat ergo ab utroque puncto ligationis circulorum punctus is, qui septemtrionali polo suppositus est, .s. gradibus, totidemque, cui australis polus superimminet. Quamobrem latum terre ad .s. usque gradus protenditur. Ex quo fit, ut in aliis locis maius fit, in aliis minus. Opitulatur autem lati terre noticia plurimum, quod ea incognita omnis inveniendis stellarum quarumlibet locis vacillat ratio, diei et noctis horarum mensura nulla colligitur, altitudinesque stellarum super orizontem ignorantur. Apparet igitur per huius rei noticiam facilem ad illa transitum parari.

[77] *Latum ergo terre est arcus circuli meridionalis, qui est inter punctum capitis et circulum recti diei.* [78/79] *A quo si imaginamur meridianum circulum per superficiem suam totum secare mundum, faciet in superficie terre circulum centrum, cuius terre centrum est, per cuius quemvis punctum linea, que de centro mundi oritur, transiens tangit punctum capitis. Arcus igitur huius circuli, qui est in superficie terre a puncto, quem tangit linea mundi usque rectam lineam, est latum terre loci eiusdem.* Et est tante hic arcus quantitatis in suo circulo, quante et ille est, qui in meridionali | circulo positum a puncto capitis usque rectum circulum significavimus in meridionali. Quod docet subiecta formula.

1 Latum] Datum (L) **16** sunt] *add.* (L) **21** ea] *ante del.* in (L) **26** quo] qua **29** est] *add.* (L)

2.3 <Longitudes and Latitudes>

Config. 4: Longitudes and Latitudes

'Terrestrial latitude' we call the distance of any place on the earth from the terrestrial equator, which lies on the surface of the earth beneath the celestial equator. For only that region is placed beneath the celestial equator which is met by the plane of that circle when it divides the entire world and produces on the surface of the earth an imaginary circle that divides the earth into two halves, one of which inclines to the north and the other one to the south. And since every great circle of the earth is divided into 360 degrees, if you imagine a circle through the poles of the world and cutting through the world, its plane produces a circle on the surface of the earth that divides the terrestrial equator into two halves and at the same position is itself divided by the latter. Thus, one complete half lies in the north and the other one in the south. But as the equator lies in the middle between the two poles, that is, remote from each pole by 90 degrees in all its parts, the circle on the earth which is produced by its plane will be distant by that same given amount of degrees from the point on the earth which is placed beneath the pole of the world on a straight line. By consequence, a circle on the earth that is produced by the plane of a circle that passes through the poles of the world will be divided into four equal quadrants at four points, two of which are beneath the poles of the world, one under the north pole and the other under the south pole, whereas the other two are the common intersections of the circles; and the points under the poles, just as the points where the circles intersect, lie diametrically opposite one another. Hence the point under the north pole is remote from each of the intersection points by 90 degrees, and by the same amount is that point remote which the south pole is looming above. The terrestrial latitude can therefore increase up to 90 degrees; thus being larger in some places and smaller in others. Knowing the terrestrial latitude is extremely helpful, because when the latitude is not known, every method of finding the positions of stars falters, the lengths of the hours of day and night cannot be determined, and the altitudes of the stars above the horizon are unknown. Thus, from knowing the former, the latter is found to be easily accessible.

[77] *The terrestrial latitude is the arc on the meridian between the zenith and the equator.* [78/79] *If we thus imagine the meridian dissecting the entire world by its plane, it will produce a circle on the surface of the earth whose centre coincides with that of the earth, and every straight line drawn from the centre of the world through any point of this circle will touch the zenith. The arc of this circle on the surface of the earth, from the point that is touched by the line of the world until the terrestrial equator, equals the terrestrial latitude of that place.*[15] And the size of this arc on its circle equals that of the arc on the meridian which we have defined as being placed from the zenith until the equator at the meridian. This is illustrated by the following diagram [Fig. 2.3].

[15] 'locus': Arab., madīnah.

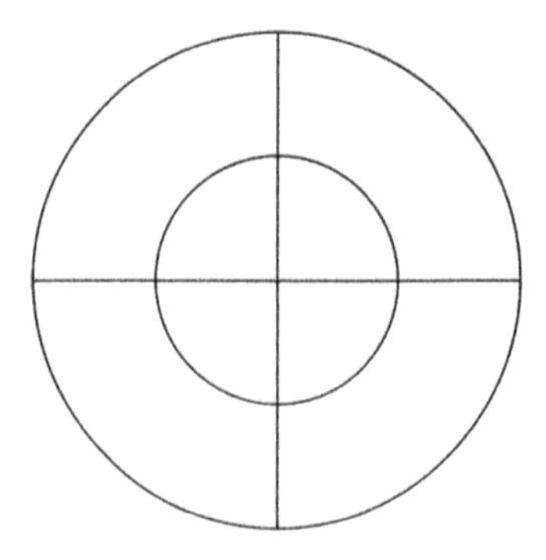

Fig. 2.3: From MS Cambrai 930, fol. 20r

[80] Et latum quidem terre, ut dictum est, a circulo recti diei in austrum aut septemtrionem consideratur, at vero longum ab eo loco quem tholum diximus in orientem aut occidentem. Quare autem ab eo loco potius et non aliunde longum intueri oporteat, ea nobis causa dici videtur quod in tholo mundi sole posito totam
5 habitationem dies illustrat et tunc est quidem meridies, sed initium vespera diei, finem autem habitationis mane introrsusque positas terras relique diei hore habent. Ab illius quovis loco sed habitationis extrema longum sumi conveniet, ut si cuiuspiam loci que sit longitudo ab initio habitationis tenebitur per eandem alterius longum celerius deprehendatur. *Est igitur longum cuiusque loci arcus circuli recti diei*
10 *qui cadit inter circulum meridiei eiusdem loci et circulum finis habitationis* sive cuiuslibet alterius loci *aut in orientem aut in occidentem. Erit autem hic arcus semper eius medietatis circuli recti diei que dividit habitationem septemtrionalem ab australi plaga,* [81] *duo vero habitationis termini hii sunt puncti in quibus sese ligant circulus recti diei et circulus orizontis* tholi mundi, *quem orizontem medie terre dicimus,* que superius
15 manifestissime explanata sunt.

[82] *Imaginamur autem circulos recto circulo paralellos, quos temporis circulos esse prediximus,* altrinsecus ad austrum et septemtrionem positos, *qui et ipsi mundum dividunt* sed non in media. Ad septemtrionem namque positi eam, in qua sunt, partem minorem, oppositam grandiorem dividendo mundum faciunt. Qui quanto a recto
20 longius absunt, tanto sue partis partem minori coartant termino. Ad austrum vergentes idem facere diligens poterit lector intueri. *Horum quisque quemvis orizontem habitationis secat in duobus punctis,* et illi quidem puncti in ipso paralello positione oppositi sunt et non in orizonte. *Partiuntur itaque orizontes in altera utra divisione*

3 et non] tamen (L) **5** illustrat] illustra (L) **10** sive] scilicet **11** hic] huius (L) **17** altrinsecus] adtrinsecus **18** positi] posita **22** duobus punctis] *corr. ex* duopunctis (L) **23** utra] itra

[80] And it has been said that the terrestrial latitude is considered from the equator towards the south or the north; the longitude, however, from the place which is called the tholus towards the east or the west. The reason why the longitude should be considered with respect to that point and not to any other seems to be that when the sun stands in the tholus of the world it is daytime in the entire inhabited part of the world, which means it is midday; but for the beginning of the inhabited region it is evening and for the end it is morning, while all places in between have the remaining hours of the day.[16] One can take the longitude from any place of the inhabited region, but it is convenient to take it from the outermost part of it, so that, if one has the distance of any place from the beginning of the inhabited region, one can quickly grasp from this distance the longitude of another place. *Thus, the longitude of any given place is the arc of the equator which falls between the meridian of that place and the circle of the end of the inhabited region,* or any other place, *to the east or the west. But this arc will always be on that half of the equator which separates the northern inhabited region from the southern plain,* [81] *whereas the two ends of the inhabited region are the points where the equator and the horizon* of the tholus of the world, *which we call the horizon of the middle of the earth, intersect,* which has been explained above in the clearest manner.

[82] *We now imagine the parallel circles of the equator, which we said are the 'time circles',* lying to the north and to the south, *which all divide the world,* but not in the middle. For those circles lying to the north will divide the world such that they make that side on which they lie smaller and the opposite side larger. And the further they are from the equator, the smaller they constrict the part on their side. The attentive reader can realise that the circles inclined southwards do the same. *Each of them intersects with any horizon of the inhabited region at two points;* and these two points are on opposite positions on the parallel circle, but not so on

[16] Cf. al-Battānī, *Ṣābiʾ Zīj*, ch.6.

omnes equales circuli paralellos in duo media. Quilibet namque orizon, cum polus in circulo recti diei positus est, per polos eiusdem transit, qui omnium eiusdem paralellorum poli sunt; radiusque mundi, qui a polo septemtrionali usque polum australem porrigitur, omnium paralellorum centra continet, ipseque a superficie orizontis, cum
5 sint eorundem altera medietas supra in orizontem, que etiam diurna dicitur, altera vero sub orizonte, que noctis est, necessaria argumentacione colligitur. Orizontium autem per ipsos particio eius eadem fit proportione, que super eisdem mundi divisione superius dicta est. Ex his igitur manifestum est omnibus his in locis, in quibus equinoctialis circulus per punctum capitis, qui orizontis polus eiusdem
10 loci est, transit, equa semper diem noctemque Libra quantitatis librari ponderari. Posito enim in Arietis Libreve capite sole terrarum ubique equinoctium esse, in quibus nox et dies alternatim variantur, nullus, quilibet ad modicum philosophie studuit, ambigere invenitur. Equinoctialis namque circulus omnes orizontes in duo media dividit, nisi et ipse orizontalis fuerit circulus. Unde quia eiusdem
15 semper medietas videtur, equalem facit diem nocti. Cum igitur paralelli circuli in supradictis orizontibus equali proportione dividantur, nulla ambiguitas, quin sole in eisdem mundo lucem fundente equalium mensurarum nox et dies semper spatio protelentur.

f. 20v | [83] ¶*Orizontium autem circuitus, a quorum polis equinoctialis in quamvis par-*
20 *tem inclinatur nec ipsi polos mundi contingunt, circulum quidem recti diei in duo media dividunt.* Dictum est enim cuiusvis spere duos magnos circulos, quantacumque seu plurima modicave inclinatione alter ab altero inclinetur, equaliter ab invicem distribui, eius vero paralellos inequaliter. *Unde et omnes eorum partes, que super orizontem sunt, inequales ab illis invenientur, que subtus.* Inclinato namque orizonte
25 in septemtrionem a circulo recti diei sepemtrionalis semper polus, australis nunquam videbitur. Ex quo fit, ut polorum radius, ex quo paralellorum centra sunt, a superficie orizontis inclinetur, nec in aliquo loco orizontis partem aliquam contingat excepto eius centro, qui et mundi et recti circuli diei centrum esse cognoscitur. Erit igitur in septemtrionalibus orizontibus ea radii medietas, que a septemtrio-
30 nali usque centrum mundi porrigitur, super orizontem, reliqua vero inferius. In australibus autem ea, que ab austro usque centrum superior, septemtrionalis inferior. Qua in re illud attendendum est, quod quanto quilibet radii punctus a centro

4 orizontis] *corr. ex* orintis (L) **6** argumentacione] augmentacione (L) **21** cuiusvis] cuius veris (L) **24** orizontem] orientem (L)

the horizon. *Thus, by mutual division, the horizons*[17] *[of the equator] divide all parallel circles into two halves. For every horizon whose pole is placed on the equator passes through the latter's poles, which are also the poles of all the parallel circles of the equator;* And the axis of the world, which extends from the north pole until the south pole, contains the centres of all parallel circles and is itself contained in the plane of the horizon, which can be concluded from due argumentation, since one of their halves, which is also called the 'diurnal half', lies above the horizon, whereas the other one, which is the 'nocturnal half', is below the horizon. But the partition of the horizons by the parallel circles corresponds to the proportion which has been mentioned above with regard to their division of the world. It is obvious from this that at all places where the equator passes through the zenith, which is also the pole of the horizon, day and night are always equally balanced on the scales of quantity. For nobody with some philosophical education will deny that when the sun is placed at the head of Aries or Libra there will occur an equinox everywhere on earth where day and night alternately change. For the equator divides all horizons into two halves, except where it is a horizon itself. Hence, as exactly half of it is always visible, it makes day and night have equal lengths. Thus when the parallel circles at the aforementioned horizons are divided by the same proportion, there is no doubt that when the sun floods the world with its light from these circles, the dimensions of night and day will always reach the same extent.

[83] *But the circuits of those horizons whose poles are inclined sidewards from the equator and which therefore do not pass through the poles of the world, still divide the equator into two halves.* For it has been said that two great circles of any sphere, irrespective of their relative inclination, be it large or small, divide each other into equal parts, whereas they divide their respective parallel circles into unequal parts. *For that reason, also all their parts which are above the horizon are found different from those beneath.* And if the horizon is inclined northwards from the equator, the north pole will always be visible, but the south pole never. In consequence, the axis of the poles, which determines the centres of the parallel circles, is inclined out of the plane of the horizon and it does not touch any part of the horizon anywhere, except the latter's centre, which is also the centre of the world and of the equator. Hence, in the northern horizons that half of the axis will be above the horizon which extends from the north until the centre of the world, while the other half will be below. In the southern horizons, however, the half from the south until the

[17] Arab. add.: 'of the equator.'

longius aberit, tanto eo, quod ipsi a parte centri proximus est, remotior superficie orizontis deprehendetur. Unde fit, ut et paralellorum, quorum idem puncti centra sunt, maiorem habeant divisionis per orizontem discordiam. Quanto enim paralelli longius seponuntur a recti diei circulo, tanto et eorum centra a superficie orizontis longius absunt, eorumque sectiones discordes amplius videntur. *Ex quo colligitur, quod paralellorum partes, qui a recto circulo inclinantur in visum polum, maiores quidem super orizontem, minores sunt eorum arcus, qui viciniores sunt equinoctiali,* quoniam et eorum centra a superficie orizontis largiori spatio semoventur. *Contra fit in eius poli parte, qui occultatur. Longius nempe quod sunt a recto remoti circuli centro, longius a superficie orizontis habent sub terra et maiores sui partes proximorum partibus circulorum a recto inferius, minores autem superius.*

[84] *Hec igitur discordia divisionis circulorum causa est discordie dierum inclinatis orizontibus. Arcus enim circulorum, qui super orizontem sunt, dicuntur arcus diei,* qui inferius, noctis. Et quoniam stellarum cursibus noctium horas earumque puncta sepe stellis necesse est exprimere, non incongruum est scire, quoniam is circulus, in quo quelibet stella volvitur, paralellus est recti diei circulo, *unde eiusdem circuli ea medietas, que orizonti superposita est, eiusdem stelle diei arcus esse dicitur, que vero subest, arcus noctis.* Sed de stellis alias, nunc de sole videamus.

[85] *Cum itaque sol quovis in puncto signiferi positus est, aliquo in circulo paralellorum recti diei volvitur; ea quidem pars eiusdem circuli, que super orizontem posita, diurna dicta est, et arcus diei solis horas diei eiusdem metietur, at inferior nocturna et arcus noctis quantitatem noctis terminant.* [86] *Et quoniam circuli, quorum <spatium> a recto circulo longius est, maiorem habent sue divisionis discordiam, siquidem hii in visum inclinantur polum, quanto longius sol ab equinoctiali paralellos discurrens pervagabitur, tanto longiori super terram videbitur mora, que diei horas longiores atque multiplices efficit,* eadem mensura noctis minuens quantitatem. Ex quo fit, ut nobis in septemtrionali habitatione positis, quoniam orizontium nostrorum circuitus in septemtrionem inclinatur, sole in septemtrionalibus pervagante signis dierum fiat augmentum multiplex.

f. 21r [87] *Recedente igitur sole ab equi-|noctiali atque ad ultima sue inclinationis in septemtrionem elapso capitis Cancri in circulo volvitur, qui signiferi inclinationem in septemtrionem terminat. In hoc itaque sol terram die illustrans maiori spatio reliquis diebus,*

---

2 et] *add.* (L) **7** arcus] arcubus **16** recti] recta (L) **23** circulo] puncto **24** inclinantur] inclinatur (L) **26** minuens] innuens **32** terminat] terminant (L)

centre will be above the horizon and the northern half below. In connection with this one has to keep in mind that the farther from the centre any point of the axis will be, the farther from it will be the point that lies closest to it on the horizontal plane from the side of the centre. As a result, also the parallel circles with those points as centres are differently divided by the horizon. For the farther from the equator the parallel circles lie, the farther remote from the horizontal plane their centres will be and the more unequal their partition will appear. *It follows from this that for parallel circles which are shifted from the equator towards the visible pole, their parts above the horizon are the larger ones, whereas the arcs of those that are closer to the equator are smaller,* because also their centres lie farther from the horizontal plane. *The opposite occurs on the side of the concealed pole. Because, as the circles are more remote from the centre, being further from the horizontal plane the circles have larger parts of themselves beneath the earth than the circles that are close to the centre, and these parts are larger the lower the circles lie on the axis, but the smaller, the higher.*

[84] *This unequal division of the circles is the cause for the unequal day lengths at the inclined horizons. And the circle arcs above the horizon are called 'day arcs',* those below 'night arcs'. And as we often need to determine the night hours by the course of the stars and also their exact moments by means of the stars, it is very convenient to know that the circle on which a star moves is parallel to the equator; *that half of this circle which is above the horizon is therefore called that star's day arc, and the half below, night arc.* But we will deal with the stars at another place; now let us consider the sun.

[85] *Hence when the sun is placed in any point of the ecliptic, it revolves on some parallel circle of the equator; then, that part of this circle which is above the horizon is called the 'day part', and the arc of the day determines the hours of that day, whereas the part below is called the 'night part', and the arc of the night defines the length of the night.* [86] *And since the circles which have a longer distance from the equator also have a greater inequality in their partition, if they are inclined to the visible pole, the further away from the equator the sun moves through the parallel circles, the longer it will be seen above the earth, thus lengthening and adding to the hours of the day* while shortening the length of the night by the same amount. As a result, for us, as we are living in the northern region and our horizons therefore incline to the north, the sun produces a manifold increase of the lengths of day when it passes through the northern signs.

[87] *And when the sun moves away from the equinox and reaches its ultimate inclination to the north, it revolves on the circle through the head of Cancer, which defines the northward inclination of the ecliptic. Hence when the sun illuminates the earth with day-*

qui in australi habitatione fuerit quolibet in loco, diem prolongat, que sequitur nocte breviorem habente quantitatem. [88] *Contra fit in omnibus his, qui in austrum inclinantur circulis.* Nam cum septemtrionalis polus in septemtrionalibus orizontibus semper videatur, qui ei ab austro oppositus est, semper occultus eadem terre subiacet mensura, qua septemtrionalis supereminet. Qua ex re datur intelligi circulum recti diei paralellum circulo, in quo caput Capricorni volvitur, eadem dividi proportione, qua et Cancri circuitus. Sed circuli capitis Cancri centrum super orizontem positum est, quantumque hoc ab orizontis superficie supereminet, tantum et centrum circuli, in quo caput Capricorni volvitur, orizonti suppositum est. Quantus igitur ille est arcus circuli Cancri, qui super orizontem est, tantus est circulus Capricorni sub terra latens, quem arcum noctis dicimus. At vero illius pars inferior huius superiori arcui est equalis. Sole igitur in australibus signis commorante dies ab equalitate defitiunt, quoniam circulorum, in quibus volvitur, partes minores super, maiores sub terra habentur. *Quanto itaque longius sol in austrum a recto circulo inclinatur, tanto dierum arcubus minutis arcus noctium augmentatur, usque dum ad eum perveniat circulum, quo in austrum amplius descendere prohibetur. Is autem est, quem diximus Capricorni capitis circuitus,* in quo eiusdem mensure noctem conficit, qua et dies est, solis radius circulo illustrato die huius noctis circularis Cancri nocti horarum certo examine respondente.

Eadem certa ratione in australibus orizontibus licet intelligi, que de septemtrionalibus dicta sunt. Quanto enim orizontes magis inclinantur in austrum, tanto magis super se polum australem vident; ex quo fit maior noctis dierumque discordia. Et quoniam in australibus orizontibus hiems, sicut iam supra diximus, quando septemtrionales partes maturat estas, atque apud nos posita hieme australia calor perurit, dumque apud nos magni volvuntur dies et breves noctes, illic contraria contingunt. Breves enim habentes dies longioribus spatiis tempora noctis metiuntur. Et inclinati quidem orizontes huiusmodi noctium et dierum atque per opposita discordiam habent.

[89] ¶*In orizontibus autem linee recte non est ita. Illic enim semper equali spatio dies et nox protelantur. Arcus enim diei et dierum eiusdem noctis et omnium noctium in suo circulo habent .tr. gradus, medium scilicet circuli, propterea quod omnes circuli paralelli*

4 qui] quam **10** qui] quod **23** hiems] hiemis

light from this circle, it prolongs the day more than on any other day, when it has been at any place in the southern hemisphere,[18] whereas the subsequent night will be the shortest. [88] *The contrary occurs in all those circles that are inclined to the south.* For as the north pole is always visible in northern horizons, its southern counterpart is always concealed and lies beneath the earth at the same angle by which the north pole stands above it. This allows one to conclude that the parallel circle on which the head of Capricorn revolves is divided by the same proportion as the circuit of Cancer. But the centre of the circle of the head of Cancer is placed above the horizon, and the length by which it stands above the horizontal plane is the same by which the centre of the circle on which the head of Capricorn revolves is placed below the horizon. Hence the arc of the circle of Cancer above the horizon equals the circle <arc> of Capricorn that is concealed under the earth and which we call the night arc. Conversely, the lower part of the first circle equals the upper arc of the latter. Therefore, when the sun stands in the southern signs the days fall short of equality, because the smaller parts of the circles on which the sun revolves lie above the earth and the larger parts below. *Hence the further the sun is inclined southwards from the equator, the more will the arcs of night increase and the arcs of day decrease, until the sun reaches that circle on which it is prevented from descending further southwards. And this, as we have said, is the circuit of the head of Capricorn,* where the sunlight will define the night to the same length as it defines the day on the illuminated part of that circle where the day of this night corresponds to the night by precise examination of the hours, namely the circuit of Cancer.

What we have said about the northern horizons can by the same certain reasoning be understood also for the southern ones. For the farther south the horizons are inclined, the closer to the zenith above them will be seen the south pole, which causes a greater difference between day and night. And as in the southern horizons it is winter when summer brings the northern regions to ripeness, as we have said above, whereas when winter is with us, the heat burns the southern world, when in our part the long days and short nights revolve, the opposite occurs there; having short days, the night times extend over very long intervals. And the inclined horizons have a corresponding difference between days and nights, but inverse.

[89] *But this is not the case for the horizons on the terrestrial equator. For day and night always have the same extent there; because the arcs of day of all days and the arcs of night of all nights comprise 180 degrees on their circle, that is, a semi-circle, since all*

[18] 'southern hemisphere': Arab., 'northern hemisphere,' meaning that only on the northern hemisphere the longest day occurs when the sun is in Cancer.

equinoctiali circulo dividuntur in duas equales medietates per sectionem orizontalium circulorum. Inde igitur fit, ut arcus, qui sunt sub orizonte, equales sint arcubus eorundem circulorum, qui super orizontis altum volvuntur. Ex quo fit diem semper equalitatem cum noctibus sperare. Mora enim solis unius diei super terram et ceterorum more dierum et sue noctis aliarumque equalis est, nisi quod propter solis altum et humilem in suo circulo varietur, de quo in suo loco dicetur, quod tamen modicum est.

[90] *Unusquisque enim paralellorum recti circuli usque ad utrumque polum secat orizontis circulum in duobus punctis,* nec est eorum quisquam vel brevissimus sub aut supra orizontem totus, aut circum eum equaliter. [91] *Unde et quecumque stella surgit in circulorum aliquo illorum, aut est in eo, necessario comprobatur* | *in uno puncto ab oriente surgere atque in alio in occidente occumbere,* quorum uterque equaliter distat a circuli equinoctialis ortu aut occasu. [92] *Arcus autem, qui est inter punctum orizontis, in quo surgit stella, et ortum recti diei circuli dicitur largitas orientalis eiusdem stelle, ille autem, qui continetur inter punctum, in quo occumbit, et occasus Arietis primi puncti aut Libre occidentalis largitas eiusdem.* In quo animadvertendum, quoniam quantacumque erit orientalis aut occidentalis largitas alicuius puncti aut circuli, tantum longitudinis arcus est meridionalis circuli, qui est a puncto capitis usque punctum meridionalis circuli, quem idem aut punctus aut circulus tangit. Ex qua datur intelligi circulum, ubi ille punctus est, rectis angulis super circulum orizontis positum.

[93] ¶Et quoniam dictum est paralellorum recti circuli fieri discordes sectiones in orizontibus, qui aut in austrum aut <in>(L) septemtrionem inclinantur, quantoque remotiores essent a recti diei circulo, tanto maiorem habere sue divisionis discordiam, memoriter teneri volumus. *Illud quoque oblitum nolumus, quod quanto propinquiores circuli sunt in viso polo, tanto maiores eorum arcus orizonti supereminent* minoribus dum sublatentibus, contraque in occulto polo contingere, ut maiores arcus circulorum sub orizonte lateant, minores vero super orizontis ferantur altitudinem. *Inde fit, ut brevis ille paralellicus recti diei circulus,* qui tot distat a polo gradibus, quoti sunt inter polum visum et orizontem eo in loco, quo meridionali sub ipso secatur circulo, aut quantum equinoctialis circulus a puncto capitis remotus est, *tangat ipsum orizontem uno in puncto,* quo scilicet meridionalis et idem iunguntur. *Est igitur totus ille circulus super orizontem.* [94] *Si qua igitur stella su-*

1 orizontalium] orientalium **9** circum eum equaliter] circulum inequaliter **25** maiores] minores

parallel circles of the equator are divided into two equal halves by the intersection with the horizons. By consequence, the arcs below the horizon are equal to the arcs of the same circle that revolve through the heaven above the horizon. As a result, the day can always rely on equality with the nights. For the sun's time above the earth[19] *on one day is equal to its time on all other days, and that of its night is equal to that of the others,* unless it changes due to the sun's higher or lower position on its circle, which we will speak about at its proper place; but this effect is small.

[90] *For every single parallel circle of the equator, until either of the poles, crosses the horizon at two points,* and not even the smallest of them is totally beneath or above the horizon or congruent with it. [91] *Therefore, every star that rises on one of those circles or is placed on it rises necessarily at one point in the east and sets at another point in the west,* which are both equally distant from the rising or setting point of the equator.[20] [92] *And the arc between the star's rising point on the horizon and the rising of the equator is called the 'amplitude of rising' of that star, whereas the arc enclosed between the point where the star sets and the setting point of the first point of Aries or Libra is called its 'amplitude of setting'.* In connection with this, one has to be aware that the amount of the amplitude of rising or setting of any point or circle equals the length of the arc on the meridian from the zenith until that point of the meridian that is touched by the point or the circle. This allows one to conclude that the circle on which that point is stands at right angles above the horizon.

[93] We also want to keep in mind what has been said about the parallel circles of the equator, that they are divided unequally at horizons which are inclined to the south or the north, and that the farther they are from the equator, the greater is the difference of their partition. *Neither should we forget that the closer the circles are to the visible pole, the greater parts of their arcs stand above the horizon* and smaller arcs are hidden beneath, and that the opposite occurs at the concealed pole, such that greater arcs of the circles are hidden under the horizon while smaller ones revolve high above the horizon. *It follows from this that the one small parallel circle of the equator* whose angular separation from the pole equals the degrees between the visible pole and the horizon at that point where the horizon is crossed by the part of the meridian that runs downwards from the pole, or whose distance from the pole equals the equator's distance from the zenith, *is tangential to that horizon in one*

[19] 'earth': the reading corresponds to Arab. Y; cf. Langermann's note to this passage.

[20] Arab. add.: 'Similarly, every point on the surface of the highest orb rises from one of these points and sets at the setting point opposite it.'

per illum circulum moveatur, semper super orizontem aspicitur nec aliquando occumbit. Statim ut orizontis punctum similis occumbenti perstrinxerit, in altum resurgit, et que putabatur occumbere, videtur oriri. *At vero in occultato semper polo contrariam inveniri rationem facile erit. Brevis enim paralellicus recti circulo diei circulus,* cuius
5 linealis circumductio tot distat ab occulto polo graduum spatio, quot et orizontis circulum ab eodem inclinari perspexeris, *non dividit illum, sed nec ab eo dividitur.* In duobus namque punctis non sese ligant invicem, *sed uno tantum sui ponunt confinia.* Unde fit, ut cum orizon, quod occultatur, a visa celi medietate dividat, atque polus ille, cui est conterminus idem circulus, semper occultetur, *idem quoque cir-*
10 *culus, cuius altior pars ad orizontem usque procedit, uno tantum in puncto videri possit, reliqua pars omnis semper occulta lateat. Inde est, ut quecumque in eius circulari motum habeant stelle, videantur quidem, cum ad punctum confinitatis conscendunt,* super orizontem surrexisse, statim in perpetuam sui devolvantur latebram.

[95] *Ex hoc igitur, quod positum est, manifestum esse credimus omnes illos, qui ultra*
15 *hos duos breviores his moventur circulos, si quidem sint intra orizontem et visum polum, semper videri* nunquamque ad occasum posse descendere, *sed omni super orizontem tempore apparere. Quo contra contingit his circulis, quorum circuitus inter occultum polum et orizontem volvuntur. Neque enim orizontis ullo confinitatis possunt puncto contingere.* Nulle igitur earum, que in eis stelle sunt, aliquatenus in eodem orizonte videri
f. 22r 20 possunt. Nunquam enim super orizontalem | circulum, sed nec ad ipsum, possunt conscendere, unde et semper in occulto latent. Opposita ratio in his invenitur sideribus, que in illis rotantur, qui circa visum positi polum nullam habent sui partem, ut dictum est, latentem, sed super orizontem perpetua feruntur circumvolutione. Unde et stelle in ipsis posite semper super orizontem inveniri possunt.

25 [96] ¶*Dictum autem* in superiori libro *recolimus omnem, que habitatur, terre partem inter duos contineri paralellos. Horum alter est circulus recti diei, quem equinoctialem diximus, transiens super punctum capitis his, qui in recta linea positi lati habitationis terminum ab austro incolunt.* Hos ultra nullos habitare homines accepimus, nec auditum nobis est aliquos sui corporis umbras, sole in Ariete aut Libre equales om-
30 nibus terris diei et noctis quantitates ponderante, in austrum deferri vidisse. Calet enim omnis ulterius terra posita, nec incoli potest ab animantibus, tum propter exuberantem calorifici solis exustionem, tum propter calentis linee, que solis incendio peruritur, affinitatem, de qua post dicemus. *At vero alter equinoctiali circulo paralellus, qui a septemtrionali terminat habitationem, is est, qui* .pf. gradibus et .le. se-

20 orizontalem] orientalem **30** ponderante] pondante (L) **31** tum] tunc (L)

single point, which of course is the same point where it meets the meridian. *Hence the entire circle is above the horizon.* [94] *If thus a star moves on that circle, it is always seen above the horizon and never sets.* Instead, as soon as it touches a point on the horizon similar to a setting star, it is elevated again, and the star that was assumed to set is seen rising. *But it will be easy to understand that at the concealed pole always the opposite will occur. For the small parallel circle of the equator* whose circular line is separated from the concealed pole by the same amount of degrees as the horizon is seen to be inclined from it *does not divide the horizon and is itself not divided by it.* For they do not intersect at two points, *but join their lines in a single point.* By consequence, as the horizon separates the concealed half of the heaven from the visible one, and as the pole to which this circle is closest is always concealed, *this circle too, whose highest point reaches the horizon, can be seen in only one point, whereas the entire rest is always hidden concealed. Therefore, all stars that move on its circular line, when they rise to the contact point* and seem to have risen above the horizon, immediately return to their eternal delitescence.

[95] *We think that it is clear from what has been said that all those circles which move beyond these two and which are smaller circles than the latter, if they lie between the horizon and the visible pole, are always visible* and can never descend to set, *but stand above the horizon at all times. The opposite occurs to those circles whose circuits revolve between the concealed pole and the horizon. For they cannot reach any point on the horizon's border.* Therefore, none of the stars that are located on them can ever be seen on the horizon, as they can never rise above, or even just until, the horizon, for which reason they are always hidden in concealment. Opposite reasoning applies to the stars which revolve on those circles which, located around the visible pole, have no such concealed part as mentioned above, but turn permanently above the horizon. Therefore, also the stars located on them can always be found above the horizon.

[96] *But we remember what has been said* in the previous book, *that the entire inhabitable region of the earth is contained between two parallel circles. One of them is the equator, which we have called the equinoctial circle, passing through the zenith for those who live on the terrestrial equator and who thus live on the southern latitudinal limit of the inhabited region.* We have learnt that people cannot live beyond these circles, and we have never heard of people who had the shadow of their bodies cast southwards when the sun stands in Aries or Libra and portions out equal lengths of day and night to all places on the earth. For all of the earth that lies beyond cannot be inhabited by animals, firstly due to the exorbitant torridity of the heating sun and, secondly, due to the proximity of the hot line, which is burnt

xagenariis *distat ab equinoctiali versus septemtrionem. Atque is circulus transit eo loco super caput illorum, a quibus finis habitationis latitudinis a septemtrione incolitur.* Ab eo autem loco usque punctum terre polo suppositum septemtrionali tanta est frigoris ubertas, ut aut nulli aut pauci certe ultra .pf. gradus et .le. sexagenarias invenian-
5 tur habitatores, atque hi quoque rari, si qui sunt, non longe positi sunt a termino habitationis, quem ponimus. Non enim viveret illic aliquod animatum propter frigoris perpetuam intolerantiam. *Hoc igitur in loco, cuius tanta est ab equinoctiali linea remotio, circulus, in quo caput Cancri volvi superius dictum est, orizontis circulum nec dividit nec ab ipso dividitur in quasvis inequales sectiones.* Cum enim sit tota inclinatio
10 capitis Cancri .lc. graduum .me. sexagenariarum, et equinoctialis circulus longe positus sit .pf. gradus et .le. sexagenarias a puncto capitis, polus quoque septemtrionalis videatur super orizontem altus totidem et gradibus et sexagenariis, ablatis .pf. gradibus et .le. sexagenariis de .s. gradibus, qui sunt quadrans circuli, medietas scilicet meridionalis circuli, qui paret super orizontem, remanebit sola inclinatio
15 Cancri sub terra, et invenitur orizontalis circulus caput Cancri eo in loco tangere, quo a meridionali circulo sub septemtrione, sic ut in eodem puncto et orizon et Cancri capitis circulus et meridionalis iungantur. Unde fit, ut sicut de brevioribus paralellis aliis diximus orizontem in uno puncto tangentibus, is quoque circulus orizonteum circulum uno in puncto perstringat. Et quoniam habet inclinationem a
20 circulo recti diei in polum, qui semper videtur, et ipse circulus semper totus super orizontem videri potest.

Contra fit in circulo, qui capitis Capricorni circuitus est. Neque enim aliquando videri potest, sed semper sub orizonte latet. Remotus est namque, ut dictum est, equinoctialis circulus a puncto capitis .pf. gradibus .le. sexagenariis, quibus subla-
25 tis de .s. remanet longitudinis .lc. graduum .me. sexagenariarum arcus meridionalis circuli super terram, tantus scilicet, quanta est et capitis Capricorni inclinatio.
f. 22v Tangit igitur orizontis circulus punctum, quo terminatur magna | inclinatio, qui est finis Sagittarii et initium Capricorni. Ex quo manifestum est circuitum capitis Capricorni non in duobus punctis, ut altera pars super terram sit, altera sub terra
30 occultetur, sed in uno tantum orizonteum tangere circulum. Unde fit, ut totus aut super orizontem videatur, aut subtus lateat. Sed quoniam et occulto semper polo proximus est australi et circulo capitis Cancri oppositus, cum ille semper videatur, hic semper occultabitur.

4 ubertas] libertas (L) **8** orizontis] *corr. ex* oriontis (L) **15** orizontalis] orientalis **26** scilicet] solis (L)

by the sun's fire, which we will speak about later. *But the other parallel circle of the equator, which terminates the inhabitable region to the north, is the one that lies at* 66;25 degrees[21] *north of the equator. It passes through the zenith for those who live on the northern latitudinal limit of the inhabitable region.* From this place until the point on the earth under the north pole there is such an excess of cold that no, or certainly just a few, inhabitants are found beyond that place; and those few, if there are any, do not live far from the above given border of the inhabitable region. For no animal could live there due to the intolerance of the permanent cold. *At this place such distant from the terrestrial equator, the circle on which the head of Cancer revolves, as has been said above, neither divides the horizon, nor is divided by the latter, into any unequal segments.* For as the total inclination of the head of Cancer is 23;35 degrees and as the equator lies 66;25 degrees from the zenith, and as also the north pole is seen above the horizon at an altitude of exactly that amount of degrees and minutes, if we subtract 66;25 degrees from 90 degrees, which equals a quadrant of a circle and thus half of the meridian arc that is seen above the horizon, the result will be precisely the inclination of Cancer below the earth, and the horizon is found to touch the head of Cancer at that point where it is touched by the meridian in the north, such that the horizon, the circle of the head of Cancer and the meridian meet in one single point. This circle therefore touches the horizon in one single point, just as we have said about other smaller parallel circles that are tangential to the horizon at one single point. And as it has its inclination from the equator towards the pole that is always visible, the circle too can always be seen in its entirety above the horizon.

The opposite occurs for the circle which is the circuit of Capricorn. For it can never be seen, but is always hidden under the horizon. For it has been said that the equator is remote from the zenith by 66;25 degrees, and if these are subtracted from 90 degrees, the remaining meridian arc above the horizon has a length of 23;35 degrees, that is, the same amount as the inclination of the head of Capricorn. Hence the horizon touches the point in which the great inclination terminates, which is the end of Sagittarius and the beginning of Capricorn. From this it is clear that the circuit of the head of Capricorn meets the horizon not at two points, such that one part would be above and another part concealed below the earth, but only in a single point. For this reason it is in its entirety either seen above the horizon or hidden below. But since it is nearer to the permanently concealed south pole and opposite to the circle of the head of Cancer, and as the latter is always

[21] '66;25 degrees': Arab., '66 parts of the parts of which 360 comprise the circle.'

[97] Et latum quidem habitationis hinc et inde secundum nonnullos his duobus paralellis circulis, ut dictum est, terminatur. *Ipsam habitationem in .g. dividunt sectiones, quas etiam climata dicunt,* [98] *ponentes initium primi climatis a recta linea, que semper equalitatem habet, ut dictum est, dierum et noctium. Ab inde vero inequantur*
5 *dierum et noctium mensure, propterea quod orizontes in septemtrionem inclinantur,* et fiunt dies suis noctibus longiores in his signis, qui inclinantur a circulo recti diei in septemtrionem, propter quod maiores circulorum arcus super orizontem sunt, sicut dictum est, minores inferius. In his autem signis, que in austrum inclinantur, solis centro posito noctes, quam in aliis passe fuerant, et ipse ingerunt diebus
10 detrimenti ab equali iniuriam.

([101]) Quidam tamen aliis partiuntur climata divisionibus, quibus tamen minime Tholomeus in astronomia magnificus et graviores philosophi assensum prebent, sicut inferius liquebit. Aiunt enim medium primi climatis .kc. horarum, secundi .kd., tercii .ke., quarti .kf., quinti .kg., sexti .kh., septimi .ki.. Multi quoque
15 sunt, qui finem primi climatis a septemtrione sub recto circulo ponunt, Tolomeum sibi volventes tutorem et auctorem assumere, quod posterius ostendemus, secundi finem, ubi .kd. hore habentur, tercii .kf., quarti .kh., quinti .l., VI .lb., VII .ld.. Tholomeus autem et ceteri, quibus sanior est intellectus, non universam habitationem in climata .g. dividunt, sed quod in ea temperantius est et frequentioribus incolitur
20 habitatoribus, non a recta linea primi climatis ponendo principium sed ab eo loco qui .kb. insignitur horis et .ne. horarum sexagenariis, medium autem illustrari dicunt .kc. horis, medium secundi .kc. et .m. sexagenariis, tercium .kd., quartum .kd. et .m. sexagenariis, quintum .ke., sextum .ke. et .m. sexagenariis, septimum .kf. horis. Quod cis aut citra superius positos habitationis terminos et eos, quibus nunc
25 ad dividenda usi sumus, climata continetur, habitatur quidem et raro inhabitatore continetur, ut predictum est, sed non inter .g. climata.

13–14 secundi] secundum (L) **15** climatis] capitis (L) **20–21** loco qui] *corr. ex* loqui (L)

visible, it will always be concealed.

[97] There are some who say that the latitude of the inhabited region is terminated from both sides by these two parallel circles, as we have said. *These people divide the inhabited region itself into seven sections, which they also call 'climates',* [98] *taking the terrestrial equator, which has been said to have always equal days and nights, as the beginning of the first climate. But from there the lengths of days and nights become unequal, because the horizons incline northwards,* and the days become longer than their nights in those signs which are inclined northwards from the equator, since greater arcs of the circles are above the horizon and smaller ones beneath, as has been said. But when the centre of the sun is placed in those signs which are inclined to the south, the nights inflict on the days the same unseemly deprivation from equality which they themselves had suffered in the other signs.

([101])[22] Others, nonetheless, use a different division to distinguish the climates; however, the great astronomer Ptolemy and other important philosophers do not agree with them, as will become clear below. For they say that the middle of the first climate had [a length of the longest day of] 13 hours, that of the second 14, that of the third 15, that of the fourth 16, that of the fifth 17, that of the sixth 18, and that of the seventh 19 hours. Many people also define the border of the first climate in the north to be under the celestial equator, claiming Ptolemy as their authority and reference, which we will show later, and the border of the second where there are 14 hours, that of the third 16, that of the fourth 18, that of the fifth 20, that of the sixth 22, and that of the seventh 24 hours. But Ptolemy and all others with a better understanding do not divide the entire inhabited region into seven climates, but only the more temperate and considerably populated part of it, assuming the beginning of the first climate not from the terrestrial equator but from that place which is indicated by [a length of the longest day of] 12 hours and 45 minutes, whereas they say that the centre of this climate was illuminated for 13 hours, the centre of the second for 13 hours and 30 minutes, the third 14, the fourth 14 hours and 30 minutes, the fifth 15, the sixth 15 hours and 30 minutes, and the seventh for 16 hours. What is contained between either of the aforementioned borders of the inhabited region and on both sides beyond the borders which we have just used to define the climates is also inhabited, though only by a few inhabitants, as has been said, but it is not among the seven climates.

[22] This paragraph replaces *Configuration*, passage 101, where a controversy about the borders of the inhabited world and the climates is vaguely indicated; cf. Langermann's note to passage 101. Stephen's treatment of the subject is discussed in Gautier Dalché, *Géographie*, 98–101.

[102] *Tholomeus vero habitationis terminos alia metitur ratione. Facit enim habitationis terminum a parte septemtrionali in lato .pc. graduum. In austrum ultra rectam extendit lineam habitationem, metiturque eam lato .kf. graduum et .le. sexagenariarum,* ubi sit latum totius habitationis ab austro in septemtrionem .qi. graduum .le. sexagenariarum. *Hec in libro, quem de habitatione dixit, scripta sunt.* Michi vero, tametsi difficillimum videatur, credendum tamen estimo eius philosophice traditioni, qua et multarum constat rerum experientiam habuisse, et antiquorum scriptis et sui temporis hominum relatione multa, que nobis incognita sunt, certo cognovisse.

[99] Et sub recta quidem linea, ut dictum est, equalibus semper permanentibus diebus et noctibus orizontes in septemtrionem inclinati dierum habent discordiam. Quanto vero plus inclinatur, tanto ampliorem faciunt alterationem. *Unde et orizon ille, cuius est* | *inclinatio* .pf. graduum .le. sexagenariarum, *habens, ut dictum est, super se totum Cancri capitis circulum, diem habet in estate .ld. horarum. Sole enim in Cancri capite morante volvitur eius centrum in eo paralello, quem dictum est esse circuitum capitis Cancri. Qui quoniam uno tantum in puncto orizontem tangit, et nulla pars eius occumbere invenitur, sed est totus super. Solis quoque centrum, dum in eo volvitur, nullum patitur sub orizonte occasum,* quia statim, ut occasurus videtur, per eundem in altum resurgit circulum. Cum igitur totam unius diei circumvolutionem super terram compleat, et dies eousque protendatur donec sol occumbat, in circulo autem .xp. gradus habeantur, quorum divisio per .ke. .ld. horas compleat, erit idem dies .ld. horarum. Totus est enim ille circulus super orizontem, et in toto illo circulo sole posito altissima spera suum complens, quem in nocte et die facit, circuitum ad occasum solem, sicut in aliis orizontibus, quorum minor est inclinatio, non potest cogere. *Recedente autem sole ab ultimo inclinationis septemtrionalis minuuntur dies et noctium detrimenta decrescunt.* At vero ingresso illo, quo signifer et recti diei circulus puncto ligantur, deficiente paululum dierum statu noctium crescit quantitas.

Sicque fit, dum sol finem transit Sagittarii et Capricorni initium, ubi magne inclinationis in austrum vergentis finis habetur. Hic punctus, sicut superius dictum est, capiti Cancri oppositus est, in quo sol positus .ld. horarum diem confecit, quoniam circulus, in quo volvebatur, totus super terram erat. *Hoc igitur in puncto, quem Capricorni initium ponimus, sol dum rotatur, in circulari eius .ld. horarum noctem*

16 Solis] Sol (L) **21** enim] *add.* (L) **26** ligantur] ligatur (L)

[102] *But Ptolemy defines the borders of the inhabited region differently. For he sets the northern end of the inhabited region at a latitude of 63 degrees. Towards the south he extends the inhabited region beyond the terrestrial equator, to a latitude of 16;25 degrees,* thus making the entire latitude of the inhabited region from south to north 79;25 degrees.[23] *This is written in the book which he entitled 'On the Inhabited World'.* Although it seems very difficult, I nonetheless consider that one should believe Ptolemy's philosophical tradition, from which it is certain that he was experienced in many fields and that through the writings of the ancients and through accounts from people of his own time he had a clear knowledge about many things that are unknown to us.

[99] And whereas at the terrestrial equator, as we have said, days and nights always remain equal, the horizons that are inclined to the north have varying lengths of the day. And the more they are inclined, the greater becomes the alteration. *The horizon with an inclination of* 66;25 degrees, *which, as has been said, has the complete circle of the head of Cancer above itself, thus has a day length of 24 hours in summer. For when the sun stands in the head of Cancer, its centre revolves on that parallel which has been said to be the circuit of the head of Cancer. As this circle touches the horizon only in a single point, none of its parts is seen to set, but the whole circle lies above the horizon. The centre of the sun too, while revolving about it, does not suffer setting under the horizon,* for as soon as it appears to be about to set, it immediately rises up again, following that circle. Thus, when it performs the total revolution of a day above the earth, and a day lasts until the sun sets, while a circle comprises 360 degrees, whose division by 15 makes 24 hours, that day will last 24 hours. For that circle lies completely above the earth, and when the sun stands in that entire circle, the revolution of the highest sphere, which takes a day and a night, cannot force the sun to set like it does in the other horizons which are less inclined. *But when the sun moves back from the northernmost inclination, the days become shorter and the deprivation of the nights becomes less.* But when it has reached the point where the ecliptic and the equator intersect, the length of the nights will become longer at the gradual expense of the days.

This happens until the sun passes the end of Sagittarius and the beginning of Capricorn, where the great inclination has its southern termination. This point, as has been said above, is opposite to the head of Cancer, in which the sun, when placed there, produced a day of 24 hours, because the circle on which it revolved was entirely above the earth. *Thus, when the sun rotates on the circuit of that point*

---

[23] *Geogr.* I,10,1.

confecit die carentem, quoniam circulus, in quo isdem punctus, caput scilicet Capricorni, volvitur, totus, ut dictum est, sub terra continetur. Huius itaque loci et similium, cuius dies estate est .ld. horarum, noctem nullam habens nox autem hieme .ld. horarum tempore protrahitur diem exhauriens.

[100] ¶*Latum est arcus meridionalis circuli, cuius quantitas est ab equinoctiali circulo usque punctum, qui super imminet recte superpositus eorum capitibus, qui in illis versantur locis. Est autem longitudo eius* .pf. graduum .le. sexagenariarum, *quemadmodum superius positum est,* complementum scilicet magne inclinationis. *Ultra vero hanc latitudinem habitationis ab equinoctiali circulo sive recta linea nullam dicunt posse esse habitationem animantium,* quoniam sempiterna frigoris congelatione nichil illic viride usui animantium utile nascitur et animantia omnia frigoris nimietate congelata exanimarentur.

([103, 104]) Illa autem loca, que ultra .pf. gradus et .le. sexagenarias posita, glaciali zona torpescunt, quanto plus ab equinoctiali versus medium glacialis, scilicet polum septemtrionalem, elongantur, tanto maiorem habent diem, .ld. horarum diem in estate, eiusdem quantitatis noctem in hieme. Hoc autem ideo fit, quod Cancri circulus, usque ad quem magna descendit inclinatio, elongatur ab orizontis circulo. Et orizonteus circulus alium de paralellis circuli recti diei, qui intra ipsum et solstitiale hostium continetur, tangit in uno puncto. Exempli si placet causa, ponimus aliquem locum, in quo circulus capitis Cancri tantum removetur ab orizonteo, ut paralellus circulus, in quo finis Geminorum volvitur, et alter, qui est initium .kf. gradus Cancri, orizontis circulum tangat in uno tantum puncto. Ad hunc itaque | circulum postquam sol ascendens in septemtrionem pervenerit, ipsum motu sue ferentis spere, de qua suo loco dicetur, pertransit, nec unquam postea ad occasum revertitur, sed semper super orizontem exaltatur, usque dum transcursis paralellis, qui sunt ab illo .ke. gradus Geminorum circulo usque solstitialem et estivum Cancri circulum, per eosdem paralellos sed non eadem via descendens, ad eundem revertitur paralellum, per quem transivit in medio signi Geminorum, nullum passus occasum. Cum igitur a septemtrionis partibus rediens, postquam transierit in medio Cancri, et occasum patitur et ortu illustratur. Opposita ratio in polo australi eodem in loco facile <est>[(L)] videri. Quot enim diebus locus ille sole in septemtrione posito sine noctibus illustratur, totidem noctium

3 est] *add.* (L) **4** exhauriens] *corr. ex* exauriens (L) **6** superpositus] suppositus **7** versantur] versatur (L) **9** latitudinem] habitudinem **17** elongatur] elongantur **21** paralellus] punctum scilicet **28** quem transivit] cuius transit (L)

which we define as the beginning of Capricorn, it will produce a night of 24 hours while not having day, because the circle on which that point revolves, that is, the head of Capricorn, lies in its entirety under the earth, as has been said. At this place, and at similar ones, where a day in summer has 24 hours without a night, in winter the night will extend to last 24 hours and exhaust the day.

[100] *The latitude is the arc on the meridian whose length is from the equator until the point which looms and is placed straight above the heads of those who live at those places. But its length, as has been said above, is* 66;25 degrees, that is, the complement to the great inclination. *It is said that beyond this latitude of the inhabited region from the equinoctial circle, or terrestrial equator, there cannot be a place of living for animals,* because due to the permanent icy freezing there cannot grow any plants for the use of animals and all animals would freeze to death in the excessive cold.

([103, 104])[24] But those places that lie beyond 66;25 degrees, torpid in the icy zone, have a longer day, the farther they are from the equator towards the centre of the ice region, that is, the north pole; namely a daylength of 24 hours in summer and a night of the same length in winter. This occurs because the circle of Cancer, until which the great inclination reaches, removes itself from the horizon. And the horizon lies tangential in a single point to another one of the parallel circles of the equator that is contained between the latter and the solstitial gate. As an example, we assume some place where the circle of the head of Cancer lies at a distance from the horizon such that the parallel circle about which the end <of the 15th degree> of Gemini and also the beginning of the 16th degree of Cancer revolve touches the horizon in a single point.[25] Hence, when the sun on its ascent northwards has reached this point, it will pass it by due to the motion of its own sphere, which we will speak about at its proper place, and it will henceforth never turn to set but be always elevated above the horizon, until it has passed through the parallel circles between that parallel of the 15th degree of Gemini and the solstitial summer circle of Cancer and, having crossed the same parallel circles, but in a different direction, returns to the same parallel circle which it had crossed at the middle of the sign of Gemini, without having experienced setting. Thus, when returning from the northern signs, after it has crossed the middle of Cancer, the sun suffers setting and appears at sunrise. It is easily seen that opposite reasoning

[24]The paragraph replaces *Configuration*, passages 103 and 104.

[25]Al-Battānī, *Ṣābiʾ Zīj*, ch. 6, contains a similar example for a distance of 30°, not 15°, from the solstice.

tenebris sine diei decore in hiemis temporibus sub australibus signis obnubilatur. Postquam enim sol paralellum, qui est circuitus mediorum Capricorni et Sagittarii, in austrum tendens transierit, semper latet sub terra, usque dum ad eundem et ab austro revertens incipit oriri et occumbere. Quemadmodum igitur dictum est, 5 quanto Cancri circulus exaltatur, tanto fit maior et prolixior dies sine nocte, atque e contrario nunc eiusdem quantitatis in opposito.

[105] *Quanto igitur maius habet latum, tanto maiores habet dies.*Latum autem eius ultra .s. gradus augmentari non poterit. Tot enim gradibus omnibus sui partibus distat equinoctialis circulus ab utroque mundi polo. Si quam igitur terram pluri- 10 bus quam .s. gradibus a circulo recti diei distantem imaginari volueris, ratio non procedet, immo pauciores, quam imaginaveris, gradus invenies. Latum enim terre esse superius positum est arcus meridionalis circuli a puncto capitis usque recti diei. Hoc autem ita intelligendum est arcum hunc esse a puncto capitis usque eam partem equinoctialis circuli, que propinquior est puncto capitis eiusdem loci. Si 15 igitur ponas aliquam terram .t. habere gradus latitudinis, falleris. Illa enim pars circuli recti, a qua latum eius metiris, sub orizonte illius loci latet, alia vero super orizontem .k. gradibus eminet, quibus sublatis de .s. remanet .r., latum scilicet loci eiusdem. *Locus igitur ille, cuius est latum in septemtrione .s. gradus, septemtrionalem polum sibi recte superpositum videt. Est igitur polus mundi illic punctus capitis,* alter 20 autem australis recte sub pedibus positus. *Unde fit, ut mundi radius, qui ab altero usque alterum dirigitur polum, ita positus sit, ut neque iacens neque in aliquam videatur inclinare partem.* Equinoctialis autem circulus, quoniam .s. distat gradibus a septemtrionali polo et polus idem in eo loco unum est cum capitis puncto, quem polum orizontis esse superius dictum est, qui orizontis circulus a polo suo totidem lon- 25 ge positus est gradibus, quot equinoctialis a suo, *invenitur idem recti diei circulus necessaria ratione eiusdem loci orizontem circumdare.* Ex hoc igitur manifestum fit supreme motu spere nullum illic duodecim signorum oriri aut occumbere, quoniam non ab inferioribus, hoc est a subterraneis, ad superiora orizontis, scilicet altum, volvitur, sed circumfertur.

30 [106] *Ex his ergo, que dicuntur, intelligi facile poterit solarem annum eo in loco in duas tantum dividi partes, altera unum diem tenet altera unam noctem. Habet autem dies iste tantum longitudinis temporis quantum et sex menses, parilique quantitate noctis*

---

1 tenebris] tenebras **19** superpositum] suppositum (L) **28** subterraneis] subtraneis (L) **31** autem] *add.* (L)

applies at the same place concerning the south pole. Because for the same duration which that place is being illuminated without nights when the sun stands in the northern signs, it is obscured by the darkness of the nights without the adornment of the day in winter when the sun stands in the southern signs. For after the sun, on its way south, has crossed the parallel circle which is the circuit of the middle of Capricorn and Sagittarius, it is always hidden under the earth, until it returns from the south to the same circle and starts rising and setting again. Hence, as has been said, the higher the circle of Cancer is elevated, the greater and longer the day will be without night, and *vice versa* with its quantity inverted.

[105] *Hence the larger its latitude, the longer days it has.* But its latitude cannot exceed 90 degrees. For this is the amount of degrees by which all parts of the equator are remote from either pole of the world. Thus, if you wanted to imagine a place on earth that lies more than 90 degrees from the equator, reasoning will not proceed; instead, you will find a smaller degree than you initially imagined. For it has been said above that the terrestrial latitude is the arc on the meridian from the zenith until the equator. But this must be interpreted such that this arc extends from the zenith until that part of the equator which lies closest to the zenith of that place. Hence, if you assume a place on earth with a latitude of 100 degrees, you will fail. For that part of the equator from which you would measure the latitude of that place will be hidden under the horizon, whereas another part will stand 10 degrees above the horizon, and if we subtract these from 90 degrees, there will remain 80 degrees, that is, the latitude of that place. *Therefore, the place whose northern latitude is 90 degrees sees the north pole placed straight above itself. One pole of the world thus coincides with the zenith there,* whereas the other, southern, pole lies straight under the feet. *Hence the axis of the world, which extends from one pole to the other, is oriented such that it does not seem to lie or to be inclined in any direction.* But since the equator lies 90 degrees from the north pole and the pole, in turn, coincides at that place with the zenith, which above has been said to be the pole of the horizon, and the horizon is distant from its pole by the same amount of degrees as the equator from its own, *by necessary consequence the celestial equator is found to enclose the horizon of that place.* It thus becomes clear from this that none of the twelve signs rises or sets there by the motion of the highest sphere, since it is not moved from the region below the horizon, that is, from beneath the earth, to the region above the horizon, that is, the altitude, but it is taken around.

[106] *From what has been said it is easily understood that the solar year is divided into two parts there, one of which comprises a single day and the other a single night. And the length of that day equals the span of six months, and the same is the length of*

tempora protrahuntur. Tanti vero diei principium est sole in caput Arietis intrante, finis dum elapsus a Virgine Libram ingreditur. Medium igitur eius est sole in capite Cancri morante. *Tamdiu igitur illic dies est quamdiu sol in septemtrionalibus signis positus nobis estatis ministrat ignes. Nox quoque que sequitur sex mensium tem-*
f. 24r 5 *pora complectitur,* et initium eius sole Libram intrante, | medium in complemento magne inclinationis que vergit in austrum, finis vero Piscium ultima solis centro relinquente. *Tamdiu igitur et nox quoque protenditur quamdiu sole in australibus signis posito nos torpentis hiemis frigora sentimus.* Cuius rei causa est, quod nulla pars signiferi motu supreme spere oritur aut occumbit, sed semper ligantes signiferum
10 et equinoctialem circulum puncti in ipsius circuli orizontalis circumductione positi sunt. Unde et australia signa semper sub terra sunt et sol eodem modo cum illa perlustrat, septemtrionalia vero super terram semper super orizontem solem monstrantia dum in ipsis rotatur. *Illo autem loco sol circa orizontem superius et inferius prope movetur, nec plus in altum aut imum mittitur quam magna zodiaci inclinatio*
15 *patitur.* Sed quoniam de lato terrarum satis dictum est, nunc de alto altitudinis videamus.

---

7 relinquente] reliquente (L)

the night. The beginning of such a day is when the sun enters into the head of Aries, and the end when it has descended from Virgo and enters Libra. Its middle is when the sun stands in the head of Cancer. *Daytime there is thus as long as the sun is placed in the northern signs and provides us with the heat of summer. And also the following night lasts six months,* and its beginning is when the sun enters Libra, its middle when the great inclination to the south has been reached, and its end when the sun's centre leaves the end of Pisces. *Hence also the night lasts as long as the sun is placed in the southern signs and we feel the cold of freezing winter.* The reason for this is that no part of the ecliptic will rise or set by the motion of the highest sphere, but the intersection points of the ecliptic and the equator are placed on the circumference of the horizon. Therefore the southern signs, and also the sun when it wanders through them, are always below the earth, whereas the northern signs above the earth always show the sun above the horizon when it rotates in them. *But at that place the sun moves closely above or below the horizon, and it is not sent higher or deeper than the greatest inclination of the zodiac permits.* But since enough has been said about the terrestrial latitude, we may now consider the altitude.

[160/161]<***A***>[(L)]*ltum igitur est arcus circuli, qui transit super punctum capitis, et est magnus in spera circulus, eiusque centrum idem est quod et mundi. Volvitur autem semper circa orizontem rotundo motu. Arcus autem huius circuli, qui continetur ab orizonteo circulo usque punctum alicuius circuli aut centrum stelle super orizontem, altum eiusdem puncti dicitur.* Imaginamur igitur in unaquaque stella sui alti designandi causa huiuscemodi circulum et transeuntem per punctum capitis et punctum medie terre et centrum medie stelle. Surgente autem stella et centrum eius super orizontem suo motu nona spera elevante crescit mensura arcus, quem altum dicimus, quantoque altius fit centrum stelle, tanto longior erit arcus etiam a centro stelle usque orizontis circulum superposite arcuatus.

Imaginamur ad designandam illius alti quantitatem lineam egredientem de centro terre et extendi per centrum stelle usque ad superficiem magne spere. Punctum igitur, quem in alta spera tetigerit eadem linea, altum stelle ponimus, estque tantus eius alti arcus, quantum ab illo puncto usque orizontem de circulo, qui transit per punctum capitis, continentur. [162] *Dum autem elevatur amplius altum stelle, et arcus quoque alti maior fit, usque dum centrum stelle motu supreme spere arcum augmentans meridionalem intret circulum. Statim fit circulus meridionalis et circulus alti idem, et est arcus alti eo loco arcus meridionalis circuli a puncto, quem linea de centro mundi exiens et per centrum stelle directa tangit in superficie none spere, usque ad orizontis circulum. Postquam autem ad meridionalem pervenerit circulum, augmentari arcus alti desistit, quia et centrum stelle altius prohibetur exaltari.* [163] *Abinde motu spere in occiduas partes remotum incipit descendere,* atque sicut paulatim a parte orientis positum promotum ad altiora alti arcum produxerat, ita sensim descendens arcum alti cogit diminui. *Quanto igitur amplius descenderit, tanto arcus eius alti minor est,* et alter arcus, qui a centro stelle usque ad punctum capitis ascendit, augmentatur. Ut autem orizontis circulum tetigerit, et alti arcum nullum habet, quoniam nec altum, et arcus a puncto capitis usque orizontis circulum productus est. *Hec que diximus de sole et omnium stellarum globis, que ortus celebrant et norunt occasus, intelligendum est.*

[164] *Verum in illis, que semper videntur aut super orizontem sunt posite, aliter se ratio habet. Paralelli namque circuli, in quibus moventur, nulla sui parte orizontis circulum attingent, sed tota earum circumductio super orizontem sempiterna volvitur rotatione.*

---

8 altum] arcum (L) **12** stelle] terre (L) **18** et] *add.* (L) **28** occasus] occassus

2.4 <The Altitude>

Config. 7: The Altitude

[160/161] *The altitude is an arc on a circle which passes through the zenith and which is a great circle on the sphere and whose centre is the same as that of the world. It always turns around the horizon with a rotating motion. The arc of this circle that is enclosed between the horizon and the point on some circle or the centre of a star above the horizon is called the 'altitude' of that point.* We imagine such a circle for every single star in order to designate its altitude, passing through the zenith and the point of the middle of the earth and through the centre of the star. Hence, when a star and its centre rise above the horizon by their motion when the ninth sphere revolves upwards, the length of the arc which we call the altitude increases; and the higher the centre of the star rises, the longer will be the arc that is spanned from the centre of the star until the horizon.

To determine the amount of its altitude, we imagine a line originating from the centre of the earth and extending through the centre of the star until the surface of the great sphere. Then we define the point on the high sphere that is touched by this line to be the altitude of the star, and the arc of its altitude is as large as that portion of the circle through the zenith which is comprised from that point until the horizon. [162] *But when the star's altitude rises further, the altitude arc also increases, until the centre of the star, by the motion of the highest sphere, prolongs its arc and enters the meridian. In that same moment, the meridian and the altitude circle become the same, and the altitude arc at that place is the arc on the meridian from the point on the surface of the ninth sphere that is touched by the line, which goes out from the centre of the world and is directed through the centre of the star, until the horizon. But after it has reached the meridian, the altitude arc ceases to grow, because also the centre of the star is restrained from rising further.* [163] *From then it is moved by the motion of the sphere towards the setting side and begins to descend,* and just like it produced the altitude arc gradually when it was elevated while being on the rising side, now steadily descending it forces the altitude arc to decrease. *And the further it descends, the smaller will be its altitude arc,* whereas the other arc increases which rises from the centre of the star until the zenith. But once the star has reached the horizon, it has no altitude arc, because it has no altitude, whereas the arc starting from the zenith has increased until the horizon. *What we have said is to be understood for the sun and the bodies of all stars that celebrate rising and experience setting.*

[164] *But for those stars which are always visible or placed above the horizon, a different reasoning applies. For the parallel circles on which they move do not touch the horizon with any of their parts, but their entire circumference rotates permanently*

f. 24v In his ergo stellis, que per illos moventur | circulos, altum quidem non habet principium, non enim oriuntur; sed neque finem, non enim occumbunt. *Variatur tamen alti arcus quantitatis* in illis locis, qui neutrum habent sibi superpositum mundi polum. Paralelli namque, in quibus moventur, non sunt iuncti in circumductionibus 5 suis paralellis orizontis, sed inclinantur ab illis et ipsos dividunt, oblique tamen neque rectis angulis aut equis divisionibus ab ipsis divisi. Meridionalis vero circulus omnes paralellos orizontis et ipsum dividit medios quippe transiens per utrumque orizontis polum, sed et paralellos equinoctialis et ipsum equa particione dividere superius positum est. Paralelli autem, in quibus moventur stelle, aliis altiores, aliis 10 vero humiliores sunt sui partibus. Altiores namque et breviores paralellos orizontis illis punctis tangunt, in quibus meridionali circulo superior eorum pars secatur; humiliores autem et latiores illis, in quibus inferior. *Cum igitur in illis partibus suorum circulorum volvuntur, que altius emicant, maius habent altum, et alti eorum arcus longior invenitur. Postquam ad eam altitudinem suorum circulorum conscenderint, qua* 15 *ipsi circuli meridionali iunguntur circulo, amplius exaltari desistunt, et arcus eorum alti maior non fit. Inde autem descendentes ad aliam et oppositam divisionem revertuntur. Meridionalis etenim circulus paralellos illos in duobus secat punctis, qui semper videntur.* Idcirco earum descensus non pervenit usque ad orizontis circulum, *sed cum ad punctum illum inferiorem descenderint, completa descensionis sue meta iterum ad al-* 20 *ta feruntur.* Hec ergo causa est eas semper altum habere, variatim tamen, ut nunc minus, nunc maius sit. Hec ergo, que dicta sunt, de arcu alti ad presens sufficiant.

[165] Sed quoniam omnis arcus cordam habet, corda huius que sit, hoc modo videndum est. *Recta ergo huius arcus alti corda est linea, que incipiens a puncto stelle aut gradus aut alicuius puncti recte defertur et cadit in suppositum superficiei orizontis* 25 *punctum.* [166] *Omnium autem arcuum corde <recte> sunt medietates cordarum duplicis arcus. Hec igitur corda recta dicitur.* [167] *Rediens autem est sagitta duplicis arcus.* De his autem apertius, si divina annuerit ac potentior maiestas, suis in locis disputabitur. Hic tantum, que sint, significasse sufficiat. *Longior autem corda huius arcus alti est corda copulamenti alti super orizontem, que est medii diei eiusdem stelle* aut so- 30 lis aut etiam puncti. Placet autem de hoc circulo et eius positione in orizontibus recte linee et inclinatis in septemtrionem pauca intueri, ut que dicta sunt superius de dierum equalitate et discordia patentiora fiant. Punctus igitur, in quo circulus volvitur in recta linea super caput, positus est in circulo equinoctiali, quippe qui supra caput transit. Si igitur sole posito in capite Arietis aut Libre imaginari vol-

---

3 alti] altum **12** inferior] inferius **18** cum] non (L) **21** que] qua

above the horizon. The altitude of stars which are moved by those circles does not have a beginning, because they do not rise; neither has it an end, because they do not set. *Nonetheless, the arc of their altitude changes* at those places that do not have one of the world's poles above themselves. For the parallel circles on which they move are not linked in their circuits to parallels of the horizon, but they are inclined from them and intersect with them; obliquely, though, and neither in right angles nor divided by the latter into equal parts. Now, the meridian, passing through both poles of the horizon, divides all parallel circles of the horizon and the horizon itself in half; and it has been said above that it does also divide the parallel circles of the equator and the equator itself into equal parts. But the parallel circles on which the stars move are higher in some of their parts and lower in others. For they are tangential to the higher and smaller parallels of the horizon at those points where their upper part is crossed by the meridian, and to the lower and wider parallels where their lower part is crossed. *When thus the stars revolve on those parts of their circles which shine higher, they have a larger altitude and the arc of their altitude is found longer. After they have risen to that altitude of their circles where the latter intersect with the meridian, they cease to rise higher and their altitude arc does not become any larger. But descending from there, they return to the other, opposite, intersection. For the meridian crosses those parallel circles in two points that are always visible.* For that reason their descent does not reach until the horizon; *rather, when they have descended to that lower point, the course of their descent has been completed and they become elevated again.* This is the reason why they always have an altitude, though changing, such that it is sometimes smaller and sometimes larger. But what has been said about the altitude arc shall be enough for the moment.

[165] But since every arc has a chord, its chord is to be considered in the following way. *The straight chord of this altitude arc is the line that starts from the point of the star, or a degree or any given point, and which is directed straight downwards and falls onto the point straight beneath it on the horizontal plane.* [166] *But the straight chords of any arcs are halves of the chords of the double arc. This chord is therefore called the 'straight chord'.* [167] *But the receding is the 'arrow' of the double arc.* But we will speak about these in more detail at their proper places, if the mighty divine majesty will concede. Here it is enough merely to have introduced their meaning. *The longest chord of this altitude arc is that which connects to the altitude above the horizon at midday of that star* or the sun or also the point. But we may look a little closer at that circle and its position at horizons on the terrestrial equator and others which are inclined northwards, in order that what we have said above about the equality and inequality of days becomes clearer. At the terrestrial equator, a point

ueris circulum alti transeuntem per centrum solis circum orizontem volvi, ratio non procedet. Non enim circumfertur, sicut supra positum est, eo die circulus is, sed idem cum equinoctiali existens, eodem quoque et ille motu movetur, atque ad ultimum alti complementum .s. gradibus distat ab orizonte, deficitque totus ille arcus circuli alti, quem superius esse diximus a centro stelle usque punctum capitis. Ex his igitur manifestum est, quod tanta erit alti circuli arcus illius quantitas, quod a centro stelle in meridie posito usque punctum capitis porrigitur, quanta est et ipsius stelle inclinatio a recto circulo. Inclinatis orizontibus non ita est. Si que enim in circulo equinoctiali stelle volvuntur, ut inclinationem non habeant, ad meridionalem circulum pervenientes tantum habent superiorem arcum, quanta est latitudo illius terre arcum sui alti complemento lati equantes. Si vero ab equinoctiali | inclinatur in occultum polum, superiori arcui latum terre et eorum inclinatio equabitur, quibus sublatis de .s. remanet arcus alti. At si in septemtrionem minus quoque, quam sit latum terre, iuncta inclinatione complemento lati habes arcum, quod restat de .s. superiorem. Si vero equalis, superior deficiet inferiori .s. gradus vendicante; si maior, lato dempto remanet superior, quo sublato de .s. restat alti arcus. In orizonte autem, qui idem est cum equinoctiali, motu none spere altum fixarum nec crescit nec minuitur, sed tantum semper altum quanta et a recto est earum inclinatio. Septem altum planetarum non per none sed per sue ferentis spere motum variatur. De circulis que dicenda videbantur et inclinatione et lato et alto altitudinis premissis ad orientalium ascensuum<que> disputationem placet accedere.

whose circle moves through the zenith is located on the celestial equator, for the latter passes along above the head. Thus, when the sun is placed in the head of Aries or Libra and you wish to imagine an altitude circle that passes through the centre of the sun and revolves around the horizon, reasoning will not procede. For on that day this circle does not revolve as described above; instead it exists identical with the equator and is moved by the same motion as the latter, and at the highest limit of altitude it would be 90° from the horizon, whereas that entire arc on the altitude circle would vanish which above we defined to be from the centre of the star until the zenith. From this it is clear that the length of that arc on the altitude circle which extends from the centre of the star until the zenith at midday is equal to the inclination of that star from the celestial equator. This is not the case at the inclined horizons. For if stars revolve on the equator and, therefore, have no inclination, when they reach the meridian their superior arc will be as large as the latitude of that place, thus adapting their altitude arc to the complement of the latitude. But if it is inclined from the equator towards the concealed pole, the terrestrial latitude plus their inclination will become equal to the superior arc, and subtracted from 90 degrees this will yield the altitude arc. But if the star is inclined to the north by less than the terrestrial latitude, you obtain a superior arc equal to 90 degrees minus the sum of the star's inclination and the complement of the latitude. But if they [i.e., the star's inclination and the terrestrial latitude] are equal, the superior arc will vanish completely, while the lower arc takes 90 degrees; and if the star's inclination is larger, subtraction of the latitude yields the superior arc, which again subtracted from 90 degrees produces the altitude arc. But where the horizon coincides with the equator, the altitude of the fixed stars neither increases nor decreases by the motion of the ninth sphere, but their altitude is always as large as their inclination from the equator. The altitude of the seven planets does not change by the motion of the ninth sphere but by that of their deferent sphere. As we have said in advance what seemed necessary to mention about the circles, the inclination, the latitude and the altitude, we may now proceed to the discussion of the ascensions and the ascendants.

[168] *<A>scensus igitur est quivis gradus circuli signorum positus ab orientibus partibus in quovis loco orizontei circuli. Occasum autem dicimus surgenti gradui oppositum in occidentis partibus super circulum orizontalem positum.* Causa eorum, que dicta sunt, hec. Circulus signorum, quem et zodiacum dicunt, magnus est spere circulus, orizon quoque in latiori parte spere circumducitur. Positum autem superius reminiscimur omnes magnos circulos, quantumcumque alter ab altero inclinetur, sese invicem partiri in duobus punctis diametrice oppositis in duas equales medietates. *Secat igitur orizontis circulum signifer in duo equalia media, et est semper medietas eius super orizontem, opposita vero subtus. In duobus enim diametrice oppositis sese mediis dividunt.* [169] *Inde ergo fit, ut oriente quovis circuli signorum gradu oppositus illi occidat.* Oriente namque Arietis capite, caput occidit Libre. Et in ceteris eadem provenit ratio. *Ad hunc itaque modum de omnibus magnis circulis in spera positis intelligendum est.*

[170] *Paralelli vero equinoctialis, in quibus volvuntur fixe et ceteri motu supreme spere, non sic. Non enim puncti, qui in orizontis circulo positi oriuntur, suos vident oppositos in occidente occumbere in omni orizonte. Nam in his orizontibus, qui polum habent in equinoctiali positum, superius data concordat regula, quoniam, ut supra dictum est, omnes ab orizontis circulo in duo dividuntur media paralelli; inclinati autem orizontes inequaliter et ipsi ab his et eos ipsi dividunt.*

[171] ¶*Orientalia autem dicimus arcum circuli spere altioris,* quam etiam sepe propter excellentiam simplici nomine speram appellamus, *qui oritur cum positis partibus sive arcu circuli signorum.* Occidentalia autem arcum circuli spere, qui occumbit cum positis partibus circuli signorum. Positas autem partes circuli signorum ideo dicimus, quoniam non eiusdem quantitatis semper habent aut orientale aut occidentale, sed mutata orizontium inclinatione mutantur et ipsa. [172] *Orientale quidem exempli causa hoc est. Punctum, quo in circulo signorum sol movetur, ponimus in orizontis circulo orientem,* illum scilicet, qui est caput Arietis. *Erit autem eiusdem temporis* athomo *aliquis punctus equinoctialis circuli in eodem orizontis circulo oriens positus.* Is autem est idem, qui et caput Arietis; in eo enim ambo se ligant et recti diei et zodiaci circuli. *Sole ergo in altum surgente motu spere movetur etiam punctus circuli recti diei in altum, usque dum tangat aliquam partem circuli, in quo est et movetur. Surrexit quoque quedam pars circuli signorum usque ad aliquem punctum,*

---

12 magnis] signis **15** in] *add.* (L) **26** hoc] hec

2.5 <The Ascendant and the Ascensions>

Config. 8: The Ascendant and the Ascensions

[168] *The 'ascendant' is the degree of the ecliptic which at a given place is placed on the eastern side of the horizon; 'descendant' we call the one that is placed opposite the rising degree, on the horizon in the west.* The reason for this is as follows. The ecliptic, which is also called the zodiac, is a great circle of a sphere, and also the horizon is drawn along the widest part of a sphere. But we remember what has been said above, that all great circles, irrespective of their inclination from one another, partition each other into two equal halves at two diametrically opposite points. *The ecliptic therefore cuts the horizon into two equal halves, and one half of it is always above the horizon and the opposite half below. For they divide each other into two opposing halves.* [169] *It follows from this that whatever degree of the ecliptic is rising, the degree opposite to it is setting.* Thus, when the head of Aries rises, the head of Libra sets. The same reasoning holds also for all other signs. *In this manner, therefore, it must be understood for all great circles on the sphere.*

[170] *But this is not the case with the parallel circles of the equator, on which the fixed stars and other bodies revolve by the motion of the highest sphere. For not at every horizon do the points that are rising at the horizon see their opposite set in the west. To those horizons which have their pole on the equator the above given rule applies, because all parallel circles are divided into two halves by the horizon, just as we have said above, but the inclined horizons are divided unequally by the parallel circles while themselves dividing the latter unequally.*[26]

[171] *'Ascension' we call the arc on the equator of the highest sphere*—which, for its eminence, we often simply call 'the sphere'—*that rises with given degrees or a given arc of the ecliptic;* 'descension', however, the arc of the equator that sets with given degrees of the ecliptic. But we say 'given' degrees of the ecliptic, because they do not always have an ascension or descension of the same size, but these quantities change with a change of the inclination of the horizon. [172] *The following is an example of the ascension. We assume the point of the ecliptic where the sun is moving to be rising at the horizon,* for example, the head of Aries. *But at the same* indivisible instant *of time, some point of the equator is rising at the same horizon.* This is also the head of Aries, for in this point both circles, that of the equator and that of the zodiac, intersect. *Hence, when the sun rises high by the motion of the sphere, the point of the equator will also rise, until it reaches a certain part of the circle on which it is and moves. And also a certain part of the ecliptic has risen until some point, and it is not the*

[26] Arab. add.: 'except for those of them which are great circles.'

nec idem est, qui fuerat tunc sole surgente ex orizontis circulo. Erit igitur arcus quidam circuli signorum a puncto solis usque punctum, quem ascensum diximus; diversi puncti in orizontem circuli utriusque positi sunt, et de singulis arcus surrexit. Illum ergo arcum circuli recti diei, qui surrexit ab ortu solis usque .ke. gradus circuli signo-
f. 25v 5 rum complentem ortum suum, dicimus orientalia | et arcum circuli signorum, qui consurrexit. Sunt autem hec in quarto climate .i. gradus .ld. sexagenarie. In aliis autem plus vel minus continetur.

[173] *Porro ille arcus circuli recti diei, quem cum circulo signorum surrexisse dicimus, equalis invenitur semper arcui paralelli, in quo sol volvetur, qui ab ortu solis usque*
10 *illud temporis momentum surrexerit.* Equali enim celeritate hic et ille volvuntur, licet breviori alter feratur semita. *In eodem itaque paralello licet orientalium, si placet, quantitatem intueri, quia et arcus eiusdem paralelli orientalia et signorum circuli dicuntur. Sunt igitur orientalia similia huius et illius equinoctialis, nec numero graduum distant.* Tametsi arcus et graduum quantitas sit diversa, eiusdem tamen minor est
15 quantitatis in suo circulo, cuius et alter in equinoctiali.

[174] *Sic igitur, ut dictum est, unaqueque pars circuli signorum habet aliquam partem circuli spere, que surgit cum ea,* et ipsa quidem ascensus, dum surgit, *illa autem orientale eius dicitur. Sunt autem orientalia diversarum partium circuli signorum,* et alia quidem partium Arietis alia partium Libre. *Quecumque tamen due partes cir-*
20 *culi signorum equaliter a quovis puncto equalitatis distant, equalem scilicet habentes inclinationem, hac in particione orientalium equalitatem servant. Exempli gratia primi decem gradus Arietis et decem ultimi Piscium equaliter distantes a puncto verne equalitatis equalia habent* numero et mensura *orientalia,* [175] *sic et omnes alie partes Arietis et Piscium. Sic partes Virginis et Libre equalia participant orientalia. Idem de*
25 *reliquorum signorum omnibus partibus intelligendum.* Cum enim .d. sint circuli signorum quadrantes, primus quidem a puncto veris usque ad punctum mutationis estatis, secundus ab illo usque punctum equalitatis autumpnalis, a quo tercius usque punctum hiemalis mutationis, quartus deinde usque primum punctum Arietis, de quo superius plenius dictum est, *quecumque erunt aliquo in loco quadrantis veris*
30 *orientalia, eadem erunt alterius quadrantis, quod in illo puncto sue equalitatis iungitur, id est hiemalis. Quotuscumque autem erit arcus orientalium totius quadrantis estivi, totus erit etiam arcus orientalium iuxtapositi quadrantis in austrum inclinati, id est autumpnalis.* Animadvertendum autem in orientalibus, quoniam quantumcumque pauciora erunt orientalia alicuius gradus in septemtrionem inclinati orientalibus eiusdem

---

1 ex] et **4–5** signorum] signi **8** quem] quidem **20** distant] distat (L) **21** hac] hec particione] particionem servant] servat (L)

same as the one when the sun was rising from the horizon. Hence there will be a certain arc of the ecliptic from the point of the sun until the point which we have called ascendant; different points of either circle are on the horizon, and an arc has risen from each one of them. And the arc of the equator which has risen since sunrise, until the 15th degree of the ecliptic completes its rising, is what we call the 'ascension', or the 'co-risen arc of the ecliptic'. In the fourth climate it is 9;24 degrees[27]; in other climates it comprises less or more.

[173] *Moreover, the arc of the equator which we say had risen with the ecliptic is always found equal to the arc of the parallel circle on which the sun revolves that has risen from sunrise until that moment.* For both circles revolve with the same speed, with one of them nonetheless being moved on a shorter course. *One therefore can also seek the length of the ascensions from that parallel circle, if desired, because also the arcs of that parallel circle are called the ascensions of the ecliptic. Hence the ascensions of this circle are according to those of the equator and do not differ in the number of degrees.* Although the size of the arc and of the degrees is different, the smaller arc nevertheless has the same size on its own circle as the other has on the equator.

[174] *In this manner, therefore, as we have said, any part of the ecliptic has a certain part of the celestial equator that rises with it,* and the former, while rising, is called ascendant, *whereas the latter is called its ascension. But there are ascensions to different parts of the ecliptic,* some for the degrees of Aries and others for the degrees of Libra. *However, any two parts of the ecliptic that are equally distant from any of the two equinoxes and, thus, have equal inclination, maintain in that partition equality of the ascensions. For example, the first ten degrees of Aries and the last ten degrees of Pisces are equally distant from the point of the spring equinox and, therefore, have equal ascensions* in number and length, [175] *and similarly do all other parts of Aries and Pisces. In this manner, the parts of Virgo and Libra share equal ascensions. The same is to be understood for all parts of the remaining signs.* For since there are four quadrants of the ecliptic, the first from the spring point until the point of the summer solstice, the second from there until the point of the autumnal equinox, from where the third extends until the point of the winter solstice, and then the fourth until the first point of Aries, which has been discussed in detail above, *whatever the ascensions to a point in the spring quadrant, they will be the same in the other quadrant which is adjacent to it at its equinoctial point, that is, in the winter quadrant. But whatever the length of the arc of the ascensions of the entire summer quadrant, of the same length will be the arc of the ascensions of the quadrant next to it inclined southwards, that is, of the autumnal*

---

[27] Unlike Stephen's examples in subsequent passages, this value is not given in the *Almagest*.

gradus in recta linea, tanto plura continebit orientalia alterius eadem quantitate in eandem partem inclinati. Verbi gratia .k. gradus Arietis, cuius est inclinatio in septemtrionem, orientalia sunt .f. gradus .kd. sexagenarie recti circuli ab equalitate orientalium recti circuli .b. gradibus et .of. sexagenariis, Virginis decimi a fine .kb. gradus .f. sexagenarie, eodem scilicet, que et .k. gradus Libre oppositi, id est Arietis. In quo mirandum, quoniam quicquid ab equalitate illorum orientalium quislibet gradus ab oriente amiserit, totum occidendo in occidentalibus resumit et opposito eius detrimentum patiente. Quicquid enim amittit .k. gradus Arietis surgendo, recuperat occumbendo, quicquid eius oppositus in ortu lucratur, in occasu perdit. Ex his manifestum est, quod occidentalia quorumvis graduum paria sint orientalibus suorum oppositorum.

[176] *Ad hunc modum consideranda orientalia* et occidentalia *in orizonte. Habent autem discordiam orientalia* et occidentalia *cuiusvis partis circuli signorum in diversis orizontibus.* Non enim eadem sunt in uno, que in alio. *Facit autem discordiam eorum latum terre discors.* In discordantibus nempe latis terre discorditer orizontes inclinantur, unde et discordia est circuli recti diei surgentis <et> circuli signorum.

f. 26r [177] *Fiet autem planius hoc modo. Orientale | Arietis totius in uno orizonte aliquas habet partes, in diverso si quidem maioris lati fiunt pauciores, si minoris plures.* Ut in IIII^{to} quidem .ki. gradus .kb. sexagenarie, in quinto .kg. gradus .mb. sexagenarie, in tercio vero .l. gradus .oc. sexagenarie. *Parili modo Piscibus contingit. Contraria est ratio* horum oppositis *Virgini et Libre.* Illorum enim orientalia in maiori lato plura, in minori pauciora esse deprehenduntur. Ex his ergo, que dicta sunt, sagaci facile est intelligi orientalia usque punctum mutationum a parte puncti equalitatis verne diminui, deinceps augmentari. *Sed et illud intelligi plenius volumus orientalia cuiusvis medietatis circuli signorum per punctos equalitatis divisi semper esse medietatem recti circuli.* Neque enim Arietis aut Libre caput oriri poterit aut occumbere, quod non puncti illi oriantur aut occumbant, in quibus uterque ligatur circulus.

---

10 sint] *corr. ex* sunt (L) **17** Orientale] Orientalem (L) **21** Virgini] Virginis **24** augmentari] augmentum (L)

quadrant. However, when dealing with the ascensions one needs to be aware that the difference by which the ascensions of any degree that is inclined north falls short of the ascensions of the same degree on the equator will be the same amount by which the ascensions of the other degree of the same quantity and inclined to the same side will be larger. For example, the 10th degree of Aries, whose inclination is to the north, has ascensions on the equator of 6;14 degrees, 2;56 degrees short of equality with the ascensions as seen from the equator, whereas the tenth degree before the end of Virgo has 12;6 degrees, that is, the same as the 10th degree of Libra, being the opposite sign of Aries.[28] It is remarkable in this that by whatever amount any degree falls short of equality of those ascensions when rising is completely recovered in the descensions when it sets, while its opposite suffers loss. For whatever the 10th degree of Aries loses when rising, it gains back when setting, whereas all which its counterpart wins during its rise, it loses when setting. From this it is clear that the descensions of any degree are equal to the ascensions of its opposite.[29]

[176] *In this manner the ascensions* and the descensions *on the horizon are to be considered. But the ascensions* and descensions *of any part of the ecliptic differ at different horizons.* For they are not the same in one horizon as they are in another. *Their difference is caused by the different terrestrial latitude.* For at different terrestrial latitudes the horizons are inclined differently, which causes a difference in the rising of the equator and the ecliptic. [177] *This will become clearer in the following way. The ascension of the entire sign of Aries has a certain amount of degrees at one horizon, at a different horizon with a higher latitude it will be less, with a lower latitude it will be more.* While thus in the fourth climate it is 19;12 degrees, it is 17;32 degrees in the fifth, but 20;53 degrees in the third.[30] *The same applies to Pisces, correspondingly. The contrary occurs* to the opposite of them, *to Virgo and Libra.* For their ascensions are found to be more at a larger latitude and less at a smaller one. From what has been said an intelligent person will easily understand that the ascensions decrease from the point of the spring equinox until the point of the solstice, and then increase. *But it should also be fully understood that the ascensions of any half of the ecliptic, if the latter is divided at the equinoctial points, is always half the equator.* For neither the head of Aries nor that of Libra can rise or set unless those points also rise or set where both circles intersect.

[28]The values correspond to Ptolemy's for the fourth climate; cf. *Alm.* II,8.

[29]Cf. al-Battānī, *Ṣābiʾ Zīj*, ch. 13.

[30]The values correspond again to Ptolemy's; cf. *Alm.* II,8.

[178] *Orizon autem, cuius latum* .pf. gradus .le. sexagenarie positum est, in quo etiam finis est habitationis, sicut proxime dictum est, *in quo et magnus dies .ld. habet horas, circulus enim Cancri totus apparet super orizontem,* et nox hieme .ld. horarum, circuitus nempe Capricorni totus sub orizonte latet, *aliter videt disposita orientalia.*
5 *In hoc enim orizonte circulus signorum iungitur circulo orizonteo omnibus sui partibus nec se quovis modo dividunt* semel *in unoquoque die duobus quandoque exceptis.* [179] *Facta vero ea, que dicta est, coniunctione motu supreme spere surgunt super orizontem sex signa simul et in momento.* Linea est enim circulus signorum circunductus. *Apparet ergo eam medietatem circuli signorum, que sic sub athomo temporis oritur, nulla ha-*
10 *bere orientalia.* Illa quoque, que hac oriente occidit, occidentalibus caret. Ut autem medietas circuli signorum altera sine orientalibus, altera carens occidentalibus in athomo temporis oritur et occumbit motu aplanos, illa quidem, que occubuit sine occidentalibus, oriri incipit habens orientalia totum rectum circulum, que surrexerat occumbit et cum ea occidentalia totus circulus. Fit huiusmodi circulorum
15 coniunctio polo signorum posito in puncto capitis, quod .lc. gradibus et .me. sexagenariis inclinatur a polo mundi et circa ipsum volvitur. Polus autem septemtrionalis altus est super terram .pf. gradibus .le. sexagenariis, quibus iunctis cum inclinatione fiunt .s.. Cum ergo polus signorum recte in meridionali positus est circulo, altior polo mundi punctum capitis tangit, et est in eo momento punctus
20 capitis Cancri in puncto septemtrionis, caput Arietis in puncto orientis, Capricorni in puncto austri, Libre occidentis in puncto. Oritur ergo simul illa pars, que est a puncto capitis Capricorni usque caput Cancri, alia simul occumbit, atque omni die fit semel duobus exceptis. Illis enim diebus bis, antequam sol iterum oriatur, circuli ambo iunguntur.

25 [180] *Orizon autem, cuius polus idem est qui et mundi, caret orientali et occidentali. Neutrum enim circulum oriri videt et occumbere.* Circulus spere semper circunfertur in circulo orizontis, circuli signorum altera medietas, que est a capite Arietis usque Libram, semper videtur, altera nunquam. [181] *In aliis autem orizontibus, qui inter hos duos habentur, quedam pars signorum non oritur, quedam non occumbit, et hec*
30 *quidem semper super hec vero sub terra est. He ergo non habent orientalia aut occidentalia.* [182] *Quod autem est inter semper visam et occultam, oritur et occumbit et habet orientalia sicut in aliis orizontibus. Quisque enim arcus earum partium oritur cum arcu*

---

22 puncto] pucto

[178] *But the horizon whose latitude is given as* 66;25 degrees, which is also the end of the inhabited region, as has just been said, *where the longest day lasts 24 hours, because the entire circle of Cancer is seen above the horizon,* and where in winter the night lasts 24 hours, because the entire circuit of Capricorn is hidden under the horizon, *sees the ascensions arranged differently. For at this horizon once every day, except for two days, the ecliptic falls together with the horizon in all its parts and the circles do not divide each other in any way.* [179] *But when this conjunction has occurred, six signs will immediately and simultaneously rise above the horizon by the motion of the highest sphere.* For the ecliptic is circumscribed as a line. *That half of the ecliptic which rises thus in an indivisible instant of time evidently has no ascensions.* And that half which sets when the former rises is deprived of descensions. But when one half of the ecliptic rises without ascensions and the other one is deprived of descensions, in an indivisible instant of time and by the motion of the aplanos, the one that set without descensions begins to rise having the entire equator as ascensions, while the one that had risen sets with a full circle as descensions. In this way a conjunction of the circles occurs when the pole of the ecliptic stands at the zenith, which is 23;35 degrees inclined from the pole of the world and revolves around the latter. But the north pole stands 66;25 degrees above the earth which, if added to the inclination, yield 90 degrees. Hence, when the pole of the ecliptic stands right on the meridian, it will be higher than the pole of the world and reach the zenith, and in that moment the point of the head of Cancer lies at the northern point, the head of Aries at the eastern point, the head of Capricorn at the southern point, and the head of Libra at the western point. Therefore, the part which is from the head of Capricorn until the head of Cancer rises at once, whereas the other part sets at once; and this happens once every day, except for two days. For on those days the two circles link twice before the sun rises again.

[180] *But the horizon whose pole is the same as that of the world lacks any ascension and descension. For it sees neither of the circles rise or set.* The celestial equator always revolves on the horizon, whereas the one half of the ecliptic from the head of Aries until Libra is always visible, and the other never. [181] *But at other horizons that are between those two, a certain part of the ecliptic does not rise and another one does not set, and one is always above the earth and the other always below. These parts therefore do not have ascensions or descensions.*[31] [182] *But what is between the always visible and the always concealed parts rises and sets and has ascensions like at other horizons.*

[31] Arab. add.: 'The corresponding part of the ecliptic is always hidden, and it [also] does not have an ascension in those places.' This sentence is also missing from all Hebrew manuscripts.

circuli recti. Hee autem, que oriuntur et occumbunt, partes divise sunt ab invicem. Altero quidem latere ea pars est, que semper videtur, altero ea que nunquam, et sunt .d.
f. 26v *arcus orientium* | *et que nunquam.* Et eorum quidem arcuum, qui non oriuntur, is qui videtur habet medium punctum mutationis estatis, alter qui semper occultus est
5 punctum mutationis hiemis. *Hii vero, qui oriuntur et occidunt, alter medius dividitur puncto equalitatis veris et ortus eius in his orizontibus conversus, occasus rectus. Alter in medio sui gestat punctum autumpnalis equalitatis,* [183] *cuius ortus in his orizontibus rectus est, occasus vero conversus.* Cuius rei causa est, quod polus circuli signorum transit punctum capitis et inclinatur in austrum. Tunc enim medietas orientium,
10 que est in parte Arietis in oriente posita, prius videt postremas sui partes oriri primis. Nimirum enim a septemtrione propinquiores sunt circulo orizontis. Surgente vero hoc arcu converso oppositus ei occumbit conversus; quo surgente recto et hic rectus occumbit. *Hec sunt, que de orientalium variatione* et ascensu *dicenda nobis ad presens videbantur.*

15 Sed quoniam hic liber in mediocrem iam pervenit sui quantitatem, in tercio de speris omnibus pauca, de sole autem et luna omnia persolvemus, que nostri mediocris tractatus, quem introducendis scripsimus, labori convenient. [184] *Quicquid autem de circulis imaginariis in superficie magne spere superius dictum est, in subiecta descriptione figuramus circulos et lineas in ea lineantes, ut ex eorum, que oculo-*
20 *rum iuditio subdita est, consideratione liceat inquirenti quecumque premissa sunt facilius imaginari.*

Explicit liber secundus; incipit tercius.

---

1 recti] signorum **2** altero] *corr. ex* altera (L) **3** oriuntur] moriuntur (L)

For every arc of these parts rises together with an arc of the equator. However, these parts that rise and set are separated from each other; on one side [between them] is the always visible part and on the other side the never visible one; thus there are four arcs, of those that rise and those that do never. And of those arcs that do not rise, the visible one has at its centre the point of the summer solstice, and the other, permanently concealed one, the point of the winter solstice. *Hence, of those that rise and set one is divided in the middle by the point of the spring equinox, and its rising in these horizons is inverted, whereas its setting is straight. The other one bears the point of the autumnal equinox in its middle,* [183] *and its rising in these horizons is straight, whereas its setting is inverted.* The reason for this is that the pole of the ecliptic goes beyond the zenith and inclines southwards. For then that half of the rising segments which lies towards Aries in the east sees its last parts rise before its first parts. This happens naturally, as they come closest to the horizon in the north. But while this arc rises inverted, the one opposite to it sets inverted, whereas when the latter rises straight, also the former sets straight. *This is what seemed necessary to us to be said for the moment, about the variation of the ascensions* and the ascendant.

But as this work has already proceeded to half of its length, in Book III we will speak only briefly about all the spheres, but about the sun and the moon as comprehensibly as possible and appropriate in our modest treatise, which we have written for novices. [184] *But all that has been said above about the imaginary circles on the surface of the great sphere is illustrated in the following diagram, in which we have drawn the circles and lines such that the enquirer may more easily imagine all that has been said in advance by considering what is presented to the judgement of the eyes.* [Fig. 2.4]

Here ends Book II, and Book III begins.

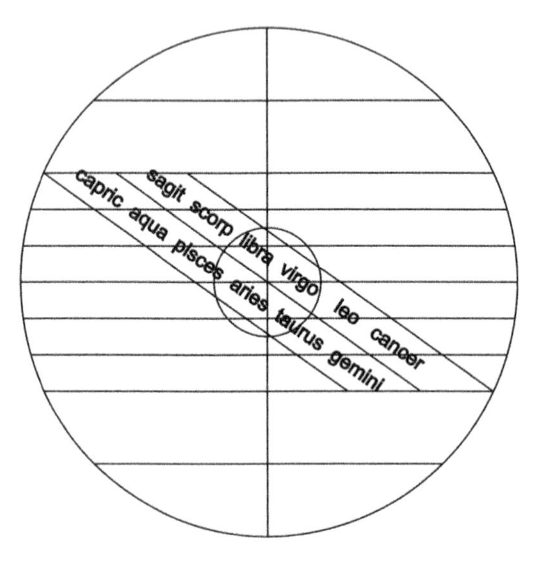

Fig. 2.4: From MS Cambrai 930, fol. 26v. The diagram, oriented south-up, shows the equator, both polar circles and the different declinations of the zodiac signs

Book III

D. Grupe, *Stephen of Pisa and Antioch: Liber Mamonis*, Sources and Studies in the History of Mathematics and Physical Sciences,
https://doi.org/10.1007/978-3-030-19234-1

<I>[(L)]am non minima propositi operis parte absoluta robustiores ad ea, que sequuntur, exurgimus, ut quem divina annuit benignitas exequendi operis transcursum, nostra non preterlabatur sine utilitate negligentia. Neque enim Epicureum aliquando dogma audivimus, sed Peripatetice potius accedimus claritati, que quamvis ad virtutem claritatem multiplici conexit rationis cathena affectum primo quidem scintillulis interlucentibus sed ex ipsius infantia defectum patientibus animum cogitantis transverberat. Post autem pleniori instantia referta circumfusis hominum vulgis lucem cogit infundere. Qua in re mirum quoddam de humani pravitate intellectus occurrit, quia hominum parte feliciori ad scintillas veritatis usquam visas tota velocitate sui ingenii currente, ut ex scintillis ignem possint iudicum excutere, pars quedam visu exterretur et tamquam ad fornacem fabri ignivomam pavet ex ignavia accedere. Pars vero ita modicum prius aspernatur lumen et quasi humile ac cito deficiens contempnit. Nonnulli vero, ne in lucem sibi ipsi prodeant, et audire fugiunt et videre. | [f. 27r] Unde fit, ut primis ad lucifluum calorem veritatis accedentibus secundi quidem ignavia torpentes ignavia iaceant; tercii autem clariora putantes ea, que manu attactant, ferali vivant immanitate aut spurca desidia. Quarti autem, ne sui ipsius iudices eorum fiant conscientie, tenebrarum malunt latibula quam in lucem prodire.

Nobis Dei gratia trium nichil instat. Hominis enim intelligens me habere celsitudinem, esse in imis refugio humanam ab horrore modestiam gravissimum facinus animadvertens mei ipsius in male contractis sepe iudiciaria sum correptus maiestate. Ex quo fit, ut otium contempnens secreta nature frequenti et multa rimari studeam investigatione. Sed quemadmodum ait Tullius temeritate G. I. Cesaris omnia suo tempore confusa in Romani imperii universitate fuisse, tametsi postea propagante altius diffusiusque Christi verissimo imperio radices sue composite sub Constantini proximeque sequentium imperatorum temporibus fuerint, nobis tamen sicut et tunc multoque durius contigit. Omnia enim apud nos quorundam temeritate, avaritia, superbia, ignavia, negligentia, falsis denique traditioni-

9 quia] qui (L) **13** humile] humilem **16** spurca] spurta **20–21** facinus] facimus (L) **21** animadvertens] *ante del.* ad (L) correptus] correptorum (L)

3.1 <Preface to Book III>

Since we have already completed a substantial part of the proposed work, we can now approach the following more vigorously, to make sure that the fulfilment of the work, though granted by divine mercy, will not fail uselessly by our own dormancy. For we have never responded to the Epicurean doctrine, but rather followed Peripatetic clarity which, by a complex chain of reasoning, connects clarity to any kind of virtue and shakes the thinker's mind, which is first affected by twinkling sparks, which nevertheless go out due to the infancy of the mind. But later, when more persistency is brought up, it makes light shine on the people around. In this, an astonishing thing occurs concerning the depravity of the human intellect. For whereas the more fortunate part of mankind, as soon as they discern the sparks of truth, run towards them with the whole swiftness of their mind, in order to strike the fire of judgement out of them, another part is frightened by the sight and languidly fears to come closer, like to a fire breathing hearth of a smith. Another part looks at the initially weak light and disregards it as something insignificant and quickly fading. And some, lest they step into the light, avoid hearing and watching. By consequence, while the first ones approach the shining[1] warmth of truth, the second, torpid by idleness, keep lying in idleness. The third, thinking that there is more clarity in objects which they can touch, live in beastlike primitivity or bawdy sluggishness. And the fourth, to prevent their consciences from becoming their judges, prefer the hideouts of darkness rather than stepping forth into the light.

By God's mercy, none of the three causes hinders us. For in the awareness that I am endowed with human celsitude and that fearful human self-restriction by hiding in lowness is a most severe offence, I am often gripped by my own judicial majesty in handling things badly. As a consequence, while despising otiosity, I am exerted to search out the secrets of nature by frequent and manifold investigation. But just as Tullius Cicero says that by Gaius Iulius Caesar's temerity everything in the whole Roman Empire was in confusion during his time,[2] and although later the truest empire of Christ extended even higher and further, with its foundations laid during the reign of Constantine and the emperors who were his immediate successors, we nevertheless suffer again from equal and even worse confusion

[1] 'lucifluum': cf. Iuvenc. 3, 294 and 4, 120.

[2] Cic. *off.* I,26: 'Declaravit id modo temeritas C. Caesaris, qui omnia iura divina et humana pervertit propter eum, quem sibi ipse opinionis errore finxerat principatum.'

bus permixta sunt, recessit pudor, fugit munditia, deest fides, sacramenta diffluxerunt. Sed de his alias. Verum hec pauca diximus, ut significaremus nostro tempore fugatis virtutibus vitiorum tumultus invaluisse.

Quare multas esse improborum cum constet turbas, proborumque pauci nu-
5 mero magni re habentur, decet invigilare ad confutandas diaboli ingenii nequitias. Solus enim, pro pudor, humanus animus deviat et a natura degenerans invenitur, cum post occasum angeli omnia naturam et indicti concordiam ordinis tueantur. Hoc lapides, metalla, plante, bruta animantia, hec omnia celestia corpora, de quibus intendimus, perpetua servant immobilitate. Que res quomodo, cur et a quibus
10 facta sit, aliud tempus disserendum. Nunc de propositis.

Hic igitur liber ordinem sperarum et de sole et luna circulos et speras, eclipsis utriusque rationes continebit. Verum artiori studio hec, et que postea sequentur, prosequemur. In duobus primis ideo diffusius pervagavi, ut lectoris intellectum admitterem. Nunc vero fundamento operis posito perpolitam volumus partem sur-
15 gere, quod et ipsa intuentium oculos magis capiat, et structure nobilitate diutius apud se animos audientium conversari compellat.

Est igitur ordo, quas superius sperarum posuimus, huiusmodi. Ultima, que nobis prima est et a ceteris omnibus clauditur, spera lune vocatur. Prima vero, que nobis ultima et ceteras omnes circumtenens, nona, que nullam habens in se stel-
20 lam omnia lucem ministrantia corpora sua celeritate in occidentem retorquet. Ab ea secunda, quam octavam dicimus, spera fixarum et signorum infra nonam clauditur. Tercia est, quam celum Saturni dicunt. Quartam autem Iovis blandum sidus irradiat. In quinta autem Martiana stella rutilat. Et de harum quidem ordine nulla est apud antiquos controversia. Unde de tribus, que restant, non eque omnes
25 sentiunt. Quidam namque, quorum unus Macrobius esse deprehenditur, solem a luna secundum ponunt, secundum Mercurium, tercium Venerem. Qua in re sibi ipse Macrobius adversari videtur. Ait enim de ordine eorum disserens: «circulus

---

14 perpolitam] per politum **24** antiquos] aliquos (L)

than in that former time. For in our days everything is corrupted by certain people's temerity, greed, arrogance, idleness, negligence or false traditions, while at the same time a sense of shame has vanished, decorum takes flight, trust is lacking and the sacraments have disappeared. But about these things elsewhere. We only have introduced this shortly to make clear that in our time, since virtues have been expelled, the cohort of vices has grown strong.

Since the crowds of the nefarious are thus undeniably many, whereas the righteous are few in number but of great value, it behoves us to watch out and fight back the devil's cunning. For, oh! shame, only the human mind goes astray and is found degenerating from nature, whereas everything else, after the fall of the angel, preserves its nature and the harmony of the enacted order. The stones, metals, plants, non-intelligent animals,[3] and all the heavenly bodies, which are our concern here, observe it in eternal consistency. How, why and by which means this is arranged must be discussed on another occasion. Now to the proposed subject.

The present book will thus contain the order of the spheres, the circles and the spheres relating to the sun and the moon, and the reasons for an eclipse of either of them. But we will pursue this and also what follows with more stringent effort. For in the first two books I have touched upon subjects in a more diffuse way to admit the reader's intellect. With the foundations of the work laid, we now want to raise the storey neatly, for then it catches the spectators' eyes more and by the gracefulness of its composition induces the minds of the audience to sojourn longer in its proximity.

The order of the spheres which we mentioned above is thus as follows. The last one, which is the first from us and which is enclosed by all others, is called the sphere of the moon. But the first one, which is the last one for us and which encloses all others, is at the ninth place, which has no stars on itself but by its velocity turns all shining bodies westwards. Enclosed by this sphere is the second one, which we call the eighth one, being the sphere of the fixed stars and the constellations, beneath the ninth. The third one is that which is called the heaven of Saturn; the fourth one is illuminated by Jupiter's charming star. In the fifth, the red star of Mars is shining. About the order of these spheres there is no dispute among the ancients. However, there is no general consent concerning the three remaining ones.[4] For some, and Macrobius is found among them,[5] put the sun in

[3] 'bruta animantia': cf. Greg. *in Iob* 10,13,23 et al.

[4] Cf. *Alm.* IX,1.

[5] *Somn.* I,19,6–10 (L).

solis a Mercurii circulo ut inferior ambitur, illum quoque superior circulus Veneris includit.» Assignans autem cuique spere propriam quantitatem Venerem Mercurii spere supposuit dicens: «quantumque a terra usque Venerem, quater tantum sit a terra usque Mercurii stellam.» Quibus verbis Venerem Mercurio et terre interpositam necessaria argumentatione iubemur credere. Sed in horum disponendo ordine insani cerebri invenitur. Quare de sole quid verissime credendum sit, videamus. Solem, ut Cicero | tradidit, inter planetas medium obtinere locum, sub Martia stella volvi occulta quidem et difficilis inventu ratio est, nec sine geometria assignari poterit. Quare rationis demonstrande si divina concedet pietas, aliud tempus veniet. Quod autem per ipsam inveniatur, dicemus.

Linea, que de centro mundi egreditur, per centrum solis directa in superficie spere signorum punctum tangit. Imaginamur autem et aliam lineam de pupilla oculi egredientem et transeuntem per idem centrum solis usque in superficiem eiusdem spere. Tangunt harum capita linearum duos punctos in diversis partibus spere, vocaturque spatium, quod inter est, discordia visus. Variatur autem hec in eadem stella, et maior mane, meridie minor, sepe nulla invenitur. Unde et in aliis signorum locis sepe planete oriri videntur, et in aliis eos ratio numerorum invenit. Est vero huiuscemodi discordia visus maior in his, que propius nobis accedunt, planetarum globis, minor in remotioribus. Quanto propinquiores nobis sunt, tanto amplius in diversa linee sese invicem secantes feruntur. Deprehensum autem a probatissimis in astronomia viris, Ptolomeo et reliquis, cognovimus maiorem hanc esse visus discordiam <in>[(L)] Mercurio quam in Venere, Veneremque maiorem habere sole. Nam luna quidem ceteris omnibus inferior longe maiorem habet. Qua ex re manifestum factum est solem quarto moveri loco. Docet hoc subiecta oculis descriptio. Recedant igitur supersticiosa, que et de ordine planetarum falsa dixerunt, et quare omnia Deus sic fecerit se putarunt invenisse. In astronomia satis est, quod factum et quomodo dispositum sit invenire, non etiam quare.

---

1 Mercurii] *add.* (L) **4** sit] fit **11** egreditur] regreditur **19** remotioribus] remotionibus (L) **21** Ptolomeo] Tptolomeo

second place after the moon, followed by Mercury, and Venus on third place. In this respect Macrobius is found to contradict himself. For when he discusses their order, he says: 'The circle of the sun is surrounded, as the lower one, by that of Mercury, and also the higher circle of Venus encloses it.'[6] But when he assigns to every sphere its particular size, he subordinates Venus to Mercury, saying: 'The distance from the earth until Mercury is four times the distance from the earth until Venus.'[7] From these words we are forced to believe by necessary consequence that Venus lies between Mercury and the earth. But in this arrangement of their order we find the product of a deranged mind. We shall therefore investigate which theory of the sun is the most plausible. That the sun, as Cicero says,[8] has the central place among the planets and revolves underneath Mars is difficult to prove and cannot be determined without geometry. Therefore, if the divine mercy concedes, we will demonstrate the argument on another occasion. Nonetheless, we will say what is found through that argument.

A line that originates from the centre of the world and is directed through the centre of the sun touches a point on the surface of the sphere of the constellations. But we imagine also another line originating from the pupil of the eye and passing through the same centre of the sun until the surface of the same sphere. Then the ends of these lines touch two points at different places of the sphere, and the distance between them is called 'parallax'. But the latter changes for the same star, being greater in the morning, smaller at noon, and often totally vanished. For the same reason, planets often seem to rise in some signs, whereas numerical calculation finds them in other signs. But the parallax of this kind is greater for the bodies of planets that come closer to us, and smaller for the more distant ones. Thus the nearer to us they are, the more diverge the intersecting lines. But from the most respected astronomers, namely Ptolemy and the others,[9] we know that this parallax is greater for Mercury than for Venus, and that Venus has a greater one than the sun. And the moon, lower than any others, has by far the greatest. From this it has been made evident that the sun moves at the fourth place. This is demonstrated to the eyes by the diagram below [Fig. 3.5]. Those who have proclaimed superstitious errors about the order of the planets, and who have come to believe to have found the reason why God created everything in that way, may therefore step back. In astronomy it is enough to find what has been created and how it is arranged, but not also why.

---

[6]*Somn.* I,19,6 (L).

[7]*Somn.* II,3,14: 'quantumque est a terra usque ad Venerem, quater tantum sit a terra usque ad Mercurii stellam.'

[8]Cic. *Somn.* 4.

[9]In *Alm.* IX,1 Ptolemy denies a noticeable parallax of the planets. In contrast, the system of contiguous spheres which he presents later, in the *Hypotheseis*, implies a considerable parallax at least for Mercury.

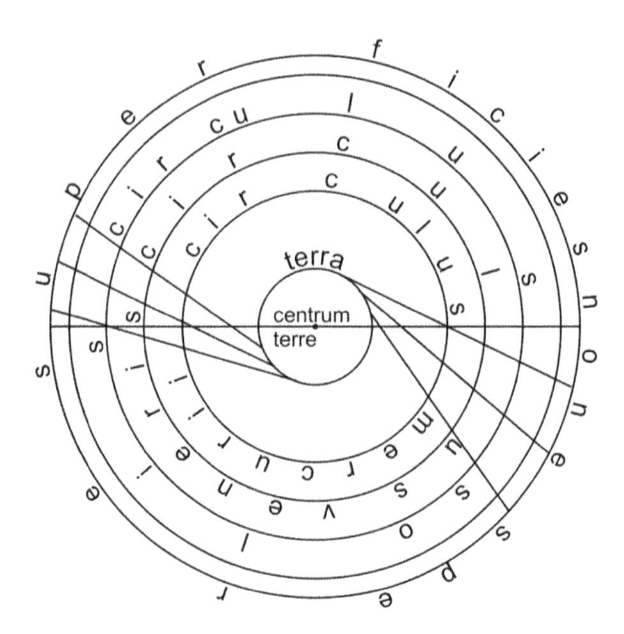

Fig. 3.5: From MS Cambrai 930, fol. 27v. Labels: "surface of the ninth sphere" (superficies none spere), "circle of the sun/Venus/Mercury" (circulus solis/Veneris/Mercurii), "the earth" (terra), and "centre of the earth" (centrum terre). In the manuscript, the label 'terra' is wrongly placed on the circle of Mercury. The diagonal lines are probably misplaced, too; to illustrate the argument, different lines should start from the same observer's position and cross the different planetary circles at their intersection points with the line through the centre

¶De ipsa quoque sperarum quantitate, utrum ea sit, quam quidam dixerunt, videre ad presens hinc licet. De earum quantitate apud Macrobium ita scriptum est: «Tantum est», inquit, «a luna usque solem, quantum a terra usque lunam, quantumque a terra usque solem, triplum est usque ad Venerem.» Que spatia cuius sint proportionis, ita videbimus. Inter terram et lunam .a. gradus esse concedamus; duplum a terra usque solem, id est .b.; triplum huius a terra ad Venerem, id est .f.; quadruplum autem huius ad Mercurium, id est .ld.; novies .ld. usque Martem, que sunt .ukf.; octies autem ducenta .kf. usque Iovem, scilicet mille septingenti .lh., qui vicesies septies multiplicati spatium a terra usque Saturnum reddunt .466g6.. De | quibus sublatis .tupa., scilicet spatio a terra usque Iovem, remanet a Iove usque Saturnum .44qpa.. Sublato de spatio Iovis spatium Martis restat a Marte .igip. usque Iovem. Sic de reliquis. Est igitur spatium a luna usque solem .a. gradus, a sole ad Venerem .d., ad Mercurium .kh., a Mercurio ad Martem .tsb., a Marte ad Iovem .igip., a Iove ad Saturnum .44qpa., cui reliquis omnibus spatiis iunctis prior surgit numerus.

11 .44qpa.] *add.* (L) .igip.] *add.* (L)

Now we may also see whether it is true what some people say about the size of the spheres. In Macrobius the following is written about their size: 'The distance from the moon until the sun is as much as the distance from the earth until the moon, and the distance until Venus is three times as much as the distance from the earth until the sun.'[10] We can see the proportions of these distances in the following way. Let us assume that there is 1 degree between the earth and the moon; the double from the earth until the sun, that is, 2 degrees; the threefold of this from the earth to Venus, that is 6; but the fourfold of this to Mercury, that is 24; nine times 24 until Mars, which is 216; but eight times 216, namely 1728, until Jupiter, which, multiplied by 27, yields 46,656 as the distance between the earth and Saturn. If we subtract from this 1728, which is the distance between the earth and Jupiter, the distance between Jupiter and Saturn remains as 44,928. If we subtract the distance of Mars from that of Jupiter, there remains a distance of 1512 from Mars until Jupiter. In the same way we proceed with the rest. Hence the distance from the moon until the sun is 1 degree, from sun to Venus 4, to Mercury 18, from Mercury to Mars 192, from Mars to Jupiter 1512, from Jupiter to Saturn 44,928. If we add all other distances to this last one, there will result the above number.

[10] *Somn.* II,3,14: 'quantum est a terra usque ad lunam, duplum sit a terra usque ad solem, quantumque est a terra usque ad solem, triplum sit a terra usque ad Venerem.'

Que spatiorum assignatio multis rationibus improbatur. Si enim est, ut idem dicit, una eademque omnium celeritas, duplo temporis sol suum peragraret circulum, quo luna suum circuit; duorum etenim circulorum si alterius diametrum duplum sit diametro alterius, et circuli sic se habent. Peragraret igitur, si vere essent assignata spatia eademque citatio, sol .b. mensibus totum zodiacum, Venus .f., Mercurius .ld., Mars .ukf., Iupiter .tnd. annis, Saturnus .Ψaad. annis. Que cum non ita sint, aut falsa est sperarum assignatio aut celeritas non erit eadem.

Celeritatem quoque non esse eandem apud eundem facile est inveniri, ubi de celesti armonia loquitur. Postquam enim dixit vocem in aere ex percussione fieri, acutam autem et gravem pro modo et impetu percussionis, ait acutiores sonos ab extremis et altioribus speris emitti quam ab infimis. Quod si verum est, non eadem celeritas. Nam si velocitas <eadem>[(L)], pari impetu aere percusso equaliter consonantes emitterent voces. Sed hee quidem graves, ille acutas. Non igitur pari feruntur velocitate. Socialis etiam cursus solis, Veneris et Mercurii non parem ipsis inesse festinantiam comprobat. Nam cum non sint in eadem spera positi, sed solis spere Veneria Veneris Mercuriana supposita sit, et he vel modicam concedantur habere quantitatem diversam, cuius brevior est circulus, paucioribus debet Venere circumduci Mercurius, pluribus autem sol, Venere medium tenente. Sed cum non sit hoc, nec illud. Non enim etiam illud concedimus, quod apud quosdam invenitur, Mercurium paucioribus Venere, Venerem sole diebus suis circummeare circulis. Toto namque hemisperio, si constaret illud, a sole secederent sicut et alii planete; non relicti quidem sicut III[es] altiores, Mars, Saturnus, Iupiter, tardi a velocitate solis, sed tardiores luna, citiores sole, usquedum opponerentur precederent, exinde consequerentur. Que non sint, nec premissa. Neque enim Venus ante aut post a sole plus quam .nh. gradibus, Mercurius autem plus quam .lf. possunt recedere. Iterum si precesserint, tardi et retrogradi, si subsequentur, velociores ad ipsum revertuntur. Quod rotunditate motu non maiorum sperarum fit iniuria. Patet ergo clarius per premissas argumentationes Macrobium falsa et sibi adversantia de ordine et spatiis et velocitate dixisse, et nunc hec, nunc illa, sicut insani capitis consuetudo est. Et de ordine quid sit, de spatiis, in sequentibus pauca, alias

---

6 annis] *suprascripsit* et b mensibus **21** secederent] secederet (L) **27** fit] *add.* (L) **29** insani] *add.* (L)

This ascription of the distances can be refuted by many arguments. For if they all have one and the same velocity, as he [*scil.* Macrobius] says,[11] it would take the sun twice as long as the moon to complete its circuit. For, if one of two circles has a diameter twice the diameter of the other, also the circles [i.e., their circumferences] have that proportion. Hence, if the ascribed distances and the statement of an equal velocity were correct, the sun would pass through the entire zodiac in 2 months, Venus in 6, Mercury in 24, Mars in 216 months, Jupiter in 144 years, Saturn in 3,888 years.[12] Since this is not the case, either the ascription of the spheres is wrong or their velocity is not equal.

And we can readily read from the same author that the velocity is not equal, where he speaks about the celestial harmony. For after saying that the sound in the air is made from impact, and high and low pitches are produced according to the type and intensity of the impact, he says that the outermost and higher spheres produce sharper sounds than the lower ones.[13] If this is true, the velocity is not the same. For if the velocity were the same, the spheres would affect the air by equal impact and thus emit equal sounds. But some sounds are heavier, others sharper. Therefore, they are not moved by the same speed. Furthermore, the cohesive course of the sun, Venus and Mercury attests that they do not have equal celerity. For, as they are not located in the same sphere, but Venus' sphere is placed below the sun's and Mercury's sphere below that of Venus, they must be admitted to have, at least minimally, different sizes, and Mercury, whose circuit is shorter, must revolve in a shorter time than Venus, and the sun in a longer time, while Venus holds the mean. But, since this is not the case, neither the other is true. Nor can we approve of that statement which is made by others, according to which Mercury revolves on its circle in fewer days than Venus, and Venus in fewer days than the sun. For if this were true, they would remove from the sun by a complete hemisphere, as do also the other planets; but not like slow bodies left behind by the speeding sun, as happens to the three higher planets, Mars, Saturn and Jupiter, but, slower than the moon, yet faster than the sun, they would precede it until they stand in opposition, and from then on would be chasing it. Since this is not the case, neither is the proposition. For Venus cannot fall behind or get in front of the sun by more than 48 degrees, and Mercury by no more than

[11] *Somn.* I,14,27 and I,21,6–7.

[12] '3,888 years': a superscript gloss reads 'and two months'.

[13] *Somn.* II,1,5–12 and (L)II,4,2–7.

plenius; de velocitate in posterioribus et suis locis disputabimus. Nunc vero, qua ratione ab oriente in occidentem ferri probentur contra nitentes, videamus.

¶In orientem planetarum .g. speras ab occidente rotari omnium est antiquorum, qui philosophie huic acutius et frequentius studuerunt, sententia. Ad cuius 5 comprobationem Macrobius tale induxit argumentum. Sole posito in capite Arietis cum occumbit, Libram videmus oriri. Lapsis autem .m. diebus in occasu solis non caput Libre sed Scorpii oritur, quia non iam sol in Ariete est, sed ad Taurum migravit. Taurus autem post Arietem oritur, et illi ab oriente est. Ex quo manifestum | est solem non in occidentem sed contra moveri. Similis est et in reliquis [f. 28v] 10 .f. ratio. Quo contra quidam se nulla videre ratione, si in orientem nitantur, quod eos ad occasum referat, cum ab aplanos corpore longe suppositi ferantur, et fieri posse, nisi firmamento superius citissime in occidentem currente superius corpora planetarum tardius volvantur. Cui rei illud quoque opitulari videtur, quod Macrobius dicit, ideo suppremam speram velocissimo cursu moveri, quia maiorem 15 habeat spiritus impetum, que terre propius accedunt spere minorem, ideoque non tanta moveri celeritate. Que cum dicunt, videntur sibi se et Macrobii et alias dissolvisse rationes, verum falso. Omnis que movetur spera celestium duos habet, ut primus liber continet, fixos polos, qui nunquam sue spere motu moventur. Si enim moverentur, non essent poli. Quod autem in duobus fixis polis moveantur, illud 20 nobis signo est, quod non fortuito et mutabili sed certo moventur motu.

Hii, ut de sole videamus, in spera solis aut polis mundi suppositi sunt, aut ab

---

3 speras] spera **11** aplanos] amplanos (L) **12** occidentem] occidente (L) **16** Macrobii] Macrobius

26 degrees. Furthermore, they always return to the sun by becoming slow and retrograde if they have preceded the sun, and by accelerating if they lie behind. This is caused by the motion of the epicyclic sphere[14] and not by a violation of the larger spheres. Hence, from the above arguments it has become totally clear that Macrobius has made wrong and contradictory statements about the arrangement, the distances and the velocity, now saying this, soon that, as is the habit of an insane mind. Concerning the actual order and the distances, we will speak briefly about this in the following, but more comprehensively elsewhere; the velocity will be discussed later at its proper place. Now let us try to explain why bodies move from east to west, although they tend in the opposite direction.

That the seven planetary spheres rotate from west to east is the opinion of all the ancients, who pursued this philosophical discipline with much astuteness and effort. To prove this, Macrobius has presented the following argument.[15] When the sun sets while standing in the head of Aries, we see Libra rise. But 30 days later, at the moment of sunset, it is not the head of Libra but that of Scorpio that is rising, because the sun stands no longer in Aries but has wandered to Taurus. Taurus, however, rises after Aries and lies east of it. It is clear from this that the sun does not move westwards, but in the opposite direction. The same applies to the remaining six planets. Some people counter that they could not see a way how the planets could move eastwards, since they lie far beneath the aplanos that takes them westwards, except that, while the firmament above moves westwards very swiftly, the planetary bodies move slower. This seems to be supported by Macrobius' statement when he says that the highest sphere moves fastest, since it has a stronger impelling breath, whereas the spheres closer to the earth have less impulsion and therefore move at a lower speed.[16] By saying so, they believe that they had reconciled Macrobius' and other theories; but, they are wrong. For, every moving heavenly sphere has two fixed poles, which are never moved by the motion of their own sphere, as is contained in Book I. For if they were moved, they would not be poles. But as they move around two fixed poles, this gives us evidence that they do not move randomly or with a changing motion, but with a regular motion.

If we consider the sun, the poles of the solar sphere can either be placed under

[14] 'epicyclic sphere': inverse to modern usage, Stephen uses the simplex, 'rotunditas', for the epicyclic sphere and a derivative, 'rotunditatis circulus', for the epicycle; cf. chs 3.2 and 4.3 of the Introduction.

[15] *Somn.* I,18,12f (L).

[16] *Somn.* II,4,4 (L).

his inclinantur. Si polis poli suppositi, non erit tanta temporum varietas ac dierum discordia, sed equinoctium. Semper enim in equinoctiali circulo sol esset, cum sit in medio sue spere positus circulus, in quo moveri habet. Si autem inclinantur et suo motu spera moveatur tantum, ratio non procedet. Nam aut suo motu movetur
5 tantum, aut alterius, aut alterius et suo. Si alterius tantum, in eodem semper parallello aut equinoctiali sol rotabitur. Si suo tantum et poli inclinentur, fit ut in eadem nocte et die .d. fiant tempora. Conatur enim esse in Capricorni capite, cum oritur suppreme spere quadrans unus, et amplius oriatur, .s. scilicet gradibus et .ke. sexagenariis. Surrexit ergo et quadrans integer circuli solaris. Que res si motu sue spere
10 facta est, cum habeat fixos polos nec moveantur, solem ad equinoctium surrexisse invenies. Surgente autem secundo quadrante spere surget secundus solis, eritque nobis positus in Cancri capite in suo occasu. Inde per noctis tempora eadem ad Capricornum relabetur ratione, sicque fieret, ut omni die estas, hiemps, ver et autumpnus nobis imminerent. Que cum non sint, suo et alterius motu movebitur.
15 Quod si alterius tardo in orientem, suo citius in occidentem, idem nobis occurret inconveniens. Si vero semel in die suppreme motu spere in occidentem, et suo similiter sed paulo minus, fieret ut ambobus coniunctis in uno circuitu suppreme geminos haberemus meridies. Motu quoque ipsius in occidentem uno gradu concesso ab Ariete non in Taurum sed in Pisces descenderet. Quibus omnibus amotis
20 restat, ut impetu suppreme ab ortu ad occasum relatus sue quidem motu spere ab occidente in orientem in anno pleno nitatur circuitu.

Similia de reliquorum planetarum globis et speris probantur, nisi quod ipse sepe videntur retrogradi. Quod non maioris spere facit motus, ferentes enim spere semper in orientem nituntur, sed rotunditatum, in quibus positi sunt, revolutione.
25 Que res in sequentibus manifestius patebit.

Hec ergo est ratio argumentatioque, quibus planetarum nisus contra firmamentum comprobantur. Ideo autem multiplices contulimus in unum probationes, ut si qui aliud de planetarum motu debacari voluerint, quibus argui possint, habeantur rationes. Et quoniam communiter de omnium planetarum speris, que dicenda
30 videbantur, breviter perstrinximus, iam nunc de singulis dicamus.

---

18 occidentem] occidente (L) **19** in] ut (L) **24** rotunditatum] rotunditatem (L)

the poles of the world or be inclined from them. If the poles were placed under the other ones, there would not be that much variation between the seasons and difference between the days, but an equinox. For the sun would always be on the equator, because the circle on which it moves lies in the middle of its sphere. But if they are inclined and the sphere moves only by its own motion, this leads to an impasse. For the sphere is necessarily moved either by its own or another sphere's motion only, or by its own and also by another sphere's motion. If it is moved only by another sphere's motion, the sun will always revolve on the same parallel circle or the equator. Whereas, if it moved only by its own motion with the poles being inclined, all four seasons would occur in the period of just one night and day. For it had to stand in the head of Capricorn when one quadrant of the highest sphere rises, and it had to rise further, namely by 90;15°.[17] Hence there has also risen a complete quadrant of the solar circle. If this happened by the motion of its own sphere, since it has fixed poles which are not moved, one would find the sun risen to the equinoctial point. And while the second quadrant of the sphere rises, the sun's second quadrant will also rise, and the sun would stand in the head of Cancer for us when it sets. From there it will correspondingly return to Capricorn during the night; thus we would be exposed to summer, winter, spring and autumn every day. Since this does not happen, it must be moved by its own and by another sphere's motion. Further, if it was moved by the other sphere's slow motion to the east and by its own fast motion to the west, we would experience the same unsuited situation. And if it revolved westwards once in a day by the motion of the highest sphere and equally, but a bit less, by its own motion, the combination of both motions would cause noon twice during one revolution of the highest sphere. And if we conceded a solar motion of one degree westwards, it would not leave Aries towards Taurus but descend to Pisces. Having excluded all these cases, there remains that the sun is taken from rising to setting by the momentum of the highest sphere, and, by its own sphere's motion, performs one complete revolution from west to east during the period of one year.

The same can be proved for the remaining planetary bodies, except that they often seem to retrograde. But this is not caused by the motion of a larger sphere, since the deferent spheres always revolve eastwards, but by the turning of the epicyclic spheres in which the planets are located. This will become clearer in the following.

This is the reason and the explanation of the planets' tending against the firmament. And we have brought together multiple proofs, so that whatever some might fantasise about the motion of the planets, there will be arguments available to refute them. And as we have shortly summarised what seemed to apply generally to the spheres of all planets, we will now discuss them individually.

[17]The fractional part provides for the sun's course on the ecliptic by about 1° per day, which corresponds to 0;15° per quadrant if a full circuit is assumed to last one day, as in the present scenario.

f. 29r [185] *Dividitur uniuscuiusque planete | spera, que claudit inferiora, in partes sperales. Ostendetur hoc de singulis suo tempore, in quot partes et quo modo dividantur. Nunc ergo a solari, que ceteris dignior et claritate ipsius notabilior est, incipiemus.* In quo illud quoque nobis erit commodo, quod et pauciores habet partes et facilior eius
5 intellectus ad reliquorum difficultatem viam parabit.

[186] ***E****st igitur spera solis sperale corpus, quam circuntenent due sperales paralelleque superficies. Earum centrum idem est, quod est mundi. Altera exterior est, altera interius habetur. Alta et exterior eius superficies iungitur intime et ime superficiei spere Martis, interior autem et ima exteriori alte superficiei spere Venerie copulatur.* [187]
10 Inter huiusmodi superficies corpus spere solaris est, de quo separatur quoddam alterius spere, in ipsa tamen spera solis et connumeratum et contentum. *Hec etiam spera, que seiungitur et fit de illa, rotunda est rotundas habens superficies, que ipsam circumterminant, altera exterius, altera vero interius. Harum centrum non idem est quod totius spere, sed punctus extra centrum mundi positus. Superficierum huius spere altera,*
15 *que alta est, tangit altam superficiem prime spere in uno puncto, interior ima interiorem et imam prime spere in uno tantummodo puncto. Ex eo igitur, quod extra centrum mundi huius spere centrum positum est, ipsa quoque spera extra centrum vocatur. Motus autem huius spere speralis circuitus est ad orientem ab occidente super duos fixos polos et fixum polorum radium. Non sunt autem hii poli aut radius poli mundi aut radius,* sed horum
20 uterque a sibi proximo mundi polo .lc. gradibus .me. sexagenariis distat. Unde fit horum radium a mundano radio in uno puncto secari. *Sed neque duo poli signorum dicendi sunt,* tametsi in radio polorum spere signifere positi sint, unde et radius eorum in radio signorum positus est.

[188] *Sol autem est corpus sperale solidum positum in corpore spere extra centrum.*

---

8 Alta] altera (L) ime] une **9** ima exteriori alte] una exterior late **15** alta] altam ima] una **16** imam] unam **24** autem] *ante del.* a (L)

[185][18] *The sphere of each planet, which encloses what is below, is divided into spherical parts. We will show this for the single ones at their time, how and into how many parts it is divided. We will now start with the solar sphere, which is the noblest and, due to the brightness of the sun, the most prominent one.* In this regard it will also be convenient for us that the sphere of the sun has the least parts, and that its easier understanding will pave the way to the complexity of the others.

3.2 <The Sphere of the Sun>

Config. 9: The Orb of the Sun

[186] *The sphere of the sun is a spherical body that is enclosed within two spherical and parallel surfaces. Their centre is the same as that of the world. One is the outer surface, and the other is considered the inner surface. Its high and outer surface is adjacent to the innermost and low surface of the sphere of Mars, whereas its inner and low surface is connected to the outer high surface of the sphere of Venus.* [187] The body of the solar sphere is placed between such surfaces, from which the body of another sphere is separated, although it is completely allocated and contained within the former. *That sphere, which is seperated from the first and originates from it, is also a round sphere with round surfaces that define it around, one outwardly and one inwardly. Their centre is not the same as that of the entire sphere, but a point that is placed outside the centre of the world. One of the surfaces of this sphere, which is the high surface, touches in one point the high surface of the first sphere, whereas its inner low surface touches the inner and low surface of the first sphere in only one point. Since the centre of this sphere is placed outside the centre of the world, it is also called the 'eccentric'*[19] *sphere. The motion of this sphere is a spherical circuit from west to east, around two fixed poles and a fixed axis of the poles. But these poles or the axis are not the poles or the axis of the world,* but each of them is distant by 23;35 degrees from that pole of the world which lies nearest to it. Hence their axis is crossed by the cosmic axis in one point. *However, the two poles should also not be called the poles of the signs,* although they are placed on the axis through the poles of the sphere that bears the signs, for which reason also their axis is placed on the axis of the ecliptic.[20]

[188] *The sun, however, is a solid spherical body that is placed within the body of the*

[18] In Arab. Y this passage is part of the previous chapter 8, whereas it is missing in K. In H1, H6 and H8 it appears separate, as a transition to the discussion of the planetary spheres, similar to Stephen's treatment of it; cf. Langermann's note to this passage.

[19] 'eccentric': lit., 'outside the centre'.

[20] The insertion clarifies a confusing formulation in the Arabic; cf. Langermann's note to this passage.

Est autem eius talis positio, quod centrum corporis solaris equaliter, id est .s. gradibus, *distat ab utroque polo spere extra centrum, et speralis quidem superficies solis ambas tangit superficies spere extra centrum in duobus punctis altero alteram. Illi ergo duo puncti duo sunt capita totius diametri solaris.* Ex quo manifestum fit longitudinem diametri solis equalem esse spisso spere extra centrum. *Solis igitur tanta est quantitas, quanta est spissitudo spere extra centrum. Quoniam autem spera hec ab occidente in orientem movetur super duos fixos,* non tamen immobiles, polos, moventur enim motu firmamenti circa polos mundi, *suo motu corpus solis tamquam inter suam positum spissitudinem et quasi sui quamdam partem movet in orientem.*

[189] *Solis igitur centrum ab occidente in orientem rotundo motu circummeans* per motum spere extra centrum, que etiam ferens dicitur, *facit* in corpore ferentis *spere imaginarium circulum,* et centrum eius centrum eiusdem spere. *Dicitur ergo et hic circulus* ab antiquis *circulus extra centrum,* ideo quia centrum eius extra centrum mundi positum est. [190] *Imaginamur autem huius circuli superficiem in eam crescere latitudinem et imanitatem, ut totum mundum in duo media dividat et secet. Facit ergo circulum in superficie* none *spere, quem circulum signorum dicimus, sicut superius iam demonstratum est,* dum de zodiaco loqueremur. *Fiunt etiam post hanc sectionem circularis superficiei duo circuli in superficiebus solaris spere, quam centrum mundi habere diximus, qui et paralelli sunt, et centrum idem cum spera, in qua sunt, et mundo sortiuntur. Horum maior,* | *qui et exterior in exteriori superficie circumducitur, circulus similis vocatur,* eo quod similis signorum circulo, qui est in superficie alte spere, et hic similis in illius superficie positus sit.

[191] *Facit etiam duos paralellos circulos in duabus superficiebus spere extra centrum, que et ferens dicta est. Hii ad invicem et ad circulum extra centrum collati paralelli inveniuntur, quia unum ferentis spere habent centrum. Horum maior similem signorum circulo tangit circulum in uno puncto, minor tangitur a paralello simili, qui est in interiori totius spere superficie, in alio puncto.*

[192] *Circinat etiam eadem sectio superficiei circuli extra centrum crescentis in unaquaque stellarum spera et aplanos, que sunt* IX *et paralelle habentes unum centrum cum mundo, duos paralellos circulos in quaque superficie* inde et inde, idem est .kh., *quorum omnium centrum idem est et mundi. Maior autem eorum, qui in superficiebus cuiusque circumducitur spere, dicitur circulus eiusdem stelle, in cuius spera positus est, similis signorum circulo. Iunguntur autem hii circuli maiores quidem sperarum minorum*

---

9 sui] *add.* (L) **16** circulum] circulorum (L) **26** qui] quod **28–29** unaquaque] unaque **32** spera] spere

eccentric sphere. Its position is such that the centre of the solar body is equally distant from each pole of the eccentric sphere, i.e. by 90 degrees, *and the sun's spherical surface touches both surfaces of the eccentric sphere in two points, each surface in one point. Hence these two points are the ends of the entire solar diameter.* It becomes clear from this that the length of the sun's diameter is equal to the thickness of the eccentric sphere. *Hence the size of the sun equals the thickness of the eccentric sphere. But since that sphere moves from west to east around two fixed,* but not unmoved, poles—for they are moved by the motion of the firmament around the poles of the world - *the solar body, embedded, as it were, in its thickness and like a part of it, moves eastwards by its motion.*

[189] *Hence the centre of the sun, revolving from west to east in a circular motion* by the motion of the eccentric sphere, which is also called the deferent sphere, *produces an imaginary circle* in the body of the deferent sphere, and the centre of that circle is the centre of that same sphere. *Therefore, also this circle is called the eccentric circle* by the ancients, because its centre is placed outside the centre of the world. [190] *But we imagine the plane of this circle to be extended to such a huge width that it cuts and divides the entire world into two halves. Thus it produces a circle on the surface of the* ninth *sphere, which we call the ecliptic, as has been demonstrated above* when we were speaking about the zodiac. *But after this dissection by a circular plane two circles are produced on the surface of the solar sphere, which we said has the centre the world; these circles are parallel and happen to have the same centre as the sphere on which they lie and also as the world. The greater one of them, which is also drawn as the outer circle on the outer surface, is called the 'similar circle',* because it is similar to the ecliptic, which is on the surface of the high sphere, which itself is placed similar in the plane of the former.

[191] *It also produces two parallel circles on the two surfaces of the eccentric sphere, which is also called the deferent sphere. These circles are found to be arranged parallel to each other and to the eccentric circle, because they have the same centre as the deferent sphere. The larger of them touches the similar circle of the ecliptic in one point, whereas the smaller one is touched at another point by a parallel to the similar circle that lies on the inner surface of the entire sphere.*

[192] *And this dissection by the extended plane of the eccentric circle draws a parallel circle on each surface of every sphere of the stars and of the aplanos, which are nine parallel spheres that have the same centre as the world;* on either side, this means 18 circles, *whose centres are all the same as that of the world. And the larger one of the circles that are drawn on the surfaces of each sphere is called the similar circle of the ecliptic of that star on whose sphere it is placed. But these larger circles of the smaller*

exterioribus, minores autem maiorum interioribus.

[193] ¶*Punctorum autem, in quibus dictum est duarum iungi sperarum superficies interius et exterius, alterum dicitur longinqua longinquitas solis a centro mundi, illud scilicet quod in exterioribus est utriusque superficiebus, alterum propinquam longinquitatem vocamus illum, quo interiores confinia superficies habent. Imaginantibus igitur nobis lineam de centro mundi egredientem, que transeat per centrum spere,* que et extra centrum et ferens spera dicitur, *atque extendi recte hinc et inde in oppositas partes tamquam diametrum, tanget duos coniunctionum sperarum, de quibus dictum est, punctos. Statim igitur ratio recte intuentis partem illam linee longiorem esse deprehendet, que a centro mundi per centrum ferentis, non solum alia, que est in oppositam partem directa, sed omnibus aliis, que de centro mundi egresse usque superficiem spere extra centrum extenduntur. Illa autem, que in oppositam partem extensa est, et hac et omnibus aliis brevior.*

[194] *Hec autem linea diametrice extensa per utrumque centrum dividit circulum extra centrum, quem per circuitum solis et motu ferentis fieri dictum est, super duos diametrice oppositos punctos. Horum qui super longiorem partem linee diametralis positus est, dicitur longinqua longinquitas centri solis, alter vero, qui supra breviorem, propinqua longinquitas. Punctus ergo ille, quem dicimus longinquam longinquitatem, altitudo solis dicitur propterea, quod dum solis centrum in eo est, longius a terra conspicitur quam in reliquis sui circuli partibus, punctus autem, qui propinqua longinquitas, sive humilitas unde solis opposita ratione.*

[195] *Atque hic punctus, quem altitudinem solis diximus, positus est in quodam puncto circuli signorum,* in .lh. gradu et paucis sexagenariis Geminorum. ([197/ 196/ 198]) *Punctus autem humilitatis solis in opposito signo est,* in .lh. gradu et paucis sexagenariis Sagittarii. *Sunt autem hii duo puncti fixi nec moventur, sicut Ptolomeus posuit in sintaxi. Aliorum autem, qui secuti sunt deinceps, astronomicorum,* quibus fuit etiam et hec et alia secretius rimari, *sententia est in orientem moveri tardo motu,* in .t. scilicet et .f. annis uno gradu. Qua in re mirum est Ptolomei deceptam fuisse sollertiam; moventur enim. Sed illi casus causa extitisse probatur quidam sui

1 maiorum] minorum **2** iungi] iugi (L) **17** propinqua] longinqua (L) **19** quod] qui (L) **26** deinceps] *corr. ex* deinces (L)

spheres coincide with outer ones, and the smaller circles of the larger ones coincide with inner ones.

[193] *One of the points where the surfaces of the two spheres were said to touch with their inner and outer sides is called the 'furthest distance' of the sun, namely that which lies on the outer surface of each sphere; the other point we call the 'closest distance', where the inner surfaces have their limits. Thus, if we imagine a line starting from the centre of the world, which passes through the centre of the sphere* called the eccentric sphere or the deferent sphere *and being extended straight in both opposite directions like a diameter, it will touch the two contact points of the above mentioned spheres. Hence any intelligent observer will immediately realise that the part of the line which is from the centre of the world through the centre of the deferent sphere is not only longer than the part that is directed in the opposite direction, but also longer than any other line starting from the centre of the world and drawn onto the surface of the eccentric sphere; whereas, the part that extends in the opposite direction is shorter than the former and also shorter than all others.*

[194] *But this line, which extends diametrically through both centres, divides the eccentric circle, which was said to be produced by the circuit of the sun and the motion of the deferent sphere, at two diametrically opposite points. That one of these points which is placed on the longer part of the diametric line is called the furthest distance of the centre of the sun, whereas the other one, on the shorter part, is called its closest distance. And the point which we call the furthest distance is also called the 'height' of the sun, for while the centre of the sun is standing in it, it is seen further from the earth than in the remaining positions of its circle; whereas, the point that is the closest distance hence is also called the 'lowness' of the sun, for the opposite reason.*

[195] *This point which we called the height of the sun is at a certain position on the ecliptic,* in the 28th degree and within a few minutes of Gemini.[21] ([197/ 196/ 198]) *And the point of the lowness of the sun is in the opposite sign,* in the 28th degree and within a few minutes of Sagittarius. *And these two points are fixed and do not move, as Ptolemy says in the Almagest. But other astronomers who followed later,* and who also examined this and other things very thoroughly, *state that it moves with a slow motion eastwards,* by one degree in 106 years.[22] In connection with this it is astonishing that this fact could evade Ptolemy's diligent attentiveness, for they do move. How-

[21]The value replaces Arab., 'whose distance from the summer solstitial point is 24 1/2 parts of the parts of which 360 make up a circle, preceding it, as Ptolemy noted in his book known as the *Almagest*.'

[22]'106 years': possibly a transmission error of al-Battānī's number, 66 years (L); see Langermann's note to the present paragraph. Cf. also below, Stephen's commentary to passage 286.

temporis precessor astronomus, cuius in querendo altitudinis solis loco falsa fuit inspectio. Illo enim in loco longinquam solis longinquitatem suo tempore dixerat, quo Ptholomeus perspicatius omnia astrorum secreta discutiens et investigans | suo tempore illam esse veracissime comperuit. Hinc ergo deceptus Ptolomeus fuisse dicitur, tametsi non ipse, nisi quod non moveri dixit. In loco enim inveniendo nichil peccavit, sed qui precesserat ipsum deceptus fuit. Non ergo Ptholomei sagacitas cecidisse arguenda est, sed illius depravanda ignorantia, qui quod nescivit scripture tradens sapienti viro cadendi causa fuit. Nunc satis quod pro illius defendenda industria diximus, ad nostra redeamus.

[199] *Sole ergo, ut dictum est, in spera ferenti et centro eius in circulo extra centrum posito movetur* ipse ab occidente in orientem *motu ferentis spere rotundum habens circuitum, et est eius centrum semper contra aliquem punctum circuli signorum suppositum;* in superficie enim illius semper volvitur. *Linea quoque, que incipit a centro mundi et transiens per centrum solis transit usque superficiem alte spere, circulum signorum in superficie eius tangit in uno puncto.* Et quoniam solis centrum quietis inscium est, *ipsa quoque cum eo semper volvitur et modo hos, modo illos, non confuse sed ordinatim punctos circuli signorum pertransit. Ad hunc modum linea illa semper volvitur, usque ad punctum circuli signorum revertatur, a quo circueundi perpetuum motum acceperat,* postquam ad primum sui circuli punctum redierit, quia ipsa semper centrum solis comitatur et nunquam excedit. *Completa linee circuitione annum quoque solis finem habere scimus.*

[200] *Solem autem in spera extra centrum semper moveri et esse supra dictum est. Spere autem alie partes nobis propinquiores sunt, alie remotiores. Est igitur sol per motum ferentis in terre partibus suppositis <modo> a remotioribus lumen ministrans, modo a magis proximis, et longius quidem a nobis recedit, quando per motum extra centrum spere in longinqua longinquitate ipsius rotatur, quam solis altitudinem esse diximus. Exinde vero descendens motu ferentis incipit fieri minus remotus, deinde propinquius. Qui cum in propinqua longinquitate pervenerit, terre propinquior lucet, nec licet ei propius accedere, quam motu ferentis pertransiens iterum ad alta renititur, donec suam tangens*

26 in] aut **28** nec] *corr. ex* n (L)

ever, there is evidence that the originator of this mistake of Ptolemy was a certain astronomer who preceded Ptolemy's time and who had made a wrong observation when searching the position of the height of the sun.[23] For he had said that the sun's furthest distance had been in his time in that place where Ptolemy, who analysed and investigated all the secrets of the stars most wisely, localised it most accurately in his own time. One therefore says that Ptolemy was wrong, although he was not wrong himself, except that he said that it did not move; for he made no mistake in determining the position, but he who preceded him was deceived. One therefore should not accuse Ptolemy's astuteness for having failed, but one has to blame the ignorance of that man who wrote something down that he did not know and thereby caused a wise man to fail.[24] Now we have said enough in defence of Ptolemy's diligence; let us return to our own concerns.

[199] *While the sun, as we have said, is thus placed in a deferent sphere and its centre on an eccentric circle, it is moved by the motion of the deferent sphere on a round circuit* from west to east, *and its centre is always placed beneath a certain point of the ecliptic;* for it always revolves in the latter's plane. *And a line which starts from the centre of the world and, passing through the centre of the sun, traverses until the surface of the high sphere touches the ecliptic on the surface of that sphere in one point.* And as the centre of the sun does not know rest, *this line too always revolves with it and passes sometimes through these points of the ecliptic, sometimes through others, not randomly but in regular order. In this manner the line always revolves, until it returns to the point of the ecliptic from where it had started that permanent circling motion,* after it has returned to the first point of its circle, because it always follows the centre of the sun and never moves away from it. *And when the line has completed a revolution, we know that also a year of the sun has ended.*

[200] *But we have said before that the sun is found, and moves, always on an eccentric sphere. However, some parts of this sphere are closer to us and others further remote. Due to the motion of the deferent sphere, the sun thus shines at the subjacent parts of the earth sometimes from more remote positions, sometimes from closer positions, and it removes furthest from us when by the motion of the eccentric sphere it revolves at its furthest distance, which we have called the height of the sun. From then on descending by the motion of the deferent sphere, it begins to become less remote, and then very close. When it has reached the closest distance, it shines on the earth from its nearest position and*

---

[23]That is, Hipparchus; cf. *Alm.* III,4.

[24]The debate has been continued in modern times; see, e.g., Manitius' critique of Ptolemy in *Almagest* (tr. Manitius), vol. I, p. 428–9, and Toomer's defence of Ptolemy in *Almagest* (tr. Toomer), p. 153, note 46.

altitudinem longius prohibetur recedere. Et quoniam sol perpetua mobilitate rotatur, sicut significamus, semper post descensum ascensu letatur, inde descensu opprimitur. Ex his que dicta sunt manifestum fit spatium, quod inter solem habetur et terram, nullam habere equalitatem mensure, sed recedente sole maius fit, accedente minus.

[201] *Unde et sol ipse in eiusdem quantitatis arcubus circuli signorum in aliis quidem percurrendum amplius, in aliis autem minus consumit temporis. Et in illa quidem medietate circuli signiferi, cuius est medius punctus altitudo solis,* que in .lh. Geminorum gradu esse diximus, *longiori moratur temporum quantitate, minus temporis consumit in discurrenda medietate opposita, cuius est medium solis humilitas punctus.*

[202] *Hinc est, quod sol duos comprobatur habere motus, unum quidem proprium, quo fertur per ferentis motum circa centrum ipsius spere et in suis fixis polis, quem etiam equalem motum dicimus; secundum autem, quod fit per designationem loci eius linee, que egreditur de centro mundi et pertransit ipsius centrum. Per hunc igitur locum, in quo loco circuli signorum sit <si>(L) queritur, invenitur. Et primus eius motus, quoniam semper idem est, equalis dictus est; hic, ex eo quod equalitatem non servat sed est discors, inequalis dicitur.*

¶Atque hoc loco res et tempus dici postulat, cur ad solis inequalem cursum inveniendum et numeri a capite Arietis collecti pars auferatur et regularium numeri crescentibus auferantur, decrescenti numero in tabula iungantur, vel minori .tr., vel maiori. Et primum quidem iam supra tetigimus, sed ut planius fiat, iterum dicemus. Duo solis | motus, de quibus diximus, nunquam eundem solis locum in circulo signorum simul designant, nisi cum utrique aut in longinqua aut in propinqua longinquitate esse inveniuntur. Tunc enim, quando solis motus equalis .xp. nec amplius gradus habet aut .tr., per hunc equalem verus locus eius in circulo signorum exprimitur. Si quid amplius habebit aut minus, invenitur adesse motus, qui dicitur inequalis. Hic sicut et alter a longinqua longinquitate habens initium finitur in .tr., a quo iterum loco incipit et finitur in .xp.. Quando autem in tabulis rectificationis eius inventio disposita est, cum in illas ad inveniendum ipsum intrare volueris, longitudinem a capite Arietis longinque longinquitatis expellere oportebit. Eadem est in reliquis .e. erraticis minuendorum numerorum ratio tenenda. Nam in luna alia est, quod in ipsius tractatu dicetur.

¶Quare autem regulares his numeris iungantur, illis auferantur, ex his que dicta

14 sit] fit **15** quod] qui **19** auferantur] auferatur **24** nec] *corr. ex* n (L) circulo] circuli (L) **31** quod] qui

cannot come any closer; passing through the closest distance by the motion of the deferent sphere, it tends upwards again, until it reaches its height and is prevented from removing any further. And since the sun revolves by perpetual mobility, as we have indicated, it will always rejoice in ascent after a descent and, subsequently, be depressed by descent. It becomes clear from what has been said that the space between the sun and the earth does not have equal extent, but becomes larger when the sun moves away and smaller when it approaches.

[201] *For this reason, in ecliptical arcs of equal lengths the sun also spends more time passing through some of them and less time passing through others. And the sun will spend the longest period of time in that half of the ecliptic at whose middle is the height of the sun,* which we said is in the 28th degree of Gemini, *and the shortest period while passing through the opposite half, whose middle is the point of the lowness of the sun.*

[202] *This proves that the sun has two motions, one of its own by the deferent sphere, which takes it around the centre of that sphere and about its fixed poles, which motion we also call the 'equal' motion, and a second motion which is described by the place of that line which originates from the centre of the world and passes through the sun's centre. By this place, therefore, one finds, if one seeks to know in what place of the ecliptic the sun stands. And whereas its first motion is called equal motion, because it is always the same, this motion is called 'unequal', because it does not preserve equality, but varies.*

At this place the order of the subject demands to say why, for finding the unequal course of the sun, we subtract a part of the number measured from the head of Aries and also subtract the rectification numbers from increasing numbers, but add them to a decreasing number in the table, depending on whether it is more or less than 180°. We have already touched on the first aspect above, but to make it clearer we will say it again. The two motions of the sun, which we spoke about, never designate the same place of the sun on the ecliptic at the same time, unless both are found at the furthest distance or the closest distance. For when the sun's equal motion has 360 degrees and not more, or 180 degrees, its true position on the ecliptic is expressed by this equal motion. If it has a bit more or a bit less, the presence of what is called the unequal motion is found. Like the other motion, it begins from the furthest distance and ends at 180 degrees, from where it begins anew and ends at 360 degrees. But since it can be found written down in the rectification tables, if you want to enter into the tables to find it, the distance between the furthest distance and the head of Aries is to be eliminated. The same reason for decreasing the numbers applies to the remaining five planets; for the moon it is different, as will be said in its discussion.

The reason why the rectifications are added to some numbers, whereas they

sunt pendet causa. Cum sol in .xp. gradibus sue ferentis est, linea que de centro mundi egreditur et que de centro ferentis, per quarum alteram equalis, alteram inequalis cursus solis invenitur, similiter iuncte eundem punctum in circulo tangunt. Sole inde per ferentis motum amoto linee centri mundi caput, per quam inequalis cursus invenitur, inter caput linee centri ferentis et punctum longinque longinquitatis est, atque ideo tunc numeri regularium auferuntur; contraria est in illa medietate ferentis ultra .tr. ratio.

Sit enim circulus signifer ABCD et in eo centrum E. Sitque circulus solis extra centrum AFGH et eius centrum I. Fiat punctus centrum A et super eum circulus solis. Circineturque alius circulus solis in puncto F, et tercius in puncto G, <quartus>[(L)] in puncto H. Egrediantur linee de centro E IIII^{or}; prima ad centrum solis A, et ipsum pertranseat II^{a} ad F et super eam B, et transeat tercia ad G, et transeat quarta quoque ad H et super eam D. Simili modo de centro I quatuor; prima ad A, secunda ad F et super eam M, tercia ad G, quarta ad H et super eam N. Linee ergo, que egrediuntur altera ab E, altera ab I et tendunt ad centrum solis A, iuncte sunt. Ille quoque que ex eisdem ad centrum solis G extenduntur iuncte. Punctus A est longinqua longinquitas solis, punctus G propinqua longinquitas. In his ergo ad inveniendum solis locum nichil aufertur aut additur. Posito sole in .S. gradibus linea EF alium punctum tangit in circulo signorum, alium linea IF. Est autem maior ab A usque lineam IF arcus circuli signorum ABM, minor ab A usque lineam EF. Linea IF equalis motus est, linea EF inequalis designans verum locum solis in circulo signorum. Ideo ergo, quod inter caput linee EF est, aufertur. Contra est in alia medietate ultra TR. Arcus ABCD, in quo finitur linea EHD, maior est arcu ABCN, in quo finitur linea IHN. Et linea EHD locus est verus in circulo signorum. Ideo ergo iungitur, quod est circuli signorum inter D et N numero circuli extra centrum, et habetur verus locus solis in circulo signorum.

8 E] est (L) **10** Circineturque] contineturque (L) **11** E IIII^{or}] hee IIII^{ta} (L) **12** II^{a} ad F] II a d f (L) **25** numero] numeri (L)

are subtracted from others, is dependent upon what has been said. When the sun stands in 360 degrees of its deferent sphere, the line that originates from the centre of the world and the line originating from the centre of the deferent sphere, one of which indicates the equal motion and the other one the unequal motion of the sun's course, fall together and touch the same point on the circle. But when from then on the sun has been moved away by the motion of the deferent sphere, the end of the line from the centre of the world, which indicates the unequal motion, is found between the end of the line from the centre of the deferent sphere and the furthest distance, and therefore the rectification numbers are to be subtracted in that moment. Contrary reasoning applies in that half of the deferent sphere which is beyond 180 degrees.

[Fig. 3.6] Let ABCD be the ecliptic, and in it the centre E. And let AFGH be the sun's eccentric circle, and its centre I. A shall become a point for the centre, and around it a circle for the sun; another circle for the sun shall be drawn at F, a third one at G, and a fourth one at H. Four lines shall originate from the centre E; the first to the centre of the sun A, and the second one shall pass through the sun's centre at F and on it be the point B, and the third shall pass through it at G, and also the fourth shall pass through it, at H, and on it be the point D. Correspondingly, we draw four lines from the centre I; the first one to A, the second one to F and on it be the point M, the third one to G, the fourth one to H and on it be the point N. Thus the lines which originate, one from E and the other from I, and which run to the sun's centre at A coincide. Also the lines that extend from the same points to the sun's centre at G coincide. But point A is the furthest distance of the sun, and point G is the closest distance. At these points, therefore, nothing is subtracted or added for finding the position of the sun. But, when the sun stands at 90 degrees, the line EF touches one point on the ecliptic, and the line IF touches a different one. And the arc ABM of the ecliptic from A until the line IF is larger than the arc from A until the line EF. The line IF is the equal motion, the line EF the unequal motion, which indicates the true position of the sun on the ecliptic. Hence the difference until the end of the line EF is subtracted. The contrary applies to the other half, beyond 180 degrees. The arc ABCD, on which the line EHD terminates, is larger than the arc ABCN, on which the line IHN terminates. But the line EHD is the true position on the ecliptic. Therefore, the part of the ecliptic between D and N is added to the number of the eccentric circle, and one obtains the true position of the sun on the ecliptic.

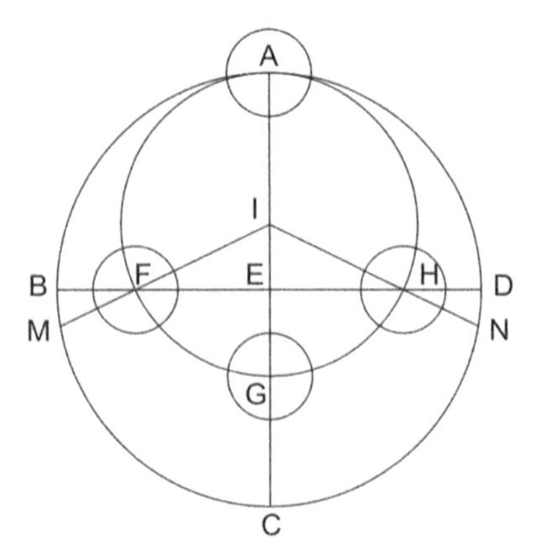

Fig. 3.6: From MS Cambrai 930, fol. 30v

[203] <**S**>(L)*ole in imis sue spere ferentis morante, quoniam in humilitate sua est, propinquior, ut dictum est, circa terram motu suppreme spere volvitur,* cum in alto, longinquior. Dictum est autem esse altitudinem eius in Geminorum .lh. gradu, humilitatem | in opposito. Quanto propinquius terre accedit, tanto maiori impetu radium in terram impellit, idemque a terra fortius repellitur; quanto longius, minor vis radii minorem patitur repulsam. *Quanto ergo terre propinquior est, supposita loca magis calefacit, quanto remotior, minus. Locus autem ille,* in quo altius fertur, septemtrionalis, *in quo propius, australis est positus, ultra, prope tamen, rectam lineam. Hic ergo locus terre suppositus puncto humilitatis solis circulus est in terra,* [204] *ideoque quod sol quando in humilitate positus est sua, motu suppreme spere movetur semel circa terram in die et nocte in eo paralello recti circuli,* quem circuitum capitis Capricorni diximus. *Est autem sub illo paralello in terra circulus illi paralello ceteris omnibus propinquior,* ceterisque circulis terre circulis paralellis superpositis.

[205] *Quando ergo imaginamur lineam egredientem de centro mundi, que porrigatur per centrum solis positum in ima parte ferentis spere, tangit ipsa linea punctum quendam in superficie terre. Movente autem* aplanos *spera solem, circa terram movetur et status linee cum centro solis et punctus eiusdem linee, qui in superficie terre positus est. Facit ergo suo motu rotundum in terre superficie circulum. Non est autem hic in terre spera magnus circulus, non enim eius centrum mundi est, sed extra. Idque idcirco, quod circulus*

---

3 .lh.] .ld. **10** movetur] morietur (L) **15** ima] una **17** positus] *corr. ex* dispositus (L) **18** terre] terra (L)

[203] *While the sun stands in the lowest point of its deferent sphere, since it is at its lowness, it revolves closest, as we have said, around the earth by the motion of the highest sphere,* and when it stands in its height, it revolves most remote. But it was said that its height is in the 28th degree of Gemini, and its lowness at the opposite position. The closer it approaches the earth, it strikes the earth with its radiation with more impact, and the stronger it is repelled by the earth; whereas, the more remote, the weaker a repulsion suffers the weaker force of the radiation. *Hence the closer it is to the earth, the more it heats up the underlying places of the earth, whereas the further remote, the less. But that place* in which it is taken furthest away lies to the north, and *that in which it is taken closest lies to the south, beyond, but still near, the terrestrial equator. Hence this place of the earth which is placed below the point of the lowness of the sun is a circle on the earth,* [204] *for when the sun is placed in its lowness, it is moved by the motion of the highest sphere once around the earth in a day and a night on that parallel circle of the equator* which we called the circuit of the head of Capricorn.[25] *But beneath that parallel circle there is a circle on the earth which is parallel and closer to it than all the remaining ones* and closer to it than to the remaining circles that lie above parallel circles on the earth.

[205] *When we thus imagine a line originating from the centre of the world and directed through the centre of the sun while the latter is placed in the low part of the deferent sphere, this line will touch some point on the surface of the earth. But as the* aplanos *sphere moves the sun, the position of the line will also move around the world together with the centre of the sun, and also the point of that line which is placed on the surface of the earth. It thus produces by its motion a round circle on the surface of the*

[25]This is approximated from the earlier stated position of the sun's perigee, at 28 degrees of Sagittarius; cf. similarly below, Fig. 3.8.

solis paralellus recti circuli, in quo volvitur, non circunductus est a centro mundi. [206] *Quando est enim sol in ima sui circuli parte, linea, que egreditur a centro mundi et ad centrum eius porrigitur, circumducta per ipsius circumvolutionem non movetur in plana superficie, sed conum facit. Ex quo manifestum fit locum terre, circa quem movetur, non*
5 *esse mediam partem terre spere. Inde est quod magnus non habetur ille circulus, quem facit in superficie terre,* sed quanto parallelus in superficie spere none minor est suo equinoctiali in quantitate ipsius spere, tanto et iste minor illo, qui circumducitur ab equinoctiali in superficie terre. [207] *Linea quoque a centro mundi incipiens et ad centrum solis directa minor est omnibus lineis que sole in aliis sui circuli partibus com-*
10 *morante ab eodem mundi centro usque ipsius centrum solis directis. In humilitate enim sui circuli rotatur, in qua positus, ut dictum est, terre spere suppositis partibus accedit. Illud autem indubitabile credimus, terra cum sit spera et centrum habeat, diametra quoque quovis modo, recto tamen, usque superficiem ipsius habere equalia utrimque directa.* Nam de alto montium superius dictum est esse quandam et parvam asperitatem
15 terree spere. *Sublata ergo illa parte linee, que a centro mundi ad centrum in imis sui circuli directa est, que est medietas diametri terre, a centro eius scilicet usque superficiem ipsius in modum diametri protensa, remanent frusta linee a superficie terre usque centrum solis brevius omnibus quarumvis linearum frustis simili modo imaginatarum sole in aliis ferentis arcubus aut punctis lucente. Ex his ergo manifesta et inviolabili colligitur*
20 *argumentatione circulum, quem in superficie terre circumductum benefitio linee dictum est, viciniorem et magis propinquum esse circulo, in quo sol in humilitate ferentis positus volvitur. Constat ergo circulum illum partemque terrarum, quam circuit, ceteris mundi partibus omnibus maiori exuri solarium radiorum incendio.*

Sed quoniam de terre calidis locis sermonem ex ipsius serie operis inserui-
25 mus, placet pauca de zonarum terre divisione summatim perstringere, atque alia rei veritate quam ponit Macrobius uti. Ille enim perustam zonam duobus para-
f. 31v lellis equinoctialis altero Cancri circulo, altero Capricorni utrobique terminans | medium eius rectam posuit lineam. Scimus autem de .e. zonis duas tantum tempe-

3 circumvolutionem] circumvolutione (L) **5** est] *add.* (L) **17** remanent] remanet (L) **24** sermonem] sermone (L) **26** rei] rem **27** altero] *corr. ex* altio (L) **28** rectam] recta (L)

earth. But this is not a great circle on the terrestrial sphere, for its centre is not the same as that of the world, but outside. This is because the sun's parallel circle of the equator on which it revolves is not drawn around the centre of the world. [206] *For when the sun is at the low part of its circle, the line that originates from the centre of the world, and which is extended to that of the sun, would not move in a plane when being turned around by the latter's revolution, but produce a cone. From this it becomes obvious that the place on the earth along which this line rotates does not represent a half of the earth's sphere. For this reason, the circle which it produces on the surface of the earth is not considered a great circle,* but the more the parallel circle on the surface of the ninth sphere is smaller than its equator in relation to its own sphere, by the same proportion this circle will be smaller than the circle which is defined on the surface of the earth by the equator. [207] *And the line starting from the centre of the world and directed to the centre of the sun is shorter than all other lines that are drawn from the same centre of the world and directed to the centre of the sun at times when the sun stands in other parts of its circle. For it rotates in the lowness of its circle, and it was said that in this position it approaches the subjacent parts of the earth. And we think that there is no doubt, since the earth is a sphere and has a centre, that also all diameters, however oriented, if only they are straight, have equal lengths in both directions until its surface.* For it has been said above that the height of the mountains is only a certain and small roughness of the terrestrial sphere. *If one thus subtracts that part which equals half the diameter of the earth—extended like a diameter from the centre to the surface—from the line between the centre of the world and that of the sun when the latter is at the lowest point of its circle, there remains a residue of the line from the surface of the earth until the centre of the sun, which will be shorter than the residue of all other lines imagined in the same manner when the sun shines at other arcs or points of the deferent sphere. From this we can conclude clearly and irrefutably that the circle which has been said to be described by means of the line on the surface of the earth is the closest and most proximate one to the circle on which the sun revolves while being placed at the lowness of the deferent sphere. It is therefore certain that this circle, and the part of the earth around which it circles, is burnt by the glowing heat of the sun's rays more than any other parts of the world.*

As we have already inserted a discussion of the hot places on the earth, which is out of the order of the subject of this work, we like to summarise in short the division of the zones of the earth, thereby following another truth of the matter than offered by Macrobius.[26] For he claims that the burnt zone is defined on both sides by two parallel circles of the equator, one of them being the circle of Cancer and

[26] *Somn.* I,15,13 and II,5–8.

ratas, trium mediam igne torreri, extremas perpetuo frigore pigrescere. Et temperatas quidem habitabiles esse, tametsi extrema earum hinc caloris, illinc patiantur iniuriam frigoris, philosophorum traditio est, frigidas et perustam inhabitabiles esse. Unde tantum illa, que a recta linea usque glacialem porrigitur septemtrionalem, habitari novimus in tantum, ut quedam insula in mari ultra Ethiopiam posita, que duobus tantum a recta linea distat gradibus et frequentibus populis habitetur et multam habeat animantium et plantarum copiam. Quod si quis dixerit non tamen illam que est a centro Cancri paralello usque rectam lineam plene habitari terram, et nos contra eam quam temperatam dicunt, ut omnino habitari <queat>, multa heremo, multa aque penuria, multa quibusdam aliis causis inculta torpere. Placet igitur meliorem quam superius posita est zonarum esse divisionem, que temperatam a recta linea ponit usque latum .pf. graduum .le. sexagenariarum, inde usque polum glacialem, a recta linea in austrum inclinari perustam. Cuius quanta sit mensura aliarumque duarum, eorum relinquo dicenda ingeniis, qui cuncta se posse comprehendere ratione autumant. Sitque hec zonarum trium figura.

---

8 plene] pleni (L) **11** quam] que

the other one the circle of Capricorn, while the terrestrial equator was its middle. We know that only two of five zones are temperate, that the central one of three zones is torrefied by fire, and that the boundary zones are paralysed by eternal cold. And it is the traditional opinion of philosophers that the temperate regions are inhabitable, although their borders suffer maltreatment by heat on one side and by cold on the other side, whereas the cold and the burnt zones are uninhabitable. However, we know that the whole region from the equator until the northern ice is inhabitable, as there is a certain island in the sea beyond Ethiopia that is only two degrees remote from the equator[27] and that is densely peopled and also bears a great mass of animals and plants. If someone says that the land between the parallel of the centre of Cancer until the equator is nonetheless not completely inhabited, we will respond that even the zone they call temperate, because it is at least in principle inhabitable, is in many of its parts left unused due to desert, lack of water or some other reasons. A better division of the zones than the one given above should therefore be considered the one which defines the temperate region as extending from the equator until a latitude of 66;25 degrees, from there to the pole the icy region, and south of the equator, the burnt zone. The extents of this zone and of the other two I leave to those minds to announce who believe that they can determine everything by rational thought. Let this be a diagram of the three zones [Fig. 3.7].

[27] Lemay considers this island to be Java or Sumatra.

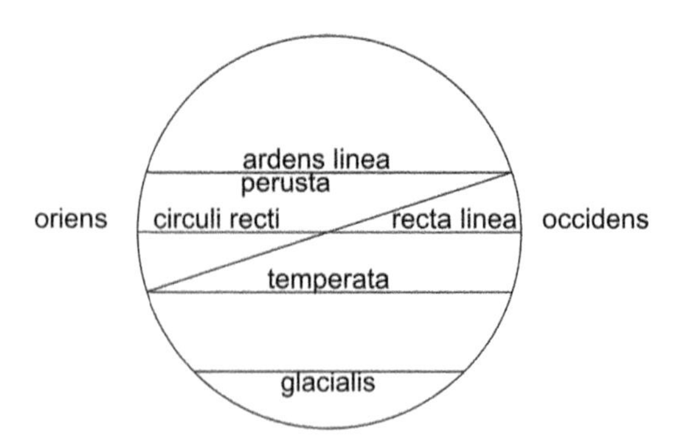

Fig. 3.7: From MS Cambrai 930, fol. 31v. Labels: "burning line" (ardens linea), "burnt/temperate/icy <zone>" (perusta/temperata/glacialis), "terrestrial equator of the celestial equator" (circuli recti recta linea), "east" (oriens), and "west" (occidens)

[208] *<**I**>*[(L)]*lle igitur circulus, quem ceteris propter nimiam solis viciniam calidiorem diximus, ardens via solis vocatur, omnes quoque terre superficiei circuli quanto ardenti circulo propinquiores sunt, tanto calidiores, qui longe remoti nichil ipsius benefitio caloris hauriunt. Recta quoque linea permultum illius calet vicinia, fitque ultra terre calorem perusta inaccessibilis. Nichil enim illic viventium posset vivere, omni aere aut terra plantis aut animantibus humoris benignam temperantiam negante.* Hec ergo causa est, que magis australem plagam torreri igne compellit credere. Expertum etenim rei veritatem nullum audivi. Verumtamen ratio ipsa suggerit sole in imis commorante perustam in austrum extendi longe et extrema etiam terrarum parumper temperari. Sed sole ad septemtrionem renitente omnia usque ad perustam naturali pruina comprimi, ultima peruste temperari.

[209] *Subscripsimus autem circulorum, quos sol in sua spera facit, figuram omniumque eorum partem non modicam, quos in aliis circunductos similes diximus.*

2 quanto] quanta (L) **5** omni aere] omnem aerem (L)

[208] *That circle which we said is hotter than the others by the extreme proximity of the sun is called the 'burning path of the sun'; and all circles on the surface of the earth are hotter, the closer they are to the burning circle, whereas the far remote ones do not receive anything of its gift of warmth. Also the terrestrial equator is very hot by its closeness, and the burnt zone beyond the heat of the earth becomes inaccessible. For no living creature could live there, as all air and earth deny plants and animals a kind and humid temperance.* This is the reason which forces one to believe that the southern part of the earth is more torrid by the fire. For I have not heard of anyone who had experienced the true reason.[28] But reason itself suggests that while the sun stands in the low positions, the burnt zone extends far to the south, and even the boundary region of the earth gets a bit more temperate. But when the sun tends back to the north, natural winter comes over all until the burnt zone, and even the border of the burnt region gets temperate.

[209] *We have added a diagram of the circles which the sun produces in its sphere and also of a large part of the circles which we said were drawn as similar circles in other spheres [Fig. 3.8].*

[28] Cf. al-Battānī, ch. 6 (L).

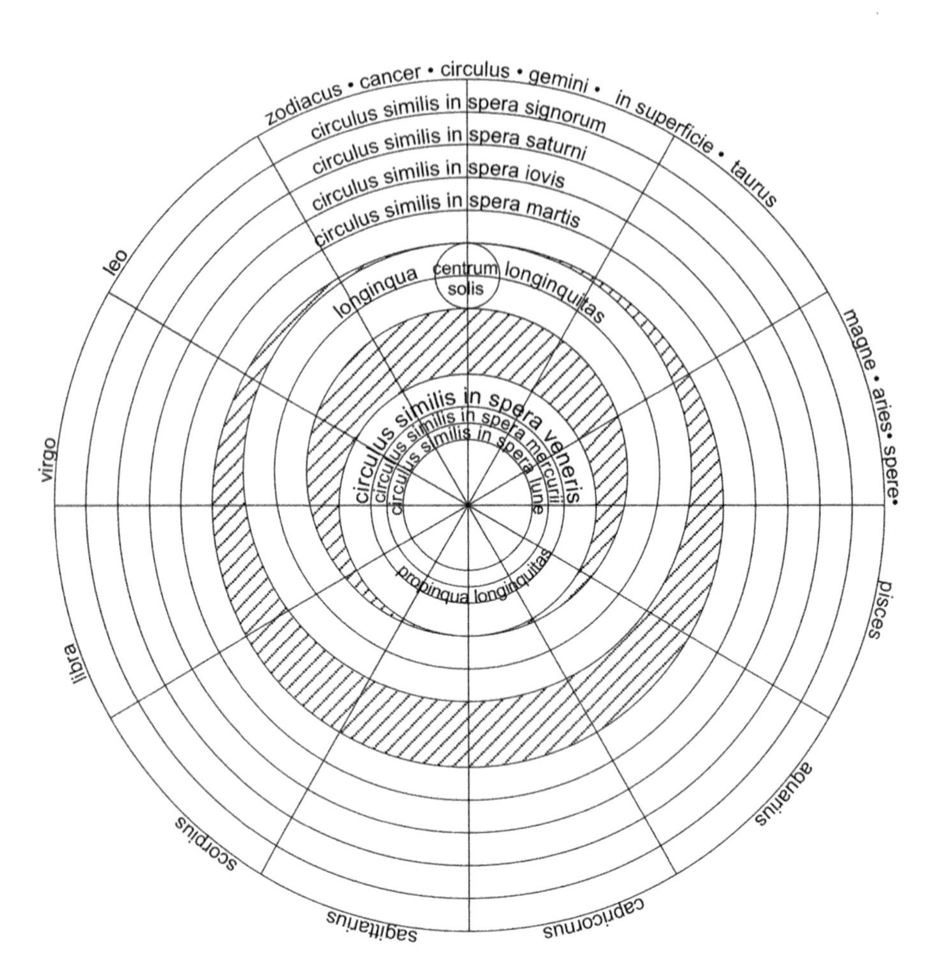

Fig. 3.8: From MS Cambrai 930, fol. 32r (hatching added). Labels along the outer circumference denote "the zodiac circle on the surface of the great sphere" (zodiacus circulus in superficie magne spere) and each of the twelve zodiac signs. Other labels: "similar circle on the sphere of the signs/Saturn/Jupiter/Mars/Venus/moon" (circulus similis in spera signorum/Saturni/Iovis/Martis/Veneris/Mercurii/lune), "centre of the sun" (centrum solis), "furthest distance" (longinqua longinquitas), and "closest distance" (propinqua longinquitas)

f. 32r [210] | <*S*>[(L)]*pera lune sperale corpus est extra intusque duabus circumducta speralibus superficiebus et paralellis. Spere et superficierum idem, quod et mundi, centrum est. Superficierum altera,* <*alta*>[(L)] *scilicet, iungitur interiori superficiei spere Mercurii, altera ima et interior exteriori superficiei spere ignis.* [211] *In alta posita circumductio* 5 *similis circuli, sicut dictum est in spera solis.* [212] *Hec autem spera lune movetur tota rotundo motu ab oriente in occidentem in duobus fixis, qui sunt suppositi polis circuli signorum in radio eorum positi,* .c. sexagenariis .k. secundis in uno die.

[213] *Dividitur hec spera in duas sperales partes, et utriusque centrum idem quod totius, et circumdatur utraque duobus paralellis superficiebus, iuncta interioris superficie* 10 *exteriori, interiori exterioris.* [214] *Et motus secunde, que propinquior est centro mundi, movetur ab oriente in occidentem* in die .ka. gradibus .i. sexagenariis .ke. secundis. *Movetur et super duos fixos polos, sed non in radio signorum aut sub alterius spere positos. Huius nomen spere inclinata.*

[215] ¶*De hac autem separatur alia, tercia scilicet, spera, quam circumdant due* 15 *sperales paralelleque superficies. Centrum earum extra centrum mundi et extra superfi-* f. 32v *ciem circuli similis signorum circulo, et longitudo eius ab* | *utroque polo inclinate spere utrimque distans equaliter. Huius spere alta superficies tangit altam superficiem inclinate in uno puncto, ima vero eiusdem inclinate imam in altero,* sicut de ferenti solis dictum est. [216] *Unde et hec spera extra centrum dicitur* et ferens lune, *que movetur ab oc-* 20 *cidente in orientem in duobus fixis polis, qui non sunt contra polos neque sub ipsis quos ante posuimus.* Motus eius in uno die equalis .ld. graduum et .lc. sexagenariarum, de quibus sublatis duarum sperarum motibus, quibus in orientem refertur, scilicet .ka. gradibus .c. et .i. sexagenariis .k. et .ke. secundis, remanet cursus huius, quo nititur in occidentem, .kc. graduum .k. sexagenariarum .me. secundarum.

---

11 movetur] moventur (L) **11–12** secundis] secundam (L) **17** inclinate] in climate (L) **18** eiusdem] *post del.* unam (L) inclinate] in climate (L) imam] *add.* (L) **22** orientem] occidentem

3.3 <The Sphere of the Moon>

Config. 10: The Orb of the Moon

[210] *The sphere of the moon is a spherical body which inwardly and outwardly is surrounded by two parallel, spherical surfaces. The centre of the sphere and of the surfaces is the same as that of the world. One of the surfaces, that is, the high one, is tangent to the inner surface of the sphere of Mercury, whereas the other, low and inner, surface is tangent to the outer surface of the sphere of fire.* [211] *On the high surface there is drawn the similar circle, corresponding to what we have said for the sphere of the sun.* [212] *This sphere of the moon, however, moves as a whole in a round motion from east to west*[29] *around two fixed poles that are placed below the poles of the ecliptic on the axis of the latter ones,* by 0;3,10 degrees per day.[30]

[213] *This sphere is divided into two spherical parts, each of which has the same centre as the entire sphere, and each is surrounded by two parallel surfaces, the surface of the inner one being tangent to the outer one and that of the outer one to the inner one.*[31] [214] *And the motion of the second sphere, which is the one closer to the centre of the world, moves from east to west* by 11;9,15 degrees per day.[32] *It also moves around two fixed poles, but these are not on the axis of the ecliptic or placed below the poles of any other sphere. This sphere is named the 'inclined sphere'.*

[215] *Separated from this one is another, i.e. a third, sphere, which is surrounded by two spherical and parallel surfaces. Their centre is placed outside the centre of the world and outside the plane of the similar circle of the ecliptic, and its distance from each pole of the inclined sphere on both sides is equal. The high surface of this sphere touches the high surface of the inclined sphere in one point, and its low surface touches the low surface of the inclined sphere in another point,* as has been said for the deferent sphere of the sun. [216] *For this reason, this sphere too is called the 'eccentric'* or 'deferent' sphere of the moon, *which moves from west to east around two fixed poles that are neither opposite to, nor identical with, the poles that we have mentioned before.* Its motion in one day equals 24;23°,[33] from which if one subtracts the motions of the two spheres by which it is taken eastwards, namely 11 degrees and 9 plus 3 minutes and 10 plus 15 seconds, there remains its course of 13;10,35 degrees, by which it tends eastwards.[34]

---

[29] Arab. add.: 'contrary to the succession of the signs of the ecliptic.'

[30] *Alm.* V,2: $0;3^{\circ/d}$; al-Battānī (ed. Nallino, II, p.75): $0;3,11^{\circ/d}$.

[31] In some of the Hebrew copies further details are included in this passage. Langermann assumes an omission of those elements from at least parts of the Arabic tradition; cf. his note to this passage.

[32] *Alm.* V,2: $11;9^{\circ/d}$; al-Battānī, ch. 30: $11;9^{\circ/d}$ (from $11;12^{\circ/d}$ - $0;3[,11]^{\circ/d}$).

[33] The value agrees with the approximation in *Alm.* V,2 and in al-Battānī, ch. 30, whereas it can be derived more precisely from *Alm.* IV,4, as 24;22,53,22,40,35,58°.

[34] The value agrees with al-Battānī (ed. Nallino, II, p.75) and closely with *Alm.* IV,3: 13;10,34,58,33,30,30°.

[217] ¶*In huius spere extra centrum corpore quedam alia spera parva posita est tante quantitatis, ut inter duas illius contineatur superficies paralellas.* Hec unam tantum habet superficiem exteriorem et centrum spere. *Longitudo centri eius ab utroque polo extra centrum spere equalis, et longitudo diametri sperule equalis spisso ferentis. Tangit* 5 *enim eius superficies in duobus suis diametrice oppositis punctis ambas ferentis superficies. Clauditur igitur totum huius corpus infra illius spissitudinis quantitatem. Hec autem spera movetur* ab oriente in occidentem .kc. gradibus .c. sexagenariis .od. secundis *in suis duobus fixis polis. Hinc* apud antiquos *nomen est rotunditas lune.* [218] *Illa quoque spera extra centrum, que ferens dicta est, ex hoc nomen adepta est, quod hanc* 10 *rotunditatem lune suo motu ferat.*

[219] ¶*Luna quoque sperale corpus est solidum et plenum positum in corpore spere, quam rotunditatem diximus, tangitque superficies illius superficiem rotunditatis in uno puncto tantum.* Et spera quidem lune tot habet partes, sex scilicet. [220] *Quando autem movetur spera extra centrum* in orientem, *in cuius spisso rotunditas circumtenetur,* 15 *super duos fixos polos, movet corpus rotunditatis quasi quamdam sui partem suo motu, movetque corpus lune, quod in ea fixum est.*

[221] In primo autem libro de imaginatione circulorum loquentes diximus, quod *in omni spera, que movetur super fixos polos, punctus quisque circulum circinat in eadem spera suo in loco imaginarium. Centrum igitur rotunditatis motu ferentis* ab 20 occidente in orientem *pleno circuitu delatum in ipsius corpore imaginarium circumducit circulum. Qui quia in eius medio positus est et ab utraque illius superficie equaliter distans, ferentis et ipsius unum est centrum.* Quia vero ferens spera extra centrum et hic circulus centrum habens extra centrum mundi ponitur, *circulus extra centrum et ferens circulus dicitur, ferens autem ideo, quia rotunditatis centrum in eo semper volvitur* 25 *nusquam inclinatum.*

[222] *Huius circuli superficiem imaginamur totum secare mundum, fierique per eius sectionem duos paralellos circulos in duabus paralellis superficiebus spere extra centrum, centrum eorum idem quod et prioris; facit etiam imaginarium magnum circulum in corpore rotunditatis, centrum eius quod ipsius sperule. Hic tangit in duobus punctis duos* 30 *priores paralellos in altero alterum.* [223] Quando autem movetur corpus lune motu rotunditatis ab oriente in occidentem, *circumducit centrum imaginarium circulum in corpore rotunditatis,* cuius centrum idem est quod et rotunditatis. *Est ergo hic et alter, qui est in superficie rotunditatis, paralelli. Ille autem, per quem discurrit centrum lune, dicitur circulus rotundus.*

---

2 duas] duos **7** .od.] .cd. **13** spera] spere **14** circumtenetur] circumtenet **21** et] *add.* (L) **32** rotunditatis] *corr. ex* rotunditas (L) **33** paralelli] *corr. ex* paralellis (L)

[217] *In the body of this eccentric sphere there is a certain other, small sphere of such size that it is contained between the former sphere's two parallel surfaces.* This sphere has only one, outer, surface and a centre of a sphere. *The distance of its centre from each pole of the eccentric sphere is equal, and the length of the diameter of this small sphere equals the thickness of the deferent sphere. For its surface touches both surfaces of the deferent sphere at two of its points that are diametrically opposite. Hence the entire body of this sphere is enclosed within the amount of thickness of that other sphere. And this sphere moves* from east to west, by 13;3,54 degrees,[35] *around its two fixed poles. Therefore,* among the ancients *its name is epicyclic sphere of the moon.* [218] *Also, the eccentric sphere, which has been called the deferent sphere, has received this name, because it carries by its motion that epicyclic sphere of the moon.*

[219] *Also the moon is a solid and massive spherical body, which lies in the body of the sphere which we called the epicyclic sphere, and its surface touches the surface of the epicyclic sphere only in one point.* The sphere of the moon has this many parts, namely six. [220] *But when the eccentric sphere, in whose thickness the epicyclic sphere is embedded, moves eastwards around the two fixed poles, it will move the body of the epicyclic sphere like a part of itself by its motion, and it moves the body of the moon which lies fixed in it.*

[221] But in Book I, when speaking about the imaginary circles, we said that *in every sphere that moves around fixed poles any point draws at its place in that sphere an imaginary circle. The centre of the epicyclic sphere, therefore, when it is taken a full revolution* from west to east *by the motion of the deferent sphere, produces an imaginary circle in the body of that sphere. As this circle lies in the middle of that sphere and equally remote from either of the latter's surfaces, its centre and that of the deferent sphere are the same.* But since the deferent sphere is eccentric, and this circle too is defined as having its centre outside the centre of the world, *it is called 'eccentric circle' or 'deferent circle'; 'deferent', because the centre of the epicyclic sphere always revolves exactly on it without any deviation.*

[222] *We imagine the plane of this circle cutting through the entire world, and that by its dissection two parallel circles are produced on the two parallel surfaces of the eccentric sphere, whose centre will be the same as that of the former; then it will also produce an imaginary great circle on the body of the epicyclic sphere, whose centre will be the same as that of this small sphere. This circle will touch the two former parallel circles in two points, each of them in one point.* [223] But when the body of the moon is moved by the motion of the epicyclic sphere from east to west, *the centre will draw an imaginary circle in the body of the epicyclic sphere* whose centre is the same as that of the epicyclic sphere. *Hence this circle and the other one, which is on the surface of the epicyclic sphere, are parallel. But the one on which the centre of the moon revolves is called 'epicycle'.*

[35] *Alm.* IV,3, IV,4 and IV,7: 13;3,53,56,17,51,59°; al-Battānī (ed. Nallino, II, p.75): 13;3,54°.

[224] *Facit quoque circuli ferentis sectio in superficie inclinate spere circulos duos, centrum quorum mundi est, poli autem eorum idem, qui et ipsius inclinate.* Ex hoc manifestum est ferentis polos esse positos in radio paralello radio polorum inclinate. f. 33r *In superficie* | *quoque alte spere prime lunarium, que claudit inclinatam,* duos *circinat* 5 *circulos paralellos circulis inclinate,* et centrum eorum mundi. *Horum alter, qui in alta superficie circumducitur, secat circulum similem signifero* in duobus diametrice oppositis punctis. [225] *Circumsecat etiam in superficie none spere* duos *circulos ex obliquo dividentes circulum signorum. Huiusmodi circulus,* cuius est obliqua positio, seu in magna spera seu in prima lunari, *dicitur inclinatus circulus lune,* [226] *et inclinatio* 10 *eius a signorum circulo tanta est, quantum latum lune esse deprehensum est,* .d. scilicet gradus et .nf. sexagenarie. Totidem ergo gradibus etiam poli inclinate spere et ferentis inclinantur a polis prime et spere signorum. *In quo aliud notandum, quod quantacumque erit latitudo lune, tanta et circuli a signifero inclinatio, propterea quod centrum lune semper in superficie huius inclinati circuli sive in austro sive in septemtrione* 15 *volvitur.*

[227] *Duos autem punctos, quibus sese ligant invicem hic et signorum circulus, duos ligantes appellamus.* Geminas quoque inclinati circuli medietates alteram, que in austrum inclinatur, australem ligantem, alteram, que in septemtrionem, septemtrionalem ligantem dicimus. *Et punctus ille ligans, qui septemtrionalis ligantis initium* 20 *est, caput draconis appellatur; ille autem, a quo australis medietas sumit principium, finis draconis sive cauda. Huius autem circuli inclinatio a signorum circulo eandem habet semper constantiam nec mutatur.*

[228] *Imaginamur autem lineam egredientem de centro mundi in modum diametri et transeuntem per centrum ferentis. Hec ergo longius extensa atque utrobique crescens in* 25 *longum, usque dum tangat circulum lune extra centrum positum,* in quo semper rotunditatis centrum moveri dictum est, *punctus eiusdem circuli, quem linee pars illa tanget, que egrediens a centro mundi transit super centrum ferentis, est remotior a terra in ferenti circulo, et dicitur altitudo lune* et longinqua longinquitas eiusdem, propterea quod centro rotunditatis in eo posito nec altius ascendere nec longius recedere rotun- 30 ditas lune a centro mundi poterit. *Punctus autem ille, quem in altera et opposito loco medietatis altera pars linee perstringit, quippe que in modum diametri utrobique posita*

9 lune] *corr. ex* linee (L) **18** ligantem] ligante **19** ligantem] ligante **21** a] aut (L)
30 quem] quoniam

[224] *But the dissection by the deferent circle also produces two circles on the surface of the inclined sphere, whose centre is that of the world, but whose poles are the same as those of the inclined sphere.* It is clear from this that the poles of the deferent sphere are placed on an axis that is parallel to the axis of the poles of the inclined sphere. *Also on the surface of the high, first one of the lunar spheres, which encloses the inclined sphere, it will draw* two[36] *circles that are parallel to the circles of the inclined sphere,* and their centre will be that of the world. *That one of them which is drawn on the high surface intersects with the similar circle of the ecliptic* at two diametrically opposite points. [225] *Also on the surface of the ninth sphere it dissects along* two *circles which divide the ecliptic obliquely. Such a circle,* whose position is oblique, be it on the great sphere or on the first lunar sphere, *is called an 'inclined circle' of the moon,* [226] *and its inclination from the ecliptic is as large as what has been found to be the latitude of the moon,* namely is 4;46 degrees. By the same degrees, therefore, also the poles of the inclined sphere and of the deferent sphere are inclined from the poles of the first [lunar] sphere and of the sphere of the ecliptic. *In connection with this, one should further notice that whatever the latitude of the moon will be, the same will be the inclination of this circle from the ecliptic, because the centre of the moon, be it in the north or in the south, revolves always in the plane of this inclined circle.*

[227] *And the two points in which this circle and the ecliptic intersect are called the two 'nodes'.*[37] Also the two equal halves of the inclined circle we call, in the case of the one which is inclined to the south, the 'southern ligant', and, in the case of the one inclined to the north, the 'northern ligant'. *And that node which is the beginning of the northern ligant is called the 'head' of the dragon, whereas the one from which the southern half takes its beginning, is called the 'end' or the 'tail' of the dragon. The inclination of this circle from the ecliptic is always constant and does not change.*

[228] *But we imagine a line originating from the centre of the world like a diameter and passing through the centre of the deferent sphere. If this line is extended further and growing in length in both directions until it touches the eccentric circle of the moon,* on which the centre of the epicyclic sphere has been said to be always moving, *the point of this circle which is touched by that part of the line which originating from the centre of the world passes through the centre of the deferent sphere is the one that is furthest remote from the earth on the deferent circle, and it is called the 'height' of the moon* or its 'furthest distance'; for, the centre of the epicyclic sphere, when standing at this point, cannot climb higher, neither can the epicyclic sphere of

[36] 'two': Arab., 'one', meaning the outer circle.

[37] 'nodes': lit., 'ligant points'.

est, propinqua dicitur longinquitas et humilitas sive imum lune. Centro enim rotunditatis in eo posito nulla prevalet propius accedere ratione. [229] *Quando autem centrum rotunditatis in puncto longinque longinquitatis volvitur, linea que de centro mundi egressa est <et> extenditur usque rotundum circulum finis eius longinqua longinquitas rotunditatis dicitur. Finis vero diametri rotundi circuli ab ea parte, que terre propinquior est, propinqua longinquitas rotunditatis lune consuevit appellari.* In his duabus longinquitatibus aut propinquitatibus centro lune posito nec propius nec longius potest aut subvehi aut proximari.

[230] ¶*Prima igitur spera lune movet super duos fixos polos, ut sic subesse polis signorum superius significavimus, qui sunt poli circuli similis signifero. Hec movet omnes speras lune, circumtenet enim et includit illas, suo motu, ab oriente scilicet in occidentem,* firmamenti cursum sequendo, sicut supra diximus. Huius cursus draconis esse dicitur, quod complet unum circuitum in .kh. annis et .upc. diebus et una hora. *Movet hic suo motu speram inclinatam et cum ea omnem superficiem inclinati circuli ab oriente in occidentem. Hic circulus inclinatus* circumducitur in superficie spere inclinate, que iungitur prime, et *in superficie alte spere dividens, ut dictum est, circulum signorum.* [231] *Unde et huius in motu moventur duo puncti* | *ligantium, draconis scilicet caput et caudam, in occidentem, atque hunc motum, ut diximus, vocant cursum draconis.*

f. 33v

[232] *Secunda autem lune spera, quam inclinatam diximus, movetur et ipsa ad occidentem in duobus fixis polis, qui inclinantur a polis signorum, et sunt poli inclinati circuli, quem superficie eiusdem spere circumductum sepius replicuimus.* Hic complet suum cursum in .lh. diebus .kf. horis .m. sexagenariis horarum. *Movet hec quoque suo motu omnes, quas in ipsa continet, speras lune ad occidentem. Movet ergo speram extra centrum suo motu et omnes superficies circuli extra centrum eodem cursu et centrum earum. Unde et longinquam longinquitatem movere comprobatur lunaris ferentis* et propinquam longinquitatem. *Est autem omnis hic motus ab oriente in occidentem circa centrum mundi.* Licet enim inclinate poli inclinentur, centrum tamen eius idem est quod et mundi. Movetur ergo inclinatus circulus circa centrum mundanum per

6 propinqua] longinqua (L) **9** lune] *corr. ex* linee (L) **11** occidentem] occidenti (L) **18** caudam] cauda motum] modum **21** complet] comprobat **23** quas] que (L)

the moon remove any further from the centre of the world. *And the point which is defined at the opposite place, in the other half, by the other part of that line which like a diameter is placed in both directions, is called the 'closest distance', and 'depth' or 'lowness', of the moon.* For when the centre of the epicyclic sphere is placed in it, it cannot by any means approach any closer. [229] *And when the centre of the epicyclic sphere revolves at the furthest distance, the end of the line which started from the centre of the world and extends until the epicycle is called the furthest distance of the epicyclic sphere. But the end of the diameter of the epicycle on that side which is closer to the earth is commonly called the closest distance of the moon's epicyclic sphere.* When the centre of the moon is placed at these two furthest or closest distances, it cannot remove or approach any further.

[230] *The first lunar sphere thus moves about two fixed poles, which we have described above as lying below the poles of the ecliptic such that they are the poles of a similar circle of the ecliptic. This sphere moves all lunar spheres—for it encloses and contains them all—by its motion, namely from east to west,* following the course of the firmament, as we have said above. Its course is called the 'course of the dragon', which completes one revolution in 18 years and 263 days and 1 hour.[38] *By its motion, the dragon moves the inclined sphere, and with it the entire plane of the inclined circle, from east to west. This inclined circle* is drawn on the surface of the inclined sphere, which is tangent to the first sphere, and it *divides the ecliptic, as we have said, on the surface of the high sphere.* [231] *For this reason, in its motion also the two nodes of the ligants, namely the head and the tail of the dragon, are moved westwards, and this motion, as we have said, is called the course of the dragon.*

[232] *Also the second sphere of the moon, which we have called the inclined sphere, moves westwards by itself around two fixed poles, which are inclined from the poles of the ecliptic and which are also the poles of the inclined circle, which we have repeatedly explained to be drawn on the surface of that sphere.* This circle completes its course in 28 days, 16 hours, 30 minutes.[39] *Also this sphere moves by its own motion all the spheres of the moon which it contains in itself, to the west. Hence, by its own motion it moves the eccentric sphere and all the planes of the eccentric circle and their centre on the same course. From this it is proven that also the furthest distance* and the closest distance *of the lunar deferent move. But this motion as a whole is from east to west around the centre of the world.* For even though the poles of the inclined sphere are

[38] The period corresponds to a daily motion of $0;3,9,40^\circ$, which differs slightly from the earlier given $0;3,10^\circ$.

[39] Again, the period does not match the earlier given motion of $11;9,15^{\circ/d}$, but corresponds to a motion of $12;32,56^{\circ/d}$.

motum speralem.

[233] ¶*Ferens etiam spera movetur in duobus fixis polis ab occidente in orientem,* et complet cursum suum in .kd. diebus .kh. horis .l. hore sexagenariis. *Hec ergo suo movet motu rotunditatem lune et centrum eius* in circuli circulari linea, que extra centrum mundi volvitur. [234] *Movet etiam per motum rotunditatis corpus lune, quod in ea fixum est, in orientem.*

[235] *Rotunditas autem movetur in suis duobus fixis polis* ab oriente in occidentem *circa suum centrum. Hec movet etiam lunarem globum* tamquam quandam partem sui *et centrum eius in rotundo circulo.* [236] *In hoc ergo centro lune posito semper, si quidem fuerit ipsa luna in medietate rotundi circuli, cuius medium est longinqua longinquitas, movebitur rotunditatis motu ab oriente in occidentem. Posita vero in inferiori arcu et medietate rotundi circuli, cuius medium est longinqua propinquitas et humilitas lune, motus eius alteratus est, movetur enim ab occidente contra.*

[237] ¶*Diametrum quoque rotundi circuli, cuius est finis ab exterioribus longinqua longinquitas,* interius autem longinqua propinquitas, *ponitur in linea, que extenditur a centro mundi et dirigitur ad punctum longinque longinquitatis ferentis, iunctum est recte diametro ipsius ferentis.* Par est ratio in longinqua propinquitate. *Quando autem movetur rotundus circulus motu sui centri, quem habet ex ferenti circulo, et recesserit centrum a diametro spere extra centrum a puncto altitudinis lune, movebitur et diametrum ipsius, cuius sunt termini longinqua longinquitas et longinqua propinquitas ipsius, nec erit eo modo positum quo prius.* [238] *Non enim erit posita longinqua propinquitas diametri aut adversus centrum mundi aut contra centrum spere, que extra mundi centrum circumvolvitur. Dico autem, quoniam si diametrum ipsum longius proteletur et recte extendatur, cum ad diametrum ferentis prius positum venerit, neque centrum mundi neque centrum ferentis tanget, sed alio in loco diametrum dividens punctum, super quem circinatur eius volubilitas, designabitur. Hic autem punctus in diametro positus est in illa parte, que a centro mundi usque longinquam propinquitatem ferentis porrigitur. Distantiam autem huius a centro mundi* antiquorum sollers indagatio repperit *tantam esse, quanta est longitudo centri extra centrum spere a centro mundi.* Caput autem diametri rotundi circuli inter illum punctum positum est. Linea vero, que de hoc puncto egreditur recte protensa usque caput diametri, circinans linea vocatur, ideo quod

---

24 ad] *bis habet* **27–28** Distantiam] Distantia **29** Caput autem diametri] Super autem diametrum **30** circinans] aranaus

inclined, its centre is nonetheless the same as that of the world. The inclined circle is therefore moved by a spherical motion around the cosmic centre.

[233] *Also the deferent sphere moves around two fixed poles from west to east,* and it completes its course in 14 days, 18 hours, 20 minutes.[40] *Thus, by its own motion, this sphere moves the epicyclic sphere of the moon and its centre* on the dirigent line of the circle which rotates outside the centre of the world.[41] [234] *By the motion of the epicyclic sphere also the body of the moon, which is fixed inside it, is moving to the east.*

[235] *But the epicyclic sphere moves about its two fixed poles* from east to west *around its centre. This again moves the lunar body and also the latter's centre on the epicycle* like a part of itself. [236] *Therefore, as the moon's centre is always located on this circle, if the moon is in that half of the epicycle whose middle is the furthest distance, it will be moved from east to west by the motion of the epicyclic sphere. But when the moon is placed in the lower arc and in that half of the epicycle whose middle is the closest distance and the lowness of the moon, its motion has changed, for it moves from west in the opposite direction.*

[237] *And the diameter of the epicycle which ends outwardly at the furthest distance* and inwardly at the closest distance, *is placed on a line that extends from the centre of the world and is directed to the point of the furthest distance of the deferent sphere <and> it falls together exactly with the diameter of its deferent.* The same applies analogously to the closest distance. *But when the epicycle is moved by the motion of its centre, which is caused by the deferent circle, and the centre moves away from the diameter of the eccentric sphere and from the point of the height of the moon, its diameter, whose ends are its furthest distance and closest distance, will also move, and it will not be placed in the same way as before.* [238] *For the closest distance of the diameter will not be oriented towards the centre of the world or towards the centre of the sphere that revolves outside the centre of the world. I mean, if this diameter is prolonged and extended straight, then when it meets the diameter of the deferent sphere defined above, it will neither touch the centre of the world nor that of the deferent sphere, but the point that divides the diameter, and around which the latter's flighty swiftness circles, will be defined at a different place. But this point is placed on that side of the diameter which extends from the centre of the world until the closest distance of the deferent sphere.* The skilful investigation of the ancients found *that the distance of this point from the centre of the world is as large as the distance of the centre of the eccentric sphere from the centre of the world.*[42] But

[40] The period corresponds to a motion of $24;23,1^{\circ/d}$, which agrees with the earlier given $24;23^{\circ/d}$.

[41] 'outside the centre of the world': Arab., 'on the circumference of the eccentric orb.'

[42] Arab. add.: 'It always preserves this alignment.'

cum diametro rotundi circuli totam circinat ferentis circuli circumductionem, et secum illo motu rotundum defert circulum.

[239] ¶*Motus autem centri rotunditatis et puncti longinque longinquitatis spere extra centrum sive altitudinis lune oppositi sunt. Alterum enim in* | *orientem, scilicet rotunditatis centrum, alter autem in occidentem, longinque longinquitatis punctus, fertur. In quorum cursu* mirandum est, *quod per inequalem motum utrimque semper a loco solis distant equaliter.* Simul enim a sole alterum in orientem, alterum autem recedit in occidentem. Ponitur enim simul esse in capite Arietis; ferens in orientem nititur in die .ld. gradibus .lc. sexagenariis, sol .oi. sexagenariis .h. secundis, prima lune spera .c. sexagenariis et .k. secundis in occidentem, inclinata eodem .ka. gradibus .i. sexagenariis .ke. secundis. Iunctis cursibus solis, prime et inclinate fiunt .kb. gradus .ka. sexagenarie .mc. secunde. Quibus omnibus sublatis de .ld. gradibus .lc. sexagenariis remanent .kb. gradus .ka. sexagenarie .lg. secunde. De tribus tantum secundis discordia est, que quadam subtilitate cursus earum evacuantur. *Manifestum ergo est, quod solis locus eorum in medio semper positus sit,* quando quidem ab ipso ad utrumque equales sunt circulorum arcus.

[240] *Patet ergo punctum longinque longinquitatis, qui circa centrum mundi* ab oriente in occidentem *volvitur, ex quo separatur a loco solis et volvitur,* hic in orientem ille in occidentem, *tantum transigi temporis, quantum mensis continet lunaris, antequam iterum coeuntes* in eundem, sed alias, hinc iste, inde ille, *obvii revertantur locum.* [241] *Centrum autem rotunditatis* nimia percurrens celeritate solare centro *volvitur, ut dictum est, in circulo extra centrum. Cuius motus initiatus a puncto longinque longinquitatis* in orientem festinans *complet temporis dimidium mensis lunaris, donec ad eundem punctum altitudinis lune redeat.* Quibus rursus separatis et in diversa properantibus, completa mensis lunaris altera medietate, in eundem simul locum conveniunt; atque ita fit, ut *centrum rotunditatis bis in mense lunari circulum extra centrum pervolet. In tempore igitur equalis sinodi et centrum rotunditatis et punctus longinque longinquitatis in uno sunt loco, finitaque equali sinodo in contraria utrisque certantibus.*

---

13 secunde] secundum **22** centro] centrum **26** bis] hec

the end of the diameter of the epicycle is placed towards that point. And the line which originates from this point and extends straight until the end of the diameter is called the 'dirigent line',[43] because it circles with the diameter of the epicycle the complete arc line of the deferent circle and by that motion takes the epicycle with itself.

[239] *But the motions of the centre of the epicyclic sphere and that of the furthest distance of the eccentric sphere, or of the height of the moon, are opposite. For, one of them, namely the centre of the epicyclic sphere, is taken eastwards, and the other one, the point of the furthest distance, westwards. In their course* it is astonishing that *in spite of unequal motion they always have the same distance from the sun on either side.* For they move away from the sun simultaneously, one to the east and the other one to the west. Assuming that they were at the head of Aries in the same moment; then the deferent sphere tends eastwards by 24;23 degrees per day, the sun by 0;59,8 degrees, the first lunar sphere by 0;3,10 degrees to the west, and the inclined sphere in the same direction by 11;9,15 degrees. If one combines the courses of the sun, the first and the inclined sphere, there will result 12;11,33 degrees. And if one subtracts all this from 24;23 degrees, there will remain 12;11,27 degrees.[44] There is disagreement by only three seconds, which are compensated for by some subtle nature of the courses. *It is therefore evident that the place of the sun is always located in the middle between them,* because the arcs of the circles in both directions from it are equal.

[240] *Hence it is obvious that the point of furthest distance—which rotates* from east to west *around the centre of the world, whereby it veers away from the place of the sun* by moving west while the latter moves east - *will take as much time as is contained in a lunar month before it meets the sun again,* at the same place but both coming back from opposite directions. [241] *But the centre of the epicyclic sphere,* which moves much faster than the centre of the sun, *has been said to revolve on the eccentric circle. Its motion, which begins at the point of furthest distance* and rushes eastwards, *takes half a lunar month until it returns to the same point of the height of the moon.* And when they have diverged again and drive in opposite directions, when the second half of the lunar month has passed, they come together again at the same place simultaneously. In this manner, *the centre of the epicyclic sphere flies through the eccentric circle twice in a lunar month. Thus, at the moment of an equal conjunction the centre of the epicyclic sphere and the point of furthest distance are at the*

[43]'dirigent'; *lit.* 'circling'.

[44]The value corresponds with *Alm.* IV,3: 12;11,26,41$^{\circ/d}$.

[242] *Cum rotunditatis centrum ad punctum longinquitatis propinque pervenit peragrata circuli extra centrum medietate, sub recto oppositi sese intuentur diametro. In opposita enim parte super diametrum positum est centrum rotunditatis. Quibus ita constitutis longinqua quoque longinquitas a centro solis motu inclinate et ferentis solis recessit* IIII*ta* *parte circuli,* .s. gradibus. Centro rotunditatis in opposito constituto *centrum solis ab utroque .s. distans gradibus ad utrumque sub quadranguli se habet figura. Manifestum ergo fit lune rotunditatis centro cum sole quadrangularem formam figurante,* sive ipsa quidem in augmento posita fuerit seu momentaneas cogat in imum diminutiones, *positum esse in longinqua propinquitate spere extra centrum.* [243] Incessanti autem utriusque motus vigilantia, *cum centrum rotunditatis lune* crescentis *plenum peragraverit circulum extra centrum, longinqua longinquitas* inclinate instantia *tantam sui discurrit circuli quantitatem, ut ab ipsa usque ad solis locum, et qua decursa est et qua restat currenda, circuli medietas copuletur, utrimque scilicet .tr. distans ab eo gradibus. Sunt igitur et centrum rotunditatis et punctus longinque longinquitatis in diametro solis posita. Ex quo manifestum fit centri rotunditatis lune equalem locum in panselinio,* quod nos plenilunium dicimus, *in puncto longinque longinquitatis spere extra centrum esse.* [244] *Est igitur locus lune semper in sinodo et panselinio in longinqua longinquitate spere ferentis, sub tetragono autem,* quando medium habet sui luminis, *in longinqua propinquitate eiusdem.*

[245] *Sed quoniam lune corpus per motum rotunditatis in orientem deferri et occidentem* diximus, *in occidentem quidem, cum in alta eius est medietate, in orientem, cum in* | *ima volvitur,* videretur fortassis aliquibus lunam orbe fieri quandoque retrogradam propter huiusmodi cursus assignationem. *Illa vero nunquam retrograditur.* Rotunditas enim lune parva est, et ferentis magna citatio. Quando igitur in occidentem movetur posita in alto rotunditatis, non videtur nec animadvertitur retrograda. *Ferens enim lune spera rotunditatem suo celeri motu in orientem movet, et fert secum lunarem globum. Motus autem ille longe citior est motu corporis lune. Hec causa est, quod luna nunquam retrograda sed semper directa videtur. Fit autem per motum rotunditatis motus lunaris corporis in circulo signorum modo tardus, sepe velotior. Et tardus quidem, cum in alta medietate eius in occidentem fertur, velocior vero, quando in ea medietate, que terre propinquior est.* [246] *Diversi autem et omnes, quos habet lunaris globus, motus sunt .f..*

---

15 panselinio] paranselinio (L) **17** panselinio] paranselinio (L) **30** alta] alto (L) velocior] velociorem (L)

same place, and when the equal conjunction has passed, both urge in opposite directions.

[242] *When the centre of the epicyclic sphere has reached the point of closest distance after traversing half the eccentric circle, they face each other diametrically opposite. For the centre of the epicyclic sphere is placed on the opposite side of the diameter. In this constellation, the furthest distance too, by the motion of the inclined [lunar] sphere and the sun's deferent sphere, has moved from the centre of the sun by a quarter of a circle,* by 90 degrees. And since the centre of the epicyclic sphere stands opposite, *the centre of the sun is 90 degrees remote from each, forming a rectangle with each of them. It thus becomes clear that when the centre of the epicyclic sphere of the moon forms a rectangle with the sun,* be it at the position of waxing moon or when it urges to gradual waning, *it is located at the closest distance of the eccentric sphere.* [243] But by the unremitting perseverance[45] of both motions, *when the centre of the epicyclic sphere* at waxing moon *has traversed the entire length of the eccentric circle, the furthest distance of the inclined sphere has* indefatigably *passed through such a portion of its circle*[46] *that its distance from the place of the sun, both the length which it has passed and also the length which is still left to be passed, equals a semi-circle, that is, it stands 180 degrees from the sun to both sides. Hence both the centre of the epicyclic sphere and the point of furthest distance are placed on the diameter of the sun. It becomes clear from this that the place of the centre of the moon's epicyclic sphere at panselinium,* which we call 'full moon', *is also at the furthest distance of the eccentric sphere.* [244] *Hence the position of the moon at conjunction and full moon is always at the furthest distance of the deferent sphere, whereas under a quadrangle [i.e., at a right angle to the sun],* when it has reached half its brightness, *it is at the closest distance of that sphere.*

[245] *But since* we said that *the body of the moon is taken eastwards and westwards by the motion of the epicyclic sphere, westwards when it stands in its high half, and eastwards when it revolves in the low one,* some people might deem from such a description of its course that the moon describes a loop and occasionally retrogrades. *But, it never retrogrades.* For the epicyclic sphere of the moon is small, and the deferent sphere moves with a high velocity. Thus, when it moves westwards, being placed in the high part of the epicyclic sphere, it is not found or perceived as retrograding. *For the moon's deferent sphere moves the epicyclic sphere eastwards by its own fast motion and takes with itself the body of the moon. And this motion is much faster than the motion of the lunar body. This is the reason why the moon is never seen retrograding but always in direct motion. Nevertheless, by the motion of the epicyclic sphere the motion of the lunar body on the ecliptic becomes sometimes slow and sometimes faster; slow, when it is taken westwards in the high half of the epicyclic sphere, and faster, when it is in that half which is closer to the earth.* [246] *However, all the different motions of the lunar body are six.*

[45] 'incessanti vigilantia': cf. Gregor ep. 4,31.

[46] 'the furthest distance...circle': Arab., 'the point of furthest distance also moves [through] a quarter of a circle.' Stephen's formulation is more accurate, as it also takes into account the simultaneous motions of the sun and the dragon.

¶*Primus est motus totius spere, que dicitur draconis cursus, ab oriente in occidentem. Secundus motus est inclinate spere in occidentem super duos alios polos, quem dicunt motum longinque longinquitatis. Tercius autem motus est spere extra centrum ab occidente in orientem, que defert secum rotunditatem lune sive in orientem seu ad austrum aut septemtrionem a circulo signorum. Quando ergo ferens movet centrum rotunditatis a circulo signorum in austrum aut septemtrionem, dicitur motus lati et inclinati circuli. Quando autem sumitur in longum, sicut est ordo signorum,* IIII^{tus} *motus dicitur et motus longitudinis. Quintus vero motus est rotunditatis* ab oriente in occidentem, et contra; *hic movet corpus lune circa centrum rotundum.* Qui quoniam modo est in orientem, sepe contra, et velocitas cursus lune per hunc alteratur, *motus discors appellatur. Sextum denique motum dicimus illum, quo diametrum rotundi circuli movetur, cuius est finis longinqua longinquitas versus positum punctum* in signorum quolibet. *Huic* a veteribus astronomis *contrarius motus nomen est impositum.*

[247] Ad designandum autem locum lune in signorum circulo et latum ab eodem *imaginamur rectam lineam, que egrediens a centro mundi ad centrum lune extenditur atque ab eo recte porrigitur usque ad superficiem magne spere et finitur in circulo inclinato. Centrum enim lune semper est in superficie inclinati circuli, nec usquam aliquando inclinatur. Magnum quoque imaginamur circulum, qui incipiens a duobus polis signorum circuli transit punctum, in quo finitur linea centri mundi in circulo inclinato.* [248] Centrum eius mundi centrum est. *Hic ergo circulus signorum circulum dividit in duas medietates. Arcus huius, qui est inter inclinatum et signiferum, latum lune dicitur. Punctus, in quo dividit zodiacum, locus est lune in circulo signorum.*

¶De rectificando lune loco, et quare nusquam prima tabula ipsius equali loco sed rotunditati superponatur, occulta est ratio. Sed hoc ideo fit, ne gravitatem

2–3 motum] *corr. ex* potum (L) **10** Sextum] Quintum (L) **17** nec] ne (L) **22** circulo] circuli (L)

The first one is the motion of the entire sphere, which is called the course of the dragon, from east to west. The second motion is that of the inclined sphere, which is westward about two other poles, and which is called the motion of the furthest distance. The third one, however, is the motion of the eccentric sphere from west to east, which takes the epicyclic sphere of the moon with itself to the east, but also southwards or northwards from the ecliptic. Thus, when the deferent sphere moves the centre of the epicyclic sphere southwards or northwards from the ecliptic, it is called the motion of the latitude and that of the inclined circle; but when it is considered in longitude, as is the order of the signs, it is called the fourth motion or the motion of the longitude.[47] *The fifth motion is that of the epicyclic sphere* from east to west, and reverse. *This one moves the body of the moon around the epicyclic centre.* Since this motion is sometimes eastwards and often westwards and as the speed of the course of the moon is varied by it, *it is called the 'anomalistic' motion. Finally, we call that motion the sixth one by which the diameter of the epicycle is moved, whose end is the furthest distance towards the assumed point* in any of the signs. *It has been given the name 'prosneusis'*[48] by the elder astronomers.

[247] But to define the position of the moon on the ecliptic and its latitude from the latter, *we imagine a straight line that originates from the centre of the world and extends to the centre of the moon and from there is prolonged straight until the surface of the great sphere and terminates on the inclined circle. For the centre of the moon is always in the plane of the inclined circle and never inclines elsewhere. We also imagine a great circle which begins at the two poles of the ecliptic and passes through the point where the line from the centre of the world terminates on the inclined circle.* [248] The centre of that circle is the centre of the world. *This circle therefore divides the ecliptic into two halves. The arc of this circle which lies between the inclined circle and the ecliptic is called the 'latitude' of the moon. The point where it divides the zodiac is the place of the moon on the ecliptic.*

The[49] rectification of the position of the moon, and why the first column[50] is in no case applied to the moon's equal position, but to the epicyclic sphere, is a

[47]Arab. defines the 'motion in latitude' as the motion of the centre of the epicyclic sphere 'with respect to the inclined circle', whereas the 'motion in longitude' as its motion 'with respect to the ecliptic'. Stephen possibly considered this misleading, as both motions are measured with regard to the ecliptic, while depending nonetheless on the moon's position on the inclined circle.

[48]'prosneusis': lit., 'contrary motion'.

[49]Having described the individual spheres of the moon and their motions, Stephen inserts here an explanation of corresponding astronomical tables in his *Regule canonis*. After that, he first deals with Ibn al-Haytham's passages 265–71 on the lunar phases, before continuing with passage 249 and the discussion of the lunar eclipses.

[50]'column': lit., 'table'.

pareret inquirenti. IIII^{te} autem addende aut auferende sic ratio colligitur. Facimus circulum in plano similem signorum circulo et super eum ABCD et centrum eius <E>$^{(L)}$. Circinamus etiam inclinatum et super eum AHP, centrum eius in sperali est. Tunc faciemus diametrum AC, et super eum ponimus centrum circuli extra centrum inter punctos A E et super hoc centrum G. Et circinamus super eum circulum extra centrum et super eum ANR, et ponimus arcum AN motum circuli rotundi, qui incepit a puncto A et cucurrit usque ad punctum N. Facimus quoque punctum N centrum rotunditatis, et super eum circumducimus rotundum circulum rotunditatis, et super hunc circulum ponimus scilicet IFL. Tunc educimus lineam de centro similis circuli ENI, et alteram de centro ferentis circuli GFNH. Erit ergo punctus rotundi circuli I longinqua longinquitas | eius, qui videtur de puncto E, scilicet centro signiferi. Punctus autem H longinqua longinquitas eiusdem circuli, qui videtur a puncto G, centro scilicet circuli ferentis. Arcus vero, qui est inter H et I, discordia est motus lune in rotundo circulo, qui positus est in prima tabula rectificationis lune. Post hec motum lunaris corporis ponimus super rotunditatem, que incipit a puncto H versus punctum I F, ponentes locum eius in rotundo circulo in puncto L, et exiet linea de centro mundi ELO. Hec tanget rotundum circulum in uno tantum puncto L. Linea autem, que egreditur a centro rotunditatis, NL, medietas est diametri ipsius rotundi circuli, et est discordia prima. Cui iungitur secunda, que invenitur per longitudinem lune a loco solis, quod est A punctus. Hoc videtur per hanc figuram, que subscripta est.

f. 35r

---

3 AHP] ABP **4** faciemus] facies (L)

complex matter. But we present it to prevent an enquirer from getting aggrieved. Whether the fourth[51] column is to be added or subtracted is understood in the following way. We draw in a plane a similar circle of the ecliptic, and on it ABCD, and its centre E. We also draw the inclined circle, and on it AHP; its centre coincides with that of the sphere. Then we will produce the diameter AC and place on it the centre of the eccentric circle, between the points A and E, and on this centre we place G. Around this centre we draw the eccentric circle, and on it ANR, and we assume the arc AN to be the motion of the epicycle, which started from point A and traversed until point N. We also make point N the centre of the epicyclic sphere, and we draw the epicycle around it, and on this circle we place IFL. Then we produce a line, ENI, from the centre of the similar circle, and another one, GFNH, from the centre of the deferent circle. Then point I of the epicycle will be the latter's furthest distance as seen from point E, that is, from the centre of the ecliptic. But point H is the furthest distance of the same circle as seen from point G, that is, from the centre of the deferent circle. But the arc between H and I is the difference of the lunar motion on the epicycle, which is written down in the first column of the rectification of the moon. After that, we assume the motion of the lunar body on the epicyclic sphere beginning from H towards I [and then to] F, and we assume its position on the epicycle at L; and there will start a line ELO from the centre of the world. This line will touch the epicycle only in one point L. But the line NL, which originates from the centre of the epicyclic sphere, is half the diameter of the epicycle itself, and it is the first anomaly. To this anomaly a second one is added, which is found from the moon's distance from the position of the sun, which is point A. This can be seen from the figure which is given below [Fig. 3.9].

[51] Although a 'fourth' column is mentioned here, Stephen refers instead consistently to the 'third' column of his table in the following discussion.

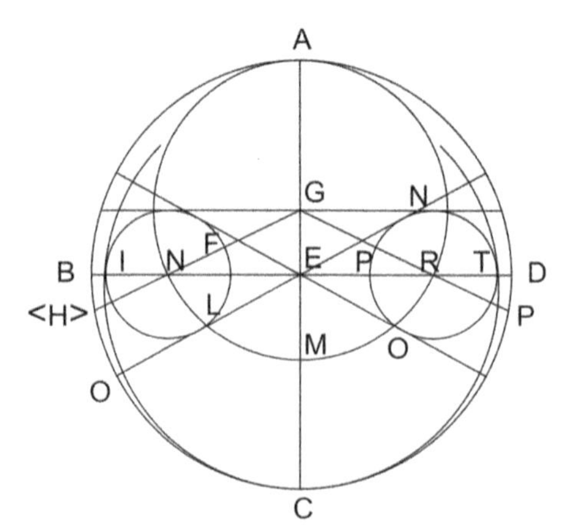

Fig. 3.9: From MS Cambrai 930, fol. 35r. The book page corresponds to the ecliptical plane, hence the inclined circle should appear as an ellipse. According to the description in the text, the inclined circle was conceived as an oval within the circle ABCD, touching the latter at the two points A and C. The illustrator apparently started to draw this as a circle tangential to C, which he then could not extend until A. Difficulties with this diagram are apparent from substantial erasure marks. The right half of the drawing is not referred to in the text, and the labels are ambiguous, as N and P are used twice

Quando vero luna volvitur in prima medietate rotunditatis, super quam sunt H I F, verus lune locus in circulo signorum minor est equali loco, atque idcirco aufertur tercia tabula de equali loco lune. Quando autem in secunda medietate, que vergit ad orientem, super quam posuimus F L H, est verus locus eius in circulo si-
5 gnorum plures continens gradus, quam equalis locus eius denuntiet. Idcirco igitur rectificacio tercie tabule equali cursui eius coadunatur, que licet in subiecta oculis descriptione animadvertere. Verum huiusmodi, quam hic ponimus descriptionem, nichil confert rectificationi, que in canone habetur; alterius enim rectificationis est.

10 [265] ***Q****uare autem luna aliquando tota clara, tociens obscura nichil habens, quod possit ministrare, luminis, modo vero plus quam sui medietas sit corporis, sepe minus habeat claritatis, et crescens atque solem insequentem prefugiens ab ea que solem recipit*
f. 35v *sui medietate, decrescens ab altera sed | que solem respiciat, lumen refundat, in promtu causam licet intueri positam.* Ab antiquis certissimo compertum est lunarem globum
15 lumine carere proprio, sed solis radiorum claritatem susceptam reverberans terris refundere. Hinc est enim, quod mediante ipsam et solem terra in recta linea luna fratris amisso luminis benefitio terre livida invidia ipsa solitum et suum represen-

2 verus] vetus (L) **4** F L H] L F H

But when the moon revolves in the first half of the epicyclic sphere, on which there are the points H I F, the true position of the moon on the ecliptic is less than the equal position, and therefore the third column is subtracted from the equal position of the moon. But when it is in the second half, which extends eastwards and on which we have placed the points F L H, its true position on the ecliptic comprises more degrees than that which its equal position indicates. Therefore, the rectification by the third column is added to its equal course, which the eyes can realise from the diagram below. But a drawing of the kind we give here does not relate to the rectification as used in the *Canon*; for it belongs to a different rectification.[52]

[265] *But the reason why the moon is sometimes full and bright, whereas just as often it is darkened without any light that it could disperse, and why it is sometimes fuller than half of its body whereas it often has less brightness, and why it shines with that half which faces the sun when the moon is waxing and fleeing ahead of the sun, which chases it, whereas when waning it shines with the other half which, nonetheless, is facing towards the sun then, will soon become perspicuous.* The ancients discovered with absolute certainty that the lunar body lacks any light of its own,[53] but that it shines back to the earth the brightness of the sun's rays which it receives and throws back.

[52]In the above simplified description, Stephen ignores the *prosneusis* and the motion of the deferent's centre in the opposite direction from the epicycle. Stephen probably simplified the problem to focus on the difference between the two lunar anomalies, one resulting from the motion of the moon on the epicycle and the other from the eccentricity of the deferent. As a consequence, the present drawing does not correspond totally to the various steps involved in the rectification procedure. This concerns the increment in epicyclic equation due to the epicycle's changing distance from the earth, which was probably the subject of the remaining two columns 2 and 3 of Stephen's table, while including the *prosneusis* would also affect the calculation of the true apogee by means of the first column.

[53]'lumine carere proprio': cf. Macr. *Somn.* I,19,8: 'luce propria caret.'

tet pallorem. Quoniam ergo a sole mutuatur lumen et ipsa spera, nisi sit obiectu terre eclipsis, *semper eius medietatem speralem sol et amplius illuminat.* Sol enim a luna longe maior est, et cum ratione probatur maiorem eius partem illustrari, que tamen variatur interiecti quantitate spatii. Quantum etenim soli propinquior et sub
5 ipso rotatur, tanto plus; quanto autem longius et per oppositionem et rotunditatem, tanto minus lumen illi confert. Contra est visus nostri ratio. Nos enim tamquam luna breviores *minus medio lune intuemur. Luminis igitur ipsius tanta variatio ex ea, quam habet adversum nos positione et solem, fit.* Cum enim, ut dictum est, plus medio ipsius illuminetur, nos minus medio videamus, si inter nos et solem posita fuerit,
10 quod illuminatur, solem respicit, quod pallet, nostro oppositum est visui, et videri solis vicinio negante non potest. [266] *Mutante autem lunari globo eam positionem sue ferentis benefitio, cum .tr. gradus medietatem sui circuli peragraverit oppositionem cum sole faciens, nobis in medio positis medietatis illius, que solis lumine irradiatur, tremulum obicit splendorem* medietatem illam, que naturali pallet livore, altera ex parte
15 occultans. *Quod, ut dictum est, fit, quia luna in panselinio soli per rectum est opposita,* sed sepe latum habens diametrum. Aliter enim omni mense suo in plenilunio fieret ecliptica. Ipsis igitur in diametri capitibus aut confinio positis *nos utrisque sumus interpositi, ut ab altera parte lunam, ab altera opposita videamus solem.* Ex quo fit, ut eam, que illuminatur a sole, medietatem umbra terre eclipticam preterlabentem
20 suspiciamus.

[267] *Pertransiens autem oppositionis et plenilunii loca fit ea medietas, que precedit, soli propinquior, que sequitur, magis remota. Et quoniam luna solis fugit diametrum, alias eius partes solaris luminis pervigil fulgor irradiat, alias vero, que ipsi latent, nos intuemur, quedam nobis et ipsi simul videntur. Non enim iam in medio positi sumus,* nec
25 que a solis centro usque lunare centrum linea dirigitur, nobis proxima est a centro mundi, sed valde remota. [268] *Que quanto a nobis magis elongantur, tanto magis*

---

15 panselinio] paranselinio (L) **23** latent] patent

Therefore, when the earth is in the middle between the moon and the sun on a straight line, the moon is deprived of the shining gift of her brother and in jealous grudge shows the earth only its natural lead-grey paleness. Hence, as it borrows the light from the sun and itself is a sphere, unless there is an eclipse by the intervening earth, *the sun will always illuminate a half of the moon's sphere, and more.* For the sun is much larger than the moon, and one can prove by reasoning that a larger part of the moon is illuminated, which nonetheless changes depending on the distance between them. For the closer to the sun and below the latter it revolves, the more light it will receive from the sun, whereas the further remote, due to opposition and the epicyclic sphere, the less. The opposite applies to our vision. For, being smaller, so to say, than the moon, *we see less than half of it. That great variation of its shining therefore results from the position which it has relative to us and to the sun.* Since, as has been said, more than half of it is illuminated, but we see less than half of it, if it stands between us and the sun, the illuminated part faces the sun while its dark side is turned to us, and the proximity of the sun makes it impossible to see that side. [266] *But when the lunar body moves away from this position by means of its deferent sphere, and when it has traversed 180 degrees, being half of its circle, and thus produces opposition to the sun, it sends to us, being in the middle, the glistening resplendence of that half which is irradiated by the light of the sun,* whereas on its other side it conceals that half which is pale by its natural lead-grey. *This occurs, as we have said, because at full moon the moon stands in direct opposition to the sun;* but its diameter often has a certain latitude. For otherwise there would be an eclipse every lunar month at full moon.[54] Thus, as they are located at the ends of the diameter or close to them, *we are placed between them to both sides, such that we see the moon on one side and the sun on the opposite side.* As a consequence of this, we should expect the half which is illuminated by the sun to avoid becoming eclipsed by the shadow of the earth.

[267] *But when passing through the places of opposition and full moon, the half which precedes gets closer to the sun, and the one that follows gets further remote. And since the moon is avoiding the diameter of the sun and some parts of it are lit by the sun's never ceasing brightness, while other parts that are concealed from the latter are seen by us, certain parts of the moon are simultaneously visible for us and the sun. For we are no longer placed in the middle,* nor is the line which is drawn from the centre of the

[54] 'but its diameter... full moon': the argument is not contained in Arab. K and Y, but it is known from the Hebrew tradition. Its appearance in the *Liber Mamonis* supports Langermann's assumption that it is an original part of Ibn al-Haytham's work.

obscuri, minus videmus clari. Decrescente ergo luna nos a medio solis et ipsius remotos medietas tota clara non despicit. Illa enim, quam nos videmus, hoc quidem in loco, qui soli propinquior, claritatem, hoc autem, qui remotior, habet obscurum.

Ut autem videamus, que pars, clara scilicet aut contrario infecta, maior sit, imaginamur lineam de centro mundi egredientem recte ad centrum lune directam. Hec quamdiu ab ea, quam nos videmus, parte claram attinget antequam centrum, claritatem, si obscuram, hanc quoque vincentem designat. Ascendente ergo ad solem luna minuitur claritas, crescit obscurum, usque dum linea de centro mundi medium tangat obscuri. [269] *Transeunte luna solem inclinatur clara medietas,* linea centri mundi mutat locum, cuius in superficie punctus, quem tangit, polus est vise medietatis circuli. *Hic ergo circulus, qui circundat visam partem <et> dimidium circulum, qui circundat claram sed oblique, quando luna apparere incipiens, partem nobis sue ostendit claritatis parvam.* [270] *Claram, quam videmus in prima, partem duo | arcus terminant eorum, quos diximus, circulorum. Quia vero circuli horum arcuum* non rectis angulis sed oblique *se secant, ipsi quoque arcus oblique positi, clare et vise partis formam in medio amplam, in capitibus subtilem arcuatamque reddunt. Atque ex hoc contingit in principio mensis et fine arcuatam habere formam.*

[271] *Crescit exinde claritas,* et linea centri terre propinquat clare parti. Que cum tetigerit circulum, qui dividit clarum ab obscura, luna in tetragono posita *nobis dimidia lucet, et facit hec linea cum ea, que extenditur ab eodem centro mundi usque centrum solis, equum angulum faciens. Tunc videmus medietatem circuli dividentis clarum quasi rectam lineam, que dividat clarum ab obscuro.* Circulus quoque vise partis hunc dividit rectis angulis, *unde medium clari, medium videmus obscuri,* sicque circulus visionis et clarum et obscurum in duo media dividit. *Exinde claritate vincente obscurum vincitur.*

1 ipsius] ipsi **12** apparere] parere **13** Claram] Clara **14** eorum] quorum **19** obscura] oscura **22** obscuro] oscuro

sun until the lunar centre closest to us from the centre of the world, but far remote. [268] *The further this line removes from us, the more we see of the dark and the less of the bright [parts of the moon]. Thus, at waning moon the bright half does not in its entirety look on us, as we have moved away from the middle between the moon and the sun. And the half which we see has brightness in that part which is closer to the sun, and darkness in that which is further remote.*

But to determine which part is larger, the bright one or the one affected by the contrary, we imagine a line originating from the centre of the world and directed straight to the centre of the moon. As long as it reaches the bright part of the part that we see, before it reaches the centre, it indicates brightness, whereas, if it reaches the dark part, it indicates this one as the winner, accordingly. Thus, while the moon ascends towards the sun, brightness wanes and the dark increases, until the line from the centre of the world touches the middle of the dark. [269] *And when the moon passes the sun, the bright half becomes inclined,*[55] [and] the line from the centre of the world changes the place, on whose surface the point that it touches is a pole of the circle around the visible half. *This circle, therefore, which encloses the visible part and also the semi-circle that encloses the bright part, but in an oblique angle, shows us a small part of its brightness when the moon begins to appear.* [270] *Two arcs of the circles that we have mentioned define the bright part that we see in the beginning.*[56] *But since the circles of these arcs intersect* not in right angles, but obliquely, *and these arcs themselves are also placed obliquely, they render the shape of the visible bright part wide in the middle, but fine and curved towards the ends.*[57] *For this reason it happens to have a curved form at the beginning and at the end of the month.*

[271] *From then on the brightness increases,* and the line from the centre of the earth approaches the bright part. When it has touched the circle that divides the bright from the dark part, the moon stands in a rectangle and *half of it shines to us, and this line produces a right angle with the line that extends from the same centre of the earth until the centre of the sun. In that moment we see the half of the circle which defines the bright part as if it was a straight line that divides the bright from the dark.* Also, the circle around the visible part divides that circle in right angles, *hence we see half of the bright and half of the dark,* thus the circle of vision divides both the bright and the dark parts into two halves.[58] *From then on, brightness is winning and the dark is*

[55] 'inclined': Following K, Langermann reads malīʾan, 'full', rather than māʾilan, as in Y. The latter seems to make better sense, though.

[56] Arab. repeats the definition of the circles.

[57] 'But since... ends': Arab., 'because this small part is circumscribed by two arcs and it is on the surface of a sphere, it is a crescent.'

[58] 'Also, the circle... halves': Arab., 'the result is that what is seen of the lit-up part is encompassed by a straight line and half the circumference of the circle which bounds the part near us. For this reason, the lit-up part of the moon which is seen in these instances is a semi-circle.'

Manifestum ergo ex his, que posita sunt, credimus, quoniam quamdiu linearum angulus, cuius finis est centrum mundi, latus erit, claritas vincet, cum latissimus, erit et plenilunium; si non erit angulus, fiet eclipsis lune. Quamdiu autem acutus, obscurum superhabundat, quanto acutior, tanto maius; si acutissimus,
5 novilunium, si nullus, eclipticum solem luna facit.

Et hec est figura eorum, que diximus.

3 Quamdiu] Quando diu (L) **5** novilunium] plenilunium (L)

being defeated.

We believe it is clear from what has been said that as long as the angle between the lines which is defined at the centre of the world is a wide angle, brightness will win, and when it is very wide, there will be full moon, and if there is no angle at all, an eclipse of the moon will occur. Whereas, as long as the angle is acute, the dark will prevail, and the more so, the more acute the angle is; and if the angle is very acute, there will be new moon, and if there is no angle at all, the moon makes the sun eclipsed.

This is an illustration of what we have said [Fig. 3.10].

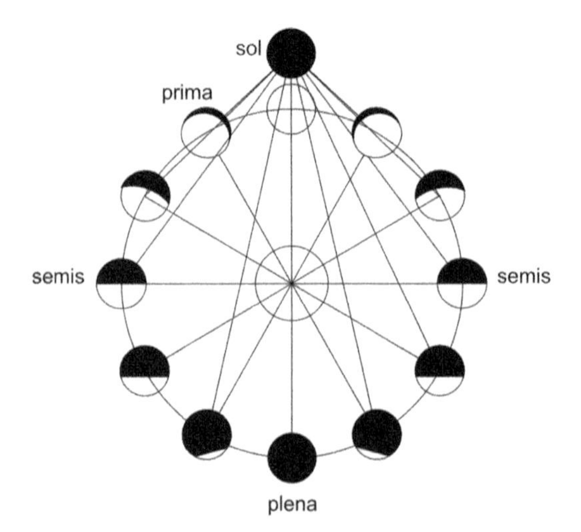

Fig. 3.10: From MS Cambrai 930, fol. 36r. Labels: "the sun" (sol), "first <visible moon>" (prima), "half <moon>" (semis), and "full <moon>" (plena)

f. 36v | Sed quoniam luna sepe fit ecliptica et ipsa quandoque defectum ingerit soli, horum quoque rationem, et quando fiat, dicere non parvam habet utilitatem. Et primo quidem de solari defectu, post lunarem eclipsim videbimus.

Qua in re linee centri mundi et circuli, qui transit per punctum alte spere, imaginatio pernecessaria teneatur, et quomodo latum lune et longum varietur videamus. Luna igitur per coadunationem, quos habet, motuum, quorum pars in orientem, pars vero nititur in occidentem, currente ipsa quidem celeritate ferentis et quandoque subsidio rotunditatis, ut dictum est, in orientem, sicut est ordo signorum.

[249] *Imaginaria quoque linea, que de centro mundi egreditur et recte directa ad centrum lune usque superficiem alte spere porrigitur et terminatur in circulo inclinato, motu lune movetur tangens nunc hos, nunc illos, inclinati punctos.* [250] *Movetur etiam*

---

3 de solari] *bis habet et correxit* (L) **8** orientem] oriente (L) **11** porrigitur] porrigatur (L)

But since the moon is often eclipsed and itself also sometimes inflicts evanescence on the sun, it is of great use to speak about the reason of these things and when it occurs. We will first deal with the solar evanescence and afterwards with the lunar eclipse.

In this matter, one must imagine a line from the centre of the world and a circle that passes through [that line's] point on the high sphere, and we should keep in mind how the moon's latitude and longitude change. For the moon moves by a combination of the motions which it has, some of which tend eastwards and some westwards, but by the speed of the deferent sphere, and sometimes enhanced by the epicyclic sphere, as we have said, it moves eastwards, as is the order of the signs.

[249] *Also the imaginary line, which originates from the centre of the world and is directed straight to the centre of the moon and is extended until the surface of the high sphere and terminates on the inclined circle, is moved by the motion of the moon, touching*

eodem motu circulus, qui transiens per polos signorum et punctum, quem tangit linea, *et inclinatum et signorum circulum in duo media dividit. Ex quo fit, ut ille punctus, in quo circulum zodiacteum secat, movetur. Hic est motus lunaris corporis, quem videmus in longitudine signorum fieri.* [251] *Motus autem ille, quo linea recte a centro mundi per centrum lune ad circulum inclinatum directa modo ad austrum modo ad septemtrionem movetur, dicitur motus lune visus in lato. Per hunc enim motum variatur arcus longinquitas, que est inter duos circulos,* inclinatum et signiferum, *et modo quidem crescens, sepe autem decrescens,* aliquando nullus invenitur. *Hinc igitur procedit, ut latum lune discors sit,* nunc maius, nunc minus, sepe nullum.

[252] *Quando motu ferentis punctus ille divisionis movetur, in quo se dividunt <circulus signorum et> signorum circulus polorum* et caput linee secans hunc et inclinatum, *procedit et lune corpus in longum. Cum autem venerit ad* circulum, qui transit super polos signorum et centrum solis, fit ille circulus, qui per centrum lune transit, cum circulo solis unus, et lineis utrisque, que egrediuntur de centro mundi, ad diversa tendentibus sed in eodem circulo finem habentibus sic, ut utreque in eadem circuli medietate a polis finiantur, *fit sinodus solis et lune. Cum autem* in altera medietate, que non habet solem sed opposita est illi, que gestat, punctum diametro solaris circuli ab eo puncto in quo sol volvitur proximum tetigerit, *panselinium fieri intuemur in eundem locum coeuntibus solis et lune ad designanda eorum loca <lineis>*[(L)] *imaginatis.*

[253] *Quando vero caput linee, que incipit a centro mundi et transit per centrum lune et movetur in superficie inclinati circuli, venerit in quemvis punctum divisionis, quo ligantur* et polorum circuli signorum circulus *et zodiacus et primus,* quorum alter caput draconis dicitur et alter secundus draconis cauda, *si quidem fiat solis et lune sinodus,* tunc inveniuntur utriusque centra solis et lune in uno eodemque loco esse circuli

---

11 caput] capitis **18** panselinium] paranselinium (L)

sometimes these, sometimes other points of the inclined circle. [250] *By the same motion also the circle is moved which runs through the poles of the ecliptic* and through the point that is touched by the line,[59] *and which divides the inclined circle and the ecliptic into two halves. It results from this that the point in which it crosses the zodiac moves. This is the motion of the lunar body which we observe to occur in longitude of the ecliptic.* [251] *But that motion by which the line that is directed straight from the centre of the world, through the centre of the moon, to the inclined circle is sometimes moved to the south and sometimes to the north, is called the observed motion of the moon in latitude. For it is by this motion that the length of the arc that is between the two circles,* namely the inclined circle and the ecliptic, *changes and is sometimes found increasing, often decreasing,* and occasionally not existent at all. *It follows from this that the latitude of the moon varies,* being sometimes larger, sometimes less, and often nil.

[252] *When by the motion of the deferent sphere that intersection point moves, in which the ecliptic and the circle through its poles* and also the end of the line which cuts through the latter and the inclined circle[60] *divide each other, the body of the moon will also move in longitudinal direction. But when it has reached the* circle that passes through the poles of the ecliptic and the centre of the sun, the circle that runs through the centre of the moon and the circle of the sun become one; and when the two lines,[61] which both originate from the centre of the world and run in different directions, but nonetheless terminate on the same circle, do so such that they both terminate in the same half of that circle as seen from the poles,[62] *a conjunction of sun and moon occurs. Whereas, when it meets the point* which is closest to that point where the sun stands on the diameter of the solar circle, but on the other half, which does not bear the sun but is opposite to the one which bears it, *we see a full moon occurring at this place, when the imaginary lines of sun and moon align, which determine the latters' positions.*

[253] *But when the end of the line which starts from the centre of the world and passes through the centre of the moon and is moved in the plane of the inclined circle reaches any of the intersection points in which* the circle through the poles of the ecliptic, *the zodiac and the first [i.e., the inclined] circle are bound to one another,* one of which is called the head of the dragon and the other, second, one the tail of the dragon, *and*

[59] The insertion is necessary for determining the circle and its motion.

[60] 'that intersection point... inclined circle': Arab., 'the point of intersection of the ecliptic circle and the circle passing through its poles, which moves on the circumference of the ecliptic circle.'

[61] 'two lines': that is, one to the sun and another to the moon.

[62] 'the circle that passes... poles': replacing Arab.: 'the point opposite the centre of the sun.' Stephen unnecessarily complicates the discussion by introducing an additional circle for the sun.

signorum, et circulus quidem unus est, *et linee, que egrediuntur de centro mundi et transeunt per utriusque centra, altera per alteram, una sunt. Eo igitur quo tempore hec fiunt, eclipsis solis esse dicitur,* idcirco quod lunaris globus nobis et solis corpori medius est interpositus. [254] *Aspectus enim nostrorum oculorum, qui extenditur ab ipsis oculorum pupillis usque ad lunare corpus, per lineam centri terre porrigitur. Qui cum ad corpus lune solidum et impenetrabile visui pervenerit, ultra dirigi prohibetur a lunari globo, nec ad solem potest venire. Sic igitur posita inter nostrum intuitum et solis globositatem lune spera solis lumen amittimus,* et umbra lune terram percutit, | que quoniam minor est terra—hoc enim etiam ipsius eclipsis comprobat, sicut in sequentibus videbitur—non ubique terrarum plenum facit solis apparere defectum. Deficit enim ipsius umbra, que per solem fit, et minor terra partem terre spere operit. In his ergo partibus, que tote umbra lune operiuntur, plenus est solis defectus; in aliis autem his quidem minor, maior autem illis. Que variatio fit propter affinitatem locorum, in quibus plenus. Nam illa loca, que non vident plenam solis, quando fit, eclipsim et sunt vicina illis, a quibus operitur totus, maiorem vident eius defectum aliis remotioribus.

[255] *Cuius rei causa est, quod spiritus visus nostri formam facit coni, cuius caput est pupilla oculi nostri, sedes autem superficies corporis, quod intuemur. In tempore ergo eclipsis solaris intuentibus nobis lune corpus radius oculi nostri dilatatur in coni mensuram, cuius caput nostri pupilla est visus, sedes autem superficialis rotunditas corporis lune.* [256] *Huius igitur superficiem imaginamur rectis lineis ab omni parte extensa crescentibus minutatim ipsius circuli protelari usque ad solare corpus, cuius superficies si totum globum solis undique concluserit, totus sol intra conum positus a nobis per lune corpus operitur. Quando ergo sic se habet coni et solis ratio, totus est eclipticus. Si vero totus intra coni non cadit superficiem, sed pars eius intus est a parte altera exterius lucente, illam solum modo lune corpus a nobis operit, que in coni cadit superficiem. Operta ergo tantummodo defectum eclipticum patiente altera nichil pati dicitur, que exterius posita terris lucem ministrat.*

3 solis] soli (L) **4** oculorum] circulorum (L) **5** oculorum] circulorum (L) pupillis] *corr. ex* pupupillis (L) **9** hoc] hec sicut] sic (L) **18** autem] aut (L) **23** concluserit] concluserint **24** operitur] aperitur (L)

if there occurs a conjunction of sun and moon, then both centres, that of the sun and that of the moon, are found in one and the same position on the ecliptic, and there results a single circle; *and also the lines which originate from the centre of the world and pass through either centres, that is, each line through one centre, become one.*[63] *In that moment when this happens, a solar eclipse is said to occur,* because the lunar body is placed in the middle between us and the body of the sun. [254] *For the sight of our eyes, which extends from our eyes' pupils until the lunar body, is directed along a line from the centre of the earth.*[64] *When it reaches the body of the moon, which is solid and impervious to sight, it is prevented by the lunar body from extending further, and it cannot reach the sun. Thus, when the sphere of the moon is located in that way between our perspective and the round body of the sun, we lose the light of the sun,* and the shadow of the moon strikes the earth; since it is smaller than the earth—this is also proven by the eclipse of the moon, as will be seen in the following—it does not cause a complete evanescence of the sun everywhere on earth. For its evanescence is caused by the moon's shadow, which is produced by the sun and, being smaller than the earth, covers only a part of the sphere of the earth. Hence in those parts which are completely covered by the moon's shadow, the sun's evanescence is total; whereas in other places, it is less in some and more in others. This variation occurs in accordance with the distance from the places where it is total. For those places which do not witness a total solar eclipse when it occurs, but are close to those places from which the sun is completely concealed, see the sun's evanescence larger than other places further remote.

[255] *The reason for this is that the breath of our vision produces the form of a cone, whose apex is the pupil of our eye and whose base is the surface of the body that we look at. Thus, when we watch the body of the moon at the time of a solar eclipse, the ray of our eye propagates in the shape of a cone whose apex is the pupil of our sight and whose base is the round surface of the body of the moon.* [256] *We therefore imagine its surface gradually projected along straight lines that extend from every part of its circle until the body of the sun; if its surface encircles the entire body of the sun on all sides, the sun is placed entirely within the cone and will be concealed from us by the body of the moon. Thus, whenever the relation of the cone and the sun is such, it is totally eclipsed. But when it does not fall into the surface of the cone in its entirety, but one part of it is within and another part is shining outside, the moon will conceal from us only that part which falls inside the surface of the cone. While thus only the concealed part suffers ecliptic*

[63] Arab. mentions only the line through the moon.

[64] Arab. add.: 'or close to it.'

[257] *Solis ergo corpore toto eclipsim patiente, postquam ad perfectionem et defectus complementum pervenerit, luna transeunte ferentis impetu velocissimo statim* ab occidentis partibus *incipit iterum videri solis claritas* subito amissa miraculo. *Hoc autem idcirco evenit, quod conus intuitus nostri eo in loco, quo solis corpus circumcingitur,* 5 *equale* ab antiquis inventus est *habere diametrum solaris corporis diametro,* quando luna in longinqua longinquitate circumvolvitur. *Quia igitur solis defectus per lunam fit, luna* ad orientis *partes festinante coni tota superficies cum ea movetur. Sic ergo ut eclipsis solaris ad complementum pervenerit, statim* ab occidente *claritas apparet, quia luna numquam in eodem perstante conus quoque totus movetur, et coopertus sol de cono* 10 *egreditur.* Ex quo manifestum est solis defectum nullo artari statu. Ut enim plenitudinis modum tetigerit, statim incipit minui, sicut et de inclinatione solis, que a puncto temporis terminatur, dictum est.

Placet autem secundum lune et solis ipsius in suis sperarum circulis positionem subtilius intueri solis defectum. In solis enim eclipsi luna, quemadmodum 15 superius dictum recolimus, in ferenti quidem circulo semper in longinqua longinquitate spere extra centrum posita, sed aliquando longinqua longinquitate sue rotunditatis quandoque in eiusdem propinqua longinquitate, sepe in mediis harum duarum locis. Ob igitur hoc per numerum rotunditatis rectificatum per primam tabulam numeros, quos de eclipsi solis, alteros quidem in longinque, alteros in 20 propinque longinquitatis, tabula colligimus, rectificare iubemur, sicut est in regulis dictum editis. Posita enim luna in longinqua longinquitate rotunditatis solis defectus huiusmodi est, quem proxime posuimus, ut postquam ad plenum pervenerit, statim minuatur, et nullam recipiat rectificacionem numerus in longinque longinquitatis tabula collectus. Quod si in longinqua propinquitate rotunditatis f. 37v 25 erit, maior fiet eclipsis, quia | luna terre propinquior, <a> sole remotior, maiorem facit in terra umbram. Quod si invenietur aliquando esse numerus ille, qui est in longinqua propinquitate, solis eclipsim pronuntiat; sin autem in horum mediis, tunc rectificacione indiget.

---

4 circumcingitur] *corr. ex* circumcingintur (L) **8** statim] statin (L) **17** propinqua] propinquitate (L) **19** numeros] numerus (L) **20** longinquitatis] longinquit

evanescence, the other part, which lies outside and spreads light to the earth, is said to be not affected in any way.

[257] *When the entire body of the sun suffers eclipse, after it has come to perfect and complete evanescence, the moon is passing through by the fast motion of its deferent sphere, and the sun's brightness,* which was suddenly and so miraculously lost, *immediately becomes visible again* from its western side. *This happens because,* as the ancients discovered, *where the cone of our vision encircles the body of the sun, it has the same diameter as the body of the sun* when the moon revolves at its furthest distance.[65] *Thus, since the evanescence of the sun is caused by the moon, when the moon speeds eastwards, the entire surface of the cone is moved with it. As soon as the solar eclipse has reached completion, therefore, the brightness will immediately reappear* from the west, *because as the moon never stands still in the same place, also the entire cone is moving, and the conceiled sun moves out of the cone.* It is clear from this that the evanescence of the sun can by no means be forced to stay. For as soon as it has reached the state of completeness, it immediately starts to diminish, as has also been said about the sun's inclination, which is determined only for an instant of time.[66]

But one should more diligently consider the evanescence of the sun according to the positions of the moon and the sun itself on the circles of their spheres. As we remember what has been said above, at a solar eclipse the moon is always located on the deferent circle at the furthest distance of the eccentric sphere, whereas in its epicyclic sphere it is sometimes at the furthest distance or at the closest distance and often between these two positions. For this reason, therefore, we must rectify the values that we have collected in the table for a solar eclipse, some at furthest distance and others at closest distance, by applying the rectified number of the epicyclic sphere from the first column, as is said in the published *Rules*. For when the moon is placed at the furthest distance of the epicyclic sphere, the evanescence of the sun will be as we have just described, namely that as soon as it has reached completion, it will immediately diminish, and the number which is recorded in the column for furthest distance should not become subject to any correction. And if it is at the closest distance of the epicyclic sphere, the eclipse will be larger, because the moon, being closer to the earth and further remote from the sun, produces a larger shadow on the earth. And if the number is once found to be that at the

[65] 'when... furthest distance': this limitation is also found in all Hebrew translations and in L1.

[66] That is, even without the fast motion of the moon the period of total eclipse would still be limited to an instant due to the sun's own motion on the ecliptic.

Solis quoque, ut dictum est, circulus extra centrum mundi positus longius propiusque quibusdam sui partibus terre accedit. In huius longinqua longinquitate aliquando sol eclipsim patitur, aliquando in propinqua longinquitate deficit, quandoque in harum mediis. Que eius positio ipsius variare eclipsis quantitatem dubium esse non debet. Si enim in longinqua longinquitate sui circuli deficiat, quocumque in sue loco rotunditatis posita sit luna, plenus est defectus, variabilis tamen ex ipsius lune situ. Si autem in longinqua propinquitate et luna in eadem in sua rotunditate, aut plenus est aut parum a pleno minor defectus. Si vero luna in utraque sui longinqua longinquitate fuerit, et rotunditatis et spere extra centrum, nequaquam plenam inferet soli eclipsim. Deprehensum est enim a quodam sollertissimo et astronomie scientia peritissimo philosopho geometricali argumentacione, quod luna et sole in locis de quibus loquimur positis, si linea, que de pupilla oculi egreditur recte, per utriusque centrum in medio eclipsi transierit, que est columpna coni, solis tantam esse quantitatem propter terre viciniam, et lune tam parvam propter maiorem sui absentiam, quod quidam circulus solis circa lunam extra coni superficiem videatur, eo quod totum non possit luna solem occulere. Hoc quidem ille ratione inventum tradidit scripture, et alter antiquorum sapientie benivolus, sicut prior rationibus probaverat, longo post tempore se idem oculis vidisse subscripsit. Iam de his, que in mediis fiunt, defectibus facile est iudicium.

[258] ¶Lune quoque eclipsis eisdem fere contingit rationibus, sed in oppositis locis. *Si enim contigerit lineam, que de centro mundi egressa ad lune centrum et inclinatum porrigitur, punctum quemvis ligantium tangere et erit centrum solis in opposito,* ut si lune quidem centrum in cauda draconis, solis in capite, si hoc in illo, hoc *in ista, tunc invenientur tria centra, solis, terre et lune, super eandem esse lineam. Ex hoc ergo fit luna defectiva,* [259] *quoniam cum proprio carens a sole lumen mutuetur, obiecta terra lumen solis amisit. Quando enim opposita est luna soli sub recto diametro, pleno, ut dictum est, orbi terre lumen de ea refunditur.*

---

7 in sua] visu **11** geometricali] geometrali **23** hoc] hec hoc] hec

closest distance, it announces a solar eclipse. But when it is between these two, it needs correction.

Also the eccentric circle of the sun, as has been said, is more remote from the earth in some of its parts, whereas it comes closer to the earth in other parts. The sun suffers an eclipse sometimes at the furthest distance of that circle, sometimes it gets evanescent at the closest distance, and sometimes between them. There should be no doubt that this position of the sun changes the extent of its eclipse. For if it becomes evanescent at the furthest distance of its circle, the evanescence will be total regardless of where on its epicyclic sphere the moon is placed, although it varies nonetheless with the position of the moon. But if it is at the closest distance and also the moon is at the closest distance of its epicyclic sphere, the evanescence is either complete or just a little less than complete. But if the moon is at both of its furthest distances, at that of the epicyclic sphere and that of the eccentric sphere, it can by no means inflict a total eclipse on the sun. For a most astute philosopher,[67] who was also very experienced in astronomy, discovered by geometric reasoning that when the moon and the sun are located at the positions we have mentioned, and if a line that originates from the pupil of the eye passes through both centres at mid-eclipse, which line is the axis of the cone, the size of the sun would be so large due to its closeness to the earth, and that of the moon would be so small due to its larger distance, that a certain circle of the sun could be seen around the moon outside the cone, because the moon could not occlude the entire sun. This man wrote down what he had found by reasoning, and much later another friend of ancient wisdom added that he had witnessed with his eyes the same that the former had proved by reasoning. Now it is easy to assess what applies to cases of evanescence that occur at positions in between.

[258] Nearly the same reasoning applies to lunar eclipses, but with opposite positions. *For if a line that originates from the centre of the world and extends to the centre of the moon and to the inclined circle happens to touch either of the nodes of the ligants, while the centre of the sun happens to be at the opposite one,* such that, if the centre of the moon is at the tail of the dragon, the centre of the sun is at its head, [that is,] if one centre is at the head and the other one at the tail, *then the three centres, that of the sun, that of the earth, and that of the moon, are found to be on the same line. As a result of this, therefore, the moon becomes eclipsed;* [259] *because, as it has no light of its own, but borrows it from the sun,*[68] *it got deprived of the light of the sun,*

[67] Cf. al-Battānī, ch. 30, and Nallino's note thereon (L).

[68] Arab. add.: 'since it [i.e. the moon] is a glossy body.'

[260] *Terra autem solidum est et spissum corpus,* et quoniam omne spissum umbram habet, *et terra. Terre autem umbra,* quoniam per solarium claritatem fit radiorum, *semper in oppositum solis dirigitur. Terrene umbre forma conus est,* qui a terra incipiens in puncto finitur. *Ratio est, quod diametrum solaris globi maius est diametro terre.* Unde et solis quantitas terream excedit magnitudinem. *Radius ergo, qui egreditur de corpore solis et percutit terram, pertransit ultra et extenditur longius,* umbraque rotunda est sicut et corpus terre, cuius est, *usque dum paulatim umbra decrescente angustantibus undique radiis et ad nichil prorsus diminuta revertuntur in se invicem radii solis. Sic umbra terre longius producta coni superficiei circumquaque radiorum superficies iungitur, et est a corpore terre usque in finem coni plena obscura umbra, que* ab ipsa antiquitate *noctis nomen sortita est.* [261] *Conus autem ille umbrarum tante longitudinis est, ut speram lune transeat* usque ad alias speras. *Cum ergo* ferentium solis et lune motibus *uterque ad oppositos* ligantium *pervenerit punctos, corpore solis et terre <et> luna <in una> positis linea, terreus globus solis et* | *lune fiat medius. Tunc ergo corpus lune cadit in conum umbrarum terre, propterea quod, ut diximus, umbre conus semper est soli oppositus, et eius longitudo speram lune pertransit. Luna ergo in umbre cono posita terraque solis et ipsius media interiacet, solarium lumen radiorum ab ipsa prohibet. Unde amissa accidentalis luminis claritate in proprie vertitur nature pallorem et videtur obscura.*

f. 38r

[262] *Quando autem centra horum trium,* terre, lune, solis, *in una sunt linea, plenus est,* et magnum habet statum, *lune defectus.* Ex quo manifestum est terre corpus maius esse lunari. Cum enim luna longe sit a terra, et sol terra maior, et umbra terre in modum coni decrescat, si luna, ut quidam autumant, equalis esset terre, nequaquam tota cono umbre clauderetur. Nunc vero cum tota circumdetur et in ea transcurrenda tota obscura duas fere consumat horas, cum longe minor sit coni circulo, quo circundatur, multo magis minor est terra. *Sepe vero fit, ut in eadem non sint posita centra trium linea, sed prope. Tunc non est centrum lune in medio coni, sed iuxta. In quo dicendum est, quod quanto linee erit propinquior, tanto maior eius obscuritas et mensura status erit.* [263] *Si igitur totum lune corpus in conum cadat, fit totum*

3 oppositum] oppositam (L) **7** paulatim] *corr. ex* pallatim (L) **17** et] *add.* (L) **18** accidentalis] occidentalis **20** centra] contra (L) **27** centra] centrum

as the earth moved in its way. For whenever the moon is opposite the sun on a straight diameter, it reflects the light to the earth with its complete disc, as has been said.

[260] *But the earth is a solid and dense body,* and since everything dense casts a shadow, *so does the earth.* But the shadow of the earth is generated by the brightness of the solar rays, *hence it will always be directed in opposition to the sun. The shape of the shadow of the earth is a cone,* which begins from the earth and ends in a point. *The reason is that the diameter of the solar body is larger than the diameter of the earth.* Consequently, also the extension of the sun exceeds the size of the earth. *Therefore, the radiation that originates from the body of the sun and hits the earth passes by and is prolonged further, and the shadow,* which is round like the body of the earth, to which it belongs, *gets gradually smaller by the converging rays on all sides, until it has vanished completely and the sun's rays come back together again. If the shadow of the earth is thus produced further, the surface of the rays coincides all around with the surface of the cone, and from the body of the earth until the end of the cone there is a completely dark shadow, which* since antiquity *has been given the name 'night'.* [261] *But that cone of shadows is that long that it penetrates through the lunar sphere,* until other spheres. *Thus, when* by the motions of the sun's and the moon's deferent spheres *either of them has arrived at opposite* nodal *points, so that the bodies of the sun, and of the earth, and the moon are placed on a line, the earthly body gets into the middle between the sun and the moon. Then the body of the moon thus falls into the cone of shadows of the earth, because, as we have said, the shadow cone is always opposite the sun and its length penetrates through the sphere of the moon. Thus, when the moon is placed in the shadow cone and the earth lies in the middle between the sun and the moon, it holds the light of the solar rays off the moon. Thus deprived of the brightness of the accidential light, it returns to its own natural paleness and appears dark.*

[262] *But when the centres of these three,* that of the earth, that of the moon and that of the sun, *are on the same line, the evanescence of the moon is total*[69] and has a long state. It is clear from this that the body of the earth is larger than the lunar one. For as the moon is far from the earth and as the sun is larger than the earth, and as the shadow of the earth converges in the manner of a cone, if the moon was of equal size as the earth, as some people believe, it could by no means be entirely enclosed by the shadow cone. However, since it is completely surrounded and since it needs nearly two hours to traverse the shadow while being completely concealed, because it is much smaller than the circle of the cone by which it is surrounded, it is much smaller than the earth. *But it often occurs that the centres of*

[69] 'the evanescence of the moon is total': Arab., 'then the moon is in the middle of this cone.'

obscurum et est tota ecliptica. Si vero pars eius intus, pars erit extra conum, que intus est, obscuratur, que extra, lumen habet, et dicitur pars lune ecliptica. [264] Illa quoque, que tota est, ecliptica varia est. *Nam quandoque centrum eius in columpna coni est, et habet eclipsis longiorem statum. Sepe vero non in columpna, sed iuxta, et quanto remo-* 5 *tior, tanto erit status eclipsis brevior, quia citius umbrarum nocte decursa festinantius de cono egreditur.*

Huius quoque eclipsim secundum utriusque luminis positionem in circulis libet intueri. Si enim sole in longinqua longinquitate posito luna in propinqua longinquitate rotunditatis erit, longior est eclipsis, nec oportet numerum longinque 10 longinquitatis tabule rectificari per rotunditatem rectam per primam tabulam. Si luna in longinqua longinquitate ecliptica, supersedendum rectificationi numerorum tabule longinque longinquitatis. Quod si sol longinquam tenet propinquitatem et luna in imo rotunditatis, est quidem magna eclipsis, sed ceteris duabus minor. Si vero in alto luna sit posita, ceteris omnibus erit eclipsis hec brevior. Causa est solis 15 et terre minor ab invicem distantia, quoniam quanto sol ab ea remotior, tanto conus umbrarum longior, quanto propinquior, quia sol terre maior, tanto breviorem efficit terre conum.

[272] *Hec sunt, que de lune speris et circulis dicenda proposuimus. Ut vero clarius et apertius fiat omne, quod dictum est, subicimus formam sperarum lune, et ponimus in ea* 20 *circulos, lineas et cetera, que diximus.* Reliquarum tractatum sperarum quartus liber complexabitur.

Explicit liber tercius; incipit quartus de retrogradatione.

---

1 extra] intra (L) **2** extra] intra (L) **2–3** Illa. . . ecliptica] *suprascripsit* (L)

the three are not placed on the same line, but nearly. Then the centre of the moon is not in the middle of the cone, but close to it. In this regard one must say that the closer it is to the axis, the larger will be its darkness and the extent of that state. [263] *Thus, if the entire body of the moon falls into the cone, the body becomes fully darkened and the entire moon is eclipsed. But if a part of it is inside the cone and a part outside, the part inside gets darkened and the part outside has light, and a part of the moon is said to be eclipsed.* [264] Also a total eclipse occurs in various forms. *For whenever its centre is on the axis of the cone, also the eclipse has the longest state. But often it is not on the axis, but close to it; and the more remote it is, the shorter will be the state of eclipse, because the night of the shadows is traversed sooner and the moon moves out of the cone faster.*

Also the eclipse of the moon is better considered according to the positions of both lights on the circles. For if the moon is at the closest distance of the epicyclic sphere while the sun is placed at the furthest distance, the eclipse lasts longest, and the value of the column for furthest distance should not be corrected by the rectified epicyclic sphere according to the first column. If the moon is eclipsed at the furthest distance, a correction of the numbers in the column for furthest distance is to be omitted. And if the sun occupies the closest distance while the moon is at the depth of the epicyclic sphere, there is still a great eclipse, but a smaller one than the other two. But if the moon is placed at its height, this eclipse will be shorter than all others. The reason is the smaller distance of sun and earth from one another, because the farther the sun is from the latter, the longer is the shadow cone, whereas the closer, since the sun is bigger than the earth, the shorter a cone of the earth it will produce.

[272] *This is what we presented as necessary to say about the spheres and the circles of the moon. But to make all that has been said clearer and more accessible, we add an illustration of the spheres of the moon and we put in it the circles, the lines and the other things that we have mentioned [Fig. A.11].* The fourth book will contain the discussion of the remaining spheres.

Here ends Book III, and Book IV, on the retrogradation, begins.[70]

[70] Book IV contains Ibn al-Haytham's complete discussion of the planets and also deals briefly with the outer spheres. However, according to the title and the following preface to the book, Stephen gives particular importance to the retrogradation of the planets.

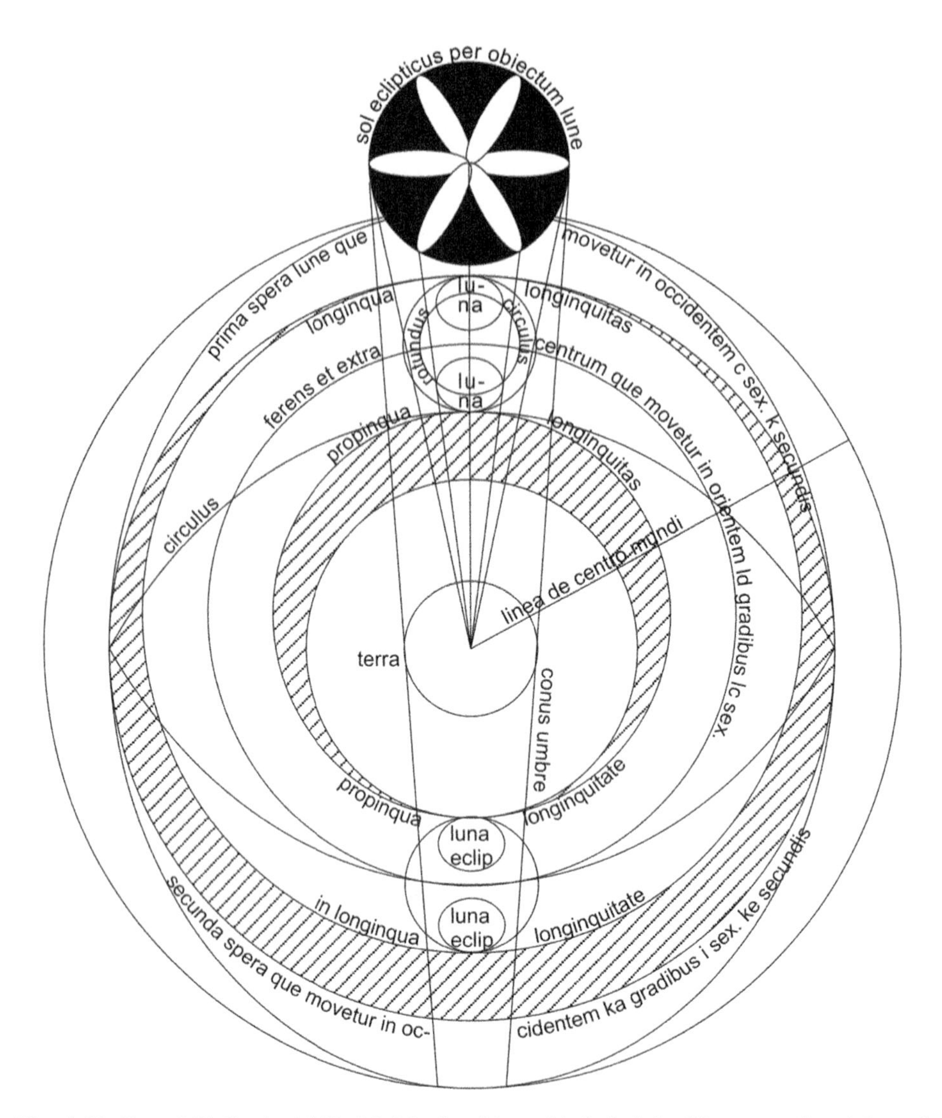

Fig. A.11: From MS Cambrai 930, fol. 38v (hatching added). Labels: "the sun, eclipsed through obstruction by the moon" (sol eclipticus per obiectum lune), "first sphere of the moon, which moves westwards by 3 arc-minutes and 10 arc-seconds" (prima spera lune que movetur in occidentem c sex. k secundis), "furthest distance" (longinqua longinquitas), "the moon" (luna), "epicycle" (rotundus circulus), "deferent and eccentric <sphere>, which moves eastwards by 24 degrees and 23 arc-minutes" (ferens et extra centrum que movetur in orientem ld gradibus lc sex.), "circle" (circulus), "closest distance" (propinqua longinquitas), "line from the centre of the world" (linea de centro mundi), "the earth" (terra), "shadow cone" (conus umbre), "eclipsed moon at closest distance/at furthest distance" (luna eclip. propinqua longinquitate/in longinqua longinquitate), and "second sphere, which moves westwards by 11 degrees, 9 arc-minutes and 15 arc-seconds" (secunda spera que movetur in occidentem ka gradibus i sex. ke secundis)

Book IV

D. Grupe, *Stephen of Pisa and Antioch: Liber Mamonis*, Sources and Studies in the History of Mathematics and Physical Sciences,
https://doi.org/10.1007/978-3-030-19234-1

Quartus hic laboris nostri decursus de .e. planetarum speris et circulis et octava, denique nona spera disserens transcurso maris alto fune anchore portus tranquillo attinget. Verum cum in aliis Arabem quendam plurimum secuti sumus, in hoc quoque permultum sequemur, licet quedam de sperarum numero et rotundi-
5 tatum invenerimus, et de circulis quidem et inclinationibus planetarum vera perstrinxit, a quibus sperarum numerus dissonat. Hoc autem suis in locis aperte monstrabitur. Ceterum cum in precedentibus ingenii vigor velocissimus pernecessarius fuerit et studii exercitium, hoc tamen in labore ultimo precipue invigilandum est propter nimiam rerum subtilitatem, que ratione tantum animi, qua differt homo a
f. 38v 10 belua, percipiantur. Verumtamen | ne pulcherrime philosophie huius simulachrum diutius sua lateat subtilitate, neve inquirentium multimodas quasi fantasticum eludat intentiones, quam aperto poterimus omnia stilo persolvemus, eotenus tamen, ne brevitatis modum possit excedere.

Non enim parva apud Latinos diutius inquievit questio, quonam modo erra-
15 ticorum .e. globi, quorum natura indictus cursus in orientem est, fiant retrogradi, et ab oriente relabantur in occiduas partes. Et hec quidem, ut verum fateamur, questio digna est et proponi et solvi, sed a nemine tamen eorum absoluta. Nec hoc mirum ducimus, cum occulta sit res et geometricalibus exquisita et approbata argumentis, quorum Latinitas inscia in divulgato diu multumque volutatur errore.
f. 39r 20 Cum enim | astutiores horum proposite rei veritatem nulla possent invenire ratione, fictum quoddam et violentum solis radiis concesserunt dicentes eorum maiori impulsu retrogrados fieri planetas, quasi possint amplius solis radii quam ipsarum, in quibus volvuntur, sempiterni cursus sperarum.

Verum id quam frivolum sit, facili patebit argumento. Saturni, Iovis ac Mar-
25 tis retrogradatio in opposito fit solis, Veneris et Mercurii in sinodo. Et Saturni retrogradatio longius ab opposito solis secundum sui tarditatem cursus incipitur

2 fune] funere **8** exercitium] exerticium **15** natura] *add.* (L) **16** hec] hoc **18** approbata] aprobata **20** proposite] *corr. ex* propoisite (L) **21** violentum] violenter radiis] radii

4.1 <Preface to Book IV>

This fourth voyage of our work will deal with the spheres and the circles of the five planets and with the eighth and, finally, the ninth spheres, and, having thus traversed the high seas, our endeavour will reach the haven with a quiet rope of the anchor.[1] But whereas in the other books we have mostly followed a certain Arab, whom we will also follow in many aspects of this book, we nevertheless found by ourselves some facts about the number of the spheres and of the epicyclic spheres. For although that man has grasped the truth behind the circles and the inclinations of the planets, the number of spheres does not harmonise with it. This will be demonstrated in a perspicuous manner at its proper places. Furthermore, just as in the preceding discussion much mental strength and astuteness and also exercise of learning were necessary, we must again be particularly attentive in this last part of our work because of the extraordinary complexity of the subject, which can only be grasped by the intellect, by which man distinguishes from beast. But to prevent the most beautiful picture of this philosophical discipline from continuing to lie concealed by its complexity or eluding like a phantasm the manifold attempts of enquirers, we will explain everything as clearly as we can, though without exceeding the appropriate concise length.

For a long time there has been a gnawing, substantial uncertainty in the Latin world about how the bodies of the five planets, whose course as ordered by nature is eastwards, can become retrograde and fall back from east to west. To admit the truth, this is a worthy problem to tackle and solve, but it has not been solved by any of them. This should not surprise us, since it is an abstract matter that is to be investigated and confirmed by geometrical proofs, which the Latin world has no knowledge of and therefore has long been enshrouded by many popular misapprehensions. Thus, as even the most intelligent among them could in no way find the true reason of the proposed matter, they allowed the sun's rays to have something imaginary and forceful, saying that the planets became retrograde by those rays' stronger impact, as if the sun's rays could be more powerful than the eternal courses of the spheres in which the planets move.

But the falsity of this will be obvious from a simple argument. Saturn, Jupiter and Mars become retrograde when they are in opposition to the sun, whereas Venus and Mercury do so when they are in conjunction with it. And the retrogra-

---

[1] Stephen resumes the picture of a journey on the sea, which he introduced in the preface to Book II.

et pluribus fit diebus, paucioribus Martis, Iove medium eorum, sicut medius est, tenente. Sed Mars soli propinquior, minus Iuppiter, remotior Saturnus. Quod si retrogradatio eorum vi radiorum solis fieri dicatur, <qui> propinquior est pluribus, tamquam cui a proximo prevalente vis inferretur, paucioribus qui remotior retrogradi deberet.

Amplius, si ab opposito sui radiorum solis violentia repellens illos facit retrogrados, Veneris et Mercurii nulla est retrogradatio. Numquam enim ad oppositum possunt ascendere. Quod si retrahi etiam a sinodo dicantur, fient et reliqui eadem ratione secundo retrogradi; nec hii duo solem aliquando precurrerent, sed citiores ad solem accederent, violentia radiorum repercussi retrocederent. Sic illis semper sepiusque repulsis nunquam cum sole sinodare liceret. Quare tamdiu inveterata cum magistro errore cadat sententia, et quod de his certissime iudicandum sit, in nostro videamus opusculo. Dissolvemus enim et huius et complurium questionis nodos, et inquirentem ab errore repressum veritatis semite inducemus.

Nunc igitur a spera Mercurii, que ceteris diutius tractandi genus desiderat, incipiemus, qui etiam a luna secundus est, deinde Veneris speras et circulos, post quam trium, quorum idem fere est tractatus pariterque omnium rectificandorum et retrogradiendi causa, exequemur. Quibus absolutis octave spere et imaginum eius mentio, quanta decet, inducetur. Novissimum huius operis stilum altissima spera, quam aplanos dicunt, vendicabit.

---

1 Iove] Iovem (L) **13** et] *add.* (L)

dation of Saturn begins farther from its opposition to the sun in accordance with its slow course and lasts for more days, whereas that of Mars lasts fewer days, and Jupiter keeps the mean between them, just as it is located in the middle. But Mars is the closest to the sun, Jupiter less, and Saturn farthest away. Hence, if their retrogradation is said to be caused by the power of the sun's rays, the closer planet would have to retrograde during more days, since the force would work on it like from an overwhelming neighbour, whereas the one more remote would have to retrograde during less.

Furthermore, if the power of the sun's rays made the planets retrograde by repelling them from the position opposite the sun, Venus and Mercury would not have retrogradation. For they can never reach until opposition. And if one said that they were dragged back also from conjunction, by the same reasoning also the other planets would become retrograde a second time; and, these two could never precede the sun, for as soon as they caught up with the sun by their faster motion, they would be thrown back by the force of the rays and follow behind. Thus, being thrown back over and over again, they could never be in conjunction with the sun. This long lasting doctrine shall therefore fall together with the error that taught it, and in our little treatise we shall find how these phenomena are to be assessed in the most certain way. For we will cut the knots of this and many other problems, and along the path of truth we will introduce the enquirer who was kept down by error.

We will now start with the sphere of Mercury, which requires longer treatment than the others and which is also the second planet after the moon, then we will discuss the spheres and the circles of Venus, after which we will continue with those of the three [outer] planets, which require a nearly identical discussion and to which the same cause of rectification and retrogradation applies. When this has been fulfilled, we will also give a description of appropriate length of the eighth sphere and its signs. The last words of this work will be demanded by the highest sphere, which is called the 'aplanos'.

[273] ¶*Spera igitur Mercurii sperale corpus est, quam dividunt due sperales et parallelle superficies, centrum quarum et ipsius mundi est. Harum alta tangit inferiorem superficiem spere Veneris, ima vero altam lune.* [274] *In alta eius superficie circulus est, quem factum diximus per motum spere solis.* [275] *Hec spera Mercurii movet circa centrum circuli signorum in duobus fixis polis suppositis ab occidente in orientem tardo motu,* sicut de solis spera dictum est.

[276] ¶*De huius* et in huius corpore *spere alia separatur circundata duabus parallelis superficiebus, quarum centrum eiusdem spere est extra centrum mundi positum et extra superficiem circuli similis. Hec movetur super suos duos fixos polos circa suum centrum, qui non sunt poli spere signorum, ab oriente in occidentem. Huic est nomen spera referens ferentem speram.*

[277] *Separatur et de hac alia spera* eadem proportione, qua ipsa de prima, *et dividunt eam due sperales paralelleque superficies, centrum quarum eiusdem spere centrum quidem et extra centrum mundi et extra centrum referentis superficiem quoque circuli similis positum est in eam partem, in qua et centrum referentis inclinatum. Et longitudo eius a centro referentis medietas longitudinis centri referentis a centro mundi. Huius alta et exterior superficies et exteriorum superficies* | *tangit altam superficiem prime spere in uno puncto,* inferior interiorem. *Hec movetur in duobus diametrice oppositis fixis polis, qui non sunt suppositi polis quarumvis duarum ab occidente in orientem, et dicitur ferens spera.*

f. 39v

[278] ¶*De hac quoque quarta separatur, que continetur inter utramque eius parallellam superficiem tota, et centrum eius equaliter distat ab utroque polo ferentis. Hec circumdatur una superficie* interius solida, cuius centrum spere est, *et tangit utramque superficiem ferentis in uno puncto, altero alteram. Ex quo constat eius diametrum equalem esse longitudine spatio ferentis, que est inter duas superficies paralellas. Motus huius est in propriis duobus fixis polis* ab occidente in orientem, *nomenque rotunditas Mercurii.*

5 occidente in orientem] oriente in occidentem **11** ferentem] referentem **19** polis] polum (L) **21** continetur] continet **22** tota] totam

4.2 <The Sphere of Mercury>

Config. 11: The Orb of Mercury

[273] *The sphere of Mercury is a spherical body which is defined by two spherical and parallel surfaces whose centre is that of the world. The high one of them is tangent to the lower surface of the sphere of Venus, and the low one to the high surface of the moon.* [274] *On its high surface there is a circle*[2] *which we have said is produced by the motion of the sphere of the sun.* [275] *This sphere of Mercury moves with a slow motion around the centre of the ecliptic about two fixed subjacent poles from west to east,* just as has been said about the sphere of the sun.

[276] *Separated from this sphere* but within its body *is another one, which is surrounded by two parallel surfaces whose centre is that of this sphere itself, which centre is placed outside the centre of the world and outside the plane of the similar circle.*[3] *This sphere moves around its centre from east to west about its two fixed poles, which are not the poles of the ecliptic. Its name is the 'referent sphere' of the deferent sphere.*[4]

[277] *From this one, another sphere is separated* by the same proportion as the former from the first, *and it is also defined by two spherical and parallel surfaces whose centre is the centre of that sphere itself, which centre is placed outside the centre of the world and outside the centre of the referent and also the plane of the similar circle, in the same direction in which also the centre of the referent is inclined. Its distance from the centre of the referent is half the distance of the referent's centre from the centre of the world. Its high and outer surface touches the high surface of the first sphere,* and its low one the inner surface of the first sphere, *in a single point [each]. It moves from west to east around two diametrically opposite fixed poles, which are not located beneath any others, and it is called the 'deferent sphere'.*

[278] *Also from this one a fourth sphere is separated, which is completely contained between the former one's two parallel surfaces, and its centre is equally distant from both poles of the deferent sphere. It is surrounded by a single surface* that is solid inside and whose centre coincides with that of the sphere, *and it touches each surface of the deferent in a single*[5] *point, each at one point. It is certain from this that its diameter is equal to the distance between the two parallel surfaces of the deferent. Its motion is around its own two fixed poles* from west to east *and its name is the 'epicyclic sphere' of Mercury.*

---

[2] Arab. add.: 'parecliptic.'

[3] Arab. add.: 'Its upper surface is tangent to the upper surface of the first orb in one common point.'

[4] Lit.: 'the sphere that carries back the deferent sphere'.

[5] K adds: 'common', which is missing from Y and all translations.

[279] ¶*Corpus autem stelle, quam Mercurium dicunt, plenum et solidum est. Forma eius speralis positumque est in corpore rotunditatis. Mercurii superficies superficiem rotunditatis tangit in uno puncto.* [280] *Quando ergo movetur ferens spera super proprios polos, movet suo motu rotunditatem, et defert eam secum* ab occidente in orientem *circa suum centrum. Facit ergo rotunditatis centrum in ferentis spere medio corpore imaginarium circulum, centrum cuius ipsius est spere. Huius circuli imaginaria circumductio non est in superficie circuli similis signorum circulo,* sed inclinatur sicut et poli eius. [281] *Imaginamur autem circuli crescere superficiem, que totum secet mundum, et facere in duabus superficiebus ferentis duos sibi et ad invicem paralellos circulos, quorum omnium centrum ferentis est spere.*

[282] *Facit autem in superficie sue prime spere, cuius centrum idem est quod et mundi, duos paralellos circulos, alterum interius, alterum extra, quorum centrum idem est quod et spere mundi. Eorum maior ille, qui in alta circumductus est superficie, dividit oblique circulum similem signorum circulo in duobus diametrice oppositis punctis, et dicitur hic circulus inclinatus.*

[283] *Circumducit etiam in superficie alte spere circulum secantem circulum signorum in duobus diametrice oppositis punctis, quos dicimus ligantes. Et hic etiam circulus in superficie alte spere inclinatus vocatur.*

[284] *Imaginamur quoque rectam lineam orientem de centro mundi, et vadit recte ad centrum referentis exinde utrobique crescens, et recte directa usque ad superficiem referentis. Huius ergo finis linee ab illa quidem parte, super quam sunt centra, est punctus tangens superficiem referentis et est longinqua longinquitas spere extra centrum magisque remotus punctus a centro mundi in tota ferenti spera Mercurii. Huic autem sub recto diametro oppositus est punctus, quem dicunt longinquam propinquitatem,* is scilicet qui ab illa parte linee tangitur, que a centro mundi directa neutrum in se centrum habet nec ferentis.

[285] ¶*Prima ergo Mercurii spera, que omnes alias claudit et circumtenet, movetur circa centrum mundi super duos diametrice oppositos polos, qui subsunt recte polis signorum spere* in eorum radio positi, *et movet secum omnes, quas claudit, speras centraque spere ferentis et referentis. Moventur etiam ab ea longinqua longinquitas et propinqua longinquitas spere ferentis et duo puncti divisionis,* in quibus ligantur signifer

---

28 polis] poli (L)

[279] *The body, however, of the star that is called Mercury is massive and solid. Its shape is spherical and it is placed within the body of the epicyclic sphere. The surface of Mercury touches the surface of the epicyclic sphere in one point.* [280] *When the deferent sphere moves around its own poles, it moves by its motion the epicyclic sphere and takes it along with itself* from west to east *around its centre. The centre of the epicyclic sphere thus produces an imaginary circle in the middle of the body of the deferent sphere, whose centre is the same as that of that sphere. The imaginary arc line of this circle is not in the plane of the similar circle of the ecliptic,* but it is inclined like its poles. [281] *But we imagine that the plane of the circle expands, such that it dissects the entire world, and that it produces two circles on the two surfaces of the deferent which are parallel to the plane and parallel to each other, the centre of all of which is the centre of the deferent sphere.*

[282] *But on the surface of its first sphere, whose centre is the same as that of the world, it produces two parallel circles, one on the inside and one on the outside, whose centre is the same as that of the sphere of the world.*[6] *The larger of them, which is drawn on the high surface, divides the similar circle of the ecliptic at two diametrically opposite points in an oblique angle, and this circle is called the 'inclined' circle.*

[283] *Also on the surface of the high sphere it produces a circle that crosses the ecliptic at two diametrically opposite points, which we call the nodes. And this circle on the surface of the high sphere is also called inclined circle.*

[284] *We also imagine a straight line which arises from the centre of the world, and it runs straight to the centre of the referent, from where it extends further in both directions, and is directed straight until the surface of the referent.*[7] *Hence the end of this line on that side where the centres lie is a point that touches the surface of the referent and it is the furthest distance of the eccentric sphere, and in the entire deferent sphere of Mercury it is the point that is most remote from the centre of the world. Straight diametrically opposite to it is the point that is called 'closest distance,'* namely the point that is touched by that part of the line which is directed from the centre of the world and has neither of the two centres on it, <hence> also not the centre of the deferent.

[285] *The first sphere of Mercury, which contains and encloses all others, thus moves around the centre of the world about two diametrically opposite poles, which are straight below the poles of the sphere of the ecliptic,* placed on the same axis; *and it moves along with itself all the spheres which it contains and also the centres of the deferent sphere and the referent sphere. Also moved by this sphere are the furthest distance and the closest*

[6]The deferent plane thus passes through the centre of the world. This will no longer be generally fulfilled with Stephen's own model of Mercury, which he presents later in this chapter.

[7]Arab. add.: 'Let the centre of the deferent orb also lie on it in the direction of the extremity of the line in which lies the centre of the dirigent orb.'

et inclinatus Mercurii circulus.

[286] *Motus autem is,* quo hec spera movetur et alias secum movet, *tardus est ab occidente in orientem complens, sicut dixit Ptolomeus in megali sintasi sua, in .t. annorum curriculis unius cursum gradus.* Aliorum sententia est in .tf. annis.

[287] *Secunda autem, referens scilicet, spera movetur circa suum centrum in duobus fixis et diametricis polis rotundo motu. Hec movet ferentem suo motu et centrum eius et quicquid in ea est rotundo suo motu <ab oriente in occidentem.* [288] *Movetur etiam ferens spera in suis duobus polis rotundo motu> ab occidente | in orientem.* [289] *Hec movet rotunditatem et centrum eius suo motu super circulum, qui dicitur ferens, in orientem,* [290] *movendoque rotunditatem movet etiam corpus stelle Mercurii in orientem.*

[291] *Rotunditas quoque Mercurii,* que quarta eius spera est, *movetur* in suo loco *circa suum centrum in duobus fixis polis rotundo motu* ab occidente in orientem, sepe contra. *Hec movet corpus stelle Mercurii suo rotundo motu,* [292] *et facit centrum stelle imaginarium circulum,* in quo semper et est et movetur, *centrum cuius ipsius rotunditatis est. Huic circulo nomen est ab antiquis impositum rotundus circulus.* [293] *In hoc rotundo circulo semper est, ut dixi, positum Mercurii centrum. Et quoniam rotunditatis motus in orientem est ab exterioribus, ab illa vero medietate sui, que nobis propinquior est, in occidentem, quando Mercurii centrum in altiori eius medietate rotatur, motus eius est ab occidente in orientem; quando autem in altera, que centro mundi propinquior est,* in qua medius est punctus, qui dicitur longinqua propinquitas Mercurii, sicut et ei oppositus in recto diametro longinqua longinquitas eiusdem, *ab oriente revertitur in occidentem,* et dicitur esse retrogradus. Posita autem sic rotunditate, ut eius longinqua longinquitas tangat longinquam longinquitatem spere extra centrum aut propinqua longinquitas, diametrum rotunditatis superpositum est diametro longinque et propinque longinquitatis spere ferentis, sic ut totum in eo sit, non ad orientem nec ad occidentem, sed <pars> ad austrum, pars ad septemtrionem inclinata.

[294] *Quando autem movetur ferens spera et movet secum rotunditatem, diametrum eius, qui fuit in linea diametri, super quam sunt centra, movetur et mutat eam, quam*

---

18 quando] quoniam

distance of the deferent sphere and the two intersection points, in which the ecliptic and the inclined circle of Mercury are bound to one another.

[286] *But that motion,* by which this sphere moves, and by which it moves the others along with itself, *is a slow motion from west to east, which completes the passage of one degree in the course of 100 years, as Ptolemy has said in his* Almagest*;* others say in 106 years.[8]

[287] *The second sphere, that is, the referent, moves with a round motion around its centre, about two fixed and diametrical poles. This sphere moves by its motion the deferent and the latter's centre and all that is inside it by its round motion <from east to west.* [288] *The deferent sphere also moves about its poles with a round motion> from west to east.* [289] *This sphere moves by its own motion the epicyclic sphere and the latter's centre eastwards, on a circle that is called the deferent circle;* [290] *and by moving the epicyclic sphere, it also moves eastwards the body of the star Mercury.*

[291] *Also the epicyclic sphere of Mercury,* which is its fourth sphere, *moves* at its place *around its centre in a round motion about two fixed poles* from west to east, and often reverse. *By its round motion this sphere moves the body of the star Mercury,* [292] *and the centre of the star produces an imaginary circle* on which it is always located and moving, *whose centre is that of the epicyclic sphere itself. This circle has been named 'epicycle' by the ancients.*[9] [293] *The centre of Mercury is always located on this epicycle, as I have said. And since the motion of the epicyclic sphere is eastwards at its outer parts, but westwards on that half which is closer to us, the motion of Mercury's centre will be from west to east when it revolves in the higher half of the epicyclic sphere; but when it is in the other half, which is closer to the centre of the world* and whose middle is the point which is called the closest distance of Mercury—just as the diametrically opposite point to it is called the furthest distance of Mercury - *it turns back from east to west* and is said to 'retrograde.' But when the epicyclic sphere is positioned such that its furthest distance or closest distance touches the furthest distance of the eccentric sphere, the diameter of the epicyclic sphere is placed on the diameter through the furthest and the closest distances of the deferent sphere, so that it lies totally in the other one, and it is not inclined eastwards or westwards but exactly to the south on one side and to the north on the other side.

[294] *But when the deferent sphere moves, and moves the epicyclic sphere along with itself, the latter's diameter, which was aligned with the diameter through the centres, is*

[8]'106 years': repeated appearance of this number, which is possibly a corruption of al-Battānī's value of 66 years; cf. above, Stephen's commentary to passages 196–198.

[9]Arab. add.: 'This circle is inclined to the plane of the eccentric orb.' Stephen discusses the orientations of the epicycles in more detail later.

habuit prius, positionem.

[295] *Est autem quidam punctus in medio linee positus, que continetur inter centrum referentis et centrum mundi, equaliter ab utroque distans. Immaginantibus autem nobis diametrum rotunditatis transire propinquam longinquitatem ipsius et crescere atque prolongari recte, cum ad secandam diametri ferentis lineam pervenerit, illum tanget punctum, quem medium centri mundi <et re>ferentis dicimus. Hec linea a diametro rotunditatis usque punctum porrecta circinans linea diametri rotunditatis dicitur. Hoc quoque ideo, quoniam cum centrum rotunditatis semper moveatur in circulo ferenti extra centrum, diametrum rotunditatis recte superpositum est illi linee, que a centro rotunditatis incipiens extenditur ad punctum duorum centrorum medium, sicut diximus. Circinat autem hec linea circulum, cuius ipse punctus centrum est, et circulus ille extra centrum aliorum,* [296] *et super tria centra in ea parte linee, que porrigitur a centro mundi usque ad longinquam longinquitatem spere extra centrum. Horum primum est punctus, cuius linea circinat diametrum rotunditatis, post hoc quod circinat referentem speram. Tercium est centrum deferentis. Et longitudo horum ab invicem equalis est a centro mundi* sicut centrum mundi primum, secundum circinantis linee, tercium referentis, IIII^{tum} ferentis. Quantumque est a centro mundi usque centrum circinantis linee, duplum est a centro mundi usque centrum referentis, triplum autem ad centrum ferentis; ita equalibus distant <s>patiis.

[297] *Quando ergo movetur spera, que circumdat intus et extra ferentem, referens, movet centrum ferentis et circumvolvit suo motu. Que cum dimidium sui cursus fecerit, centrum ferentis in centro ponit circinantis linee, ut sint unum. Postea pertransiens centrum eius in complemento sui cursus refert in | eundem, a quo eduxerit, locum.*

[298] *Motus autem ferentis et referentis contrarii sunt, sicut diximus. Referens enim in occidentem movetur in unoquoque solis anno uno circuitu, cum qua volvitur diametrum, quod transit super centrum ferentis,* in quo dictum est esse centra sperarum, *mo-*

3 referentis] ferentis **14** referentem] ferentem

moved and changes the position that it had before.

[295] *However, there is a certain point placed in the middle of the line between the centre of the referent and the centre of the world, equally distant from either of the latter. If we imagine the diameter of the epicyclic sphere to pass through the latter's closest distance*[10] *and to grow and be prolonged straight, when it comes to cross the diameter of the deferent sphere, it will touch that point which we say is the middle between the centre of the world and that of the referent. This line, which is drawn from the diameter of the epicyclic sphere until that point, is called the 'dirigent line' of the diameter of the epicyclic sphere. This is because, as the centre of the epicyclic sphere always moves on the eccentric deferent circle, the diameter of the epicyclic sphere lies right on that line which starts from the centre of the epicyclic sphere and extends until the point in the middle between the two centres, as we have said. But this line produces a circle whose centre is just that point, and that circle is not concentric with any other circles,*[11] [296] *and it passes through three centres on that part of the line which extends from the centre of the world until the furthest distance of the eccentric sphere. The first of them is the point whose line is dirigent of the diameter of the epicyclic sphere; then the centre dirigent of the referent sphere. The third is the centre of the deferent. Starting from the centre of the world, their distances from one another are equal,* so that the first centre is that of the world, the second one that of the dirigent line, the third one that of the referent, the fourth one that of the deferent. And the distance from the centre of the world until the centre of the dirigent line, when doubled, equals the distance from the centre of the world until the centre of the referent, and, when tripled, until the centre of the deferent; in this way they are spaced by equal distances.

[297] *When thus the referent sphere moves, which surrounds the deferent inwardly and outwardly, it will move and take around by its motion the centre of the deferent. And when it has made half of its course, it places the centre of the deferent at the centre of the dirigent line, so that both become one. Then moving on and completing its course, it brings the centre of the deferent back to the same place from where it had taken it away.*

[298] *But the motions of the deferent and the referent are in opposite directions, as we have said. For the referent moves westwards by one revolution every solar year, and with it revolves the diameter which passes through the centre of the deferent* and on which the

[10] Arab. add.: 'after its separation from the furthest distance', referring to the departure of the epicyclic sphere from its furthest distance on the deferent. Stephen seems to refer the phrase to the diameter of the epicyclic sphere and thus had to change the wording to the 'closest distance', i.e. that of the epicyclic sphere.

[11] 'But this line... circles': Arab., 'The circle which is described by the extremity of this line is also called the eccentric orb.'

venturque cum eo longinqua longinquitas et propinqua longinquitas spere extra centrum, et redit in pristinum statum semel in anno solis.

[299] *Ferens vero, que fert rotunditatem, movetur ab occidente contra orientem in anno solis bis, et linea, que centrum rotunditatis circinat, tangit lineam, que inter centra ferentis et referentis est, in anno solari bis,* [300] *semel quidem, cum centrum rotunditatis est in longinqua longinquitate, secundo, cum in propinqua longinquitate ferentis.* [301] *Per hunc igitur motum movetur centrum rotunditatis, et ex quo separatur ab aliquo puncto inclinati circuli, priusquam ad eundem redeat, pertransit annus solaris. Sic fit, ut in anno solis semel volvatur. Super circulum autem extra centrum bis in anno volvitur,* sicut et ferens spera.

[302] *Centrum autem rotunditatis est in eodem loco circuli signorum semper per motum ferentis, in quo et centrum <solis>. Hec ergo causa est, ut* in rectificatione sua *inveniatur abesse modo ante, modo retro, a solis loco, equaliter in suo cursu.* [303] *Atque is motus, qui sic variatur, est proprius rotunditatis,* qui continetur prima tabula. [304] *Motus quoque stelle, qui fit per rotunditatem, equaliter illam ante et retro elongat aut longinqua longinquitate aut propinqua rotunditatis, equali scilicet spatio a diametro, quod dividit medium directi aut retrogradi cursus stelle Mercurii. Sicque fit, ut diametrum, quod iungitur circinanti, subsit semper centro solis.*

¶Quare autem numerus de collecto numero, scilicet qui est Mercurii cursus, maior auferatur quam ad rectificandum solis locum, causa est, quod longinqua longinquitas ferentis Mercurii remotior est ab Ariete quam longinqua longinquitas solis. Auferendi vero numerum eadem ratio est in hoc et .d. reliquis, etiam et in sole.

[305] ¶*Corpus autem stelle movetur circa centrum rotunditatis, et quando est in alta medietate,* cuius est medium longinqua longinquitas rotunditatis, *et motu ferentis et motu rotunditatis movetur in orientem, et est citus motus eius utroque coniuncto. Quando*

13 cursu] cursu figit **17** quod] qui **19** scilicet] solis (L) **26** coniuncto] *corr. ex* coniunctos (L) Quando] Quoniam

centres of the spheres were said to be, *and with the diameter move the furthest distance and the closest distance of the eccentric sphere, and once in a solar year it returns to its initial state.*

[299] *But the deferent, which carries the epicyclic sphere, moves twice per solar year from west to east, and the line which is dirigent of the centre of the epicyclic sphere is tangential to the line between the centres of the deferent and the referent twice in a solar year;* [300] *once, when the centre of the epicyclic sphere is at the furthest distance of the deferent, and for a second time, when at the closest distance.*[12] [301] *By this motion, therefore, moves the centre of the epicyclic sphere, and from the moment when it departs from any point of the inclined circle, until it returns to the same point, there passes one solar year. In this way, it makes one revolution per solar year.* But just like the deferent sphere, *it makes two revolutions per year on the eccentric circle.*

[302] *However, by the motion of the deferent the centre of the epicyclic sphere is always at the same position on the ecliptic as the centre <of the sun>. This is the reason why,* in its rectification, *it is sometimes found ahead and sometimes behind the position of the sun, by equal distances.* [303] *And the motion that varies in that way is the epicyclic sphere's own motion,* which is contained in the first column. [304] *Also the motion of the star which is produced by the epicyclic sphere takes the star equally ahead or behind the furthest distance or the closest distance, that is, by an equal distance from the diameter that runs through the middle of the straight or the retrograding course of the star Mercury. In this way, the diameter which coincides with the dirigent line is always under the centre of the sun.*

But the reason why we subtract a greater number from the accumulated course, that is, from the course of Mercury, than for rectifying the position of the sun is that the furthest distance of Mercury's deferent is farther remote from Aries than the furthest distance of the sun. But the reason why the number is subtracted is the same in this case as in those of the other four planets, and it is the same as in the case of the sun.

[305][13] *But the body of the star is moved around the centre of the epicyclic sphere,*[14] *and when it is in the high half,* whose middle is the furthest distance of the epicyclic sphere, *it is moved eastwards by the motions of the deferent and of the epicyclic sphere,*

[12] Arab. add.: 'similar to the situation with the orb of the moon.'

[13] Stephen has made substantial modifications to the following three passages. Stephen emphasises the cyclic change from direct motion to station, retrogradation and back to direct motion, whereas Ibn al-Haytham does not discuss stations, but instead pays more attention to the direct and the retrograde motions as distinct states.

[14] Arab. add.: 'It faces in its motion, as [we] shall explain, parts of the ecliptic circle.'

autem est in ima parte rotunditatis, cuius est medium propinqua longinquitas eiusdem, *movet ab ea in occidentem.* Quamdiu ergo motus rotunditatis in occidentem motum ferentis in orientem equare non poterit, movetur corpus stelle cum ferenti. Ubi vero hic in occidentem, ille in orientem equalis est, stationaria videtur. [306]
5 *Cum prevolverit et maior erit in occidentem motus, retrograda est stella.* Quo iterum decrescente et adequato motui ferentis, nichil procedit stella et est in secunda statione. Ut vero diminuitur et minor est in occidentem, directa fit. Cum vero motus rotunditatis in occidentem totus defecerit, stella pleno motu ferentis movetur in orientem. Ea autem posita in ea parte, que in orientem movetur, utroque motu in
10 orientem citior solito festinat. [307] *Sic ergo ex rotunditatis motu et ferentis Mercurii citior cursus directus efficitur;* tardior autem, cum hec in occidentem minus ferenti est; nullus, cum equaliter hec in occidentem, illa in orientem fertur; retrogradus, cum huius in occidentem prevalet citatio.

[308] ¶De lato autem stelle Mercurii subtile est iudicium, quia et per rotundi-
15 tatem et per inclinatum fit circulum. *Inclinatus ergo circulus parum inclinatur a circulo signorum* dividens ipsum in duobus diametrice oppositis punctis, estque eius sectio cum illo in duabus longinquitatibus medianis circuli extra centrum. *Huius*
f. 41r *inclinatio non est fixa,* | *sed mutatur huius circuli superficiei positio et propinquat superficiei circuli signorum, usque dum eisdem iungatur et sit una tantum. Deinde separantur, et*
20 *ea quidem medietas, que in septemtrionem inclinabatur prius, in austrum incipit inclinari, que in austrum prius, in septemtrionem descendit. Et tantumdem inclinatur ad austrum, quantum et in septemtrionem inclinata fuerat; idem facit altera in septemtrionem. Post redit, et iterum iungitur, et inclinatur, atque id semper facit.* Et hoc quidem sub rerum et experimenti veritate compertum est ab antiquis superficiem inclinati circuli, quod
25 fit per circulum extra centrum, ita alterari, modo se iungendo superficiei circuli signorum, modo inclinando, nunc ad austrum, nunc ad septemtrionem. Verum per supradictarum numerum sperarum nulla id posse fieri ratione videtur.

3 corpus] tempus (L) **4** stationaria] statio varia (L) **5** occidentem] occidente (L) **11** hec] hoc **12** hec] hoc **14** quia] qui (L) **22** inclinata] inclina

and by the combination of these two its motion will be fast. But when it is in the low part of the epicyclic sphere, whose middle is the latter's closest distance, *it moves westwards by the latter.*[15] Thus, as long as the westward motion of the epicyclic sphere fails to become equal to the eastward motion of the deferent, the body of the star moves together with the deferent. But when the westward motion and the eastward motion are equal, the star appears stationary. [306] *And when the westward motion prevails and is faster, the star retrogrades.* When it decreases again and becomes equal to the motion of the deferent, the star does not move on, but has reached the second stationary position. But as soon as the westward motion decreases and is slower, the star will move forwards again. But when the westward motion of the epicyclic sphere comes totally to a halt, the star moves eastwards by the full motion of the deferent. And when it is located in that part which moves eastwards, it speeds eastwards faster than normal due to both motions. [307] *In this way, therefore, from the motion of the epicyclic sphere and that of the deferent of Mercury, the fastest forward motion is produced;* however, it is slower when the former to the west is less than that of the deferent; and there is no motion at all, when the former is equally taken to the east as the latter to the west; and the motion is retrograde, when the westward speed of the former prevails.

[308] The latitude of the star Mercury is difficult to assess, because it is produced by the epicyclic sphere as well as by the inclined circle.[16] *The inclined circle inclines slightly from the ecliptic,* dividing the latter at two diametrically opposite points, and its intersection with the latter is at the two mean distances of the eccentric circle. *Its inclination is not constant; rather, the plane of this circle changes its position and approaches the plane of the ecliptic, until it falls together with the latter and both become one. Then they diverge again, and that half which before inclined to the north begins to incline to the south, whereas the one previously to the south moves down to the north. And it will incline southwards by the same amount by which it had been inclined to the north; the same does the other half to the north. Subsequently it returns, and coincides again, and inclines, and this it does perpetually.* The ancients realised truly and from accurate observation that the plane of the inclined circle, as it is produced by the eccentric circle, changes in this way, sometimes coinciding with the plane of the ecliptic and sometimes inclining, alternately southwards and northwards. But it is evident that this can in no way be effected by the number of spheres as aforementioned.

---

[15] Arab. add.: 'contrary to the situation of the moon.'

[16] Stephen emphasises here the inclination of the epicyclic sphere, which he had omitted in his translation of passage 292. He thus postponed the subject, to give now a comprehensive discussion of the latitudes.

¶Prima spera est, cuius motus in orientem est tardo motu; secunda referens in occidentem; tercia ferens in orientem. Prime poli suppositi polis signorum, centrum quod mundi. Secunde poli inclinaturi, et centrum aliud. Tercie poli ab his inclinantur, centrum ut idem. Harum numerum et ordinem de Arabico transtulimus. 5 Que si sufficiant, ita videndum est.

Prima spera tardo motu movetur, omnes eodem movet et centrum referentis. Secunda movetur in anno semel in occidentem, et movet, quas claudit, et centrum ferentis. Si secunde polorum radium imaginemur paralellum esse prime radio, et ferentis his paralellus, quem diximus inclinatum, non inclinatur circulus, sed est 10 in superficie zodiaci. Cum igitur non inclinetur, nec variabitur. Si radius ferentis non erit prime radio paralellus, sed sese dividunt, et referentis radio paralellus erit ferentis, eadem semper erit inclinati inclinatio. Centrum enim ferentis et poli circumferuntur semper motu referentis, et huius radius illius fit radio paralellus, tercius oblique secundum dividit, eadem est semper positio, nec mutatur inclinatio, 15 sicut nec in luna.

Si nec secundus primo nec secundo tercius, et tercius quidem aliquando primo per motum secundi, aliquando non, semel in anno tantum paralellus fiet, eodemque post disiunctionem inclinabitur queque medietas quo primo. Movetur enim radius secunde tardo motu prime. Secunda semel et radium tercie secum 20 circumvolvit et semel in anno prime paralellum facit.

Ponantur inclinati poli secunde duobus gradibus, alter in circulo capitis a septemtrione positus, alter in circulo eodem inter caput Capricorni et polum australem signorum, sed in eadem medietate spere, tamquam quorum centrum extra mundi est centrum, in eadem, quia tardo moventur motu prime, diu prestant; harum spera 25 circumvolvitur semel. Inclinentur ferentis poli tantumdem ab his et radius eorum tantumdem. Volvitur prima, volvit radium secunde.

---

4 transtulimus] trastulimus (L) **6** referentis] refentis (L) **7** Secunda] *corr. ex* Sca (L) **9** paralellus] paralellos **11** radio] radios (L) **13** referentis] refentis (L) **16** primo] prime (L)

The first sphere is the one whose motion is eastwards in a slow motion, the second one is the referent sphere, moving westwards, and the third one is the deferent sphere, moving eastwards. The poles of the first one are placed below the poles of the ecliptic, and its centre is that of the world. The poles of the second one are about to incline, and also its centre is a different one. The poles of the third one become inclined by the former ones, as does its centre. I have translated their number and arrangement from Arabic; whether they are sufficient must be considered in the following way.

The first sphere moves with a slow motion, and by the same motion it moves all other spheres and also the centre of the referent. The second one moves westwards once per year, and it moves the spheres which it encloses and also the centre of the deferent. If we imagine the axis of the poles of the second sphere to be parallel to the axis of the first one, and that of the deferent sphere to be parallel to these two, the circle that we called the 'inclined' circle does not incline, but it is in the plane of the zodiac. Hence, since it does not incline, it will not change either. If the axis of the deferent sphere is not parallel to the axis of the first sphere, but both divide one another, and the axis of the deferent is parallel to that of the referent, the inclination of the inclined circle will always be the same. For, the centre and the poles of the deferent are always carried around by the motion of the referent and its axis becomes parallel to that of the former, while the third axis [i.e., that of the first sphere] divides the second axis in an oblique angle, and the position is always the same, and the inclination does not change, as happens neither in the case of the moon.

If neither the second axis is parallel to the first one nor the third axis to the second one, and if the third axis, by the motion of the second axis, becomes sometimes parallel to the first and sometimes not, it will be parallel only once per year and, after disjunction, each half will incline again to the same side as in the beginning. For, the axis of the second sphere is moved by the slow motion of the first sphere; and the second sphere takes the axis of the third one with itself on one revolution and makes it parallel to that of the first once per year.

Let us assume the poles of the second sphere inclined by two degrees, one placed on the circle through the zenith from north and the other, on the same circle, between the head of Capricorn and the south pole of the ecliptic, but both in the same half of the <first> sphere, as their centre lies outside the centre of the world, and they remain in that half for a long time, as they are moved by the slow motion of the first sphere; whereas, their own sphere revolves once. Let the poles

Imaginemur radios tres in superficies binas extendi ita, ut superficies una alteram equis angulis dividat. Extendatur superficies prima sic, ut non dividat radium secundum, sed paralella illi sit, et secunda super hanc rectis angulis. Eodem modo secundi prima superficies prime primi et secunda sicut secunda primi, pariterque prima tercii paralella duabus primis primi et secundi, secunda super hanc equis angulis sicut, ut primi primam superficiem rectis angulis et secundo obliquis dividat eadem cum secunda primi.

Sic erit radius tercius primo paralellus et tanget superficies secundi, tercii et primi centra primi, tercii et secundi simul. Volvitur secundi utraque suo motu, volvuntur terciam, fit secundi prima motu suo non paralella dividens primam <primi>, sed et tercii prima, quia secundi paralella.

8 tercius] secundus secundi] secunda **10** volvuntur] voluntur **11** secundi] secunde

of the deferent sphere and also their axis be inclined from these poles by the same amount. When the first sphere turns, it turns the axis of the second sphere.

Let us imagine the three axes to be extended into two planes each, such that one plane divides the other one at right angles.[17] Let the first plane <of the first axis> be extended such that it does not divide the second axis, but extends parallel to it, and let the second plane be at right angles to the first one. In the same way, the first plane of the second axis shall be parallel to the first plane of the first axis, and the second plane be oriented like the second plane of the first axis [i.e., perpendicular to its first plane]; likewise, the first plane of the third axis shall be parallel to the first planes of the first and the second axes, and the second plane be at right angles to the first one, such that it divides the first plane of the first axis at right angles, whereas for the second axis it crosses at oblique angles in the same way as the second plane of the first axis.

In this way the third axis will be parallel to the first one, and the <first> plane of each of the second, the third and the first axes will simultaneously pass through the centres of the first, the third and the second axes. When both planes of the second axis revolve by the motion of the axis, they move the third sphere, and due to its motion the first plane of the second axis will no longer be parallel to the first plane <of the first axis>, but will divide it; and so will the first plane of the third axis, because it is parallel to that of the second axis.

[17] Cf. Fig. 4.2 on p. 59 of the Introduction.

Volvuntur prima et secunda dimidio secunde circulo, volvuntur tercii utramque; de quarum motu nichil | dicimus, quia cum secundus ad suum, et tercius similiter ad eundem revertitur locum radius.

Ponamus enim, ut dictum est, superficies se habere, erit longinqua longinquitas tangens superficiem prime spere. Moveatur secunda spera uno quadrante, mota est tercia dimidio. Prima superficies secundi radii non paralella, sed ipsam est secans, et secunda paralella. Tercii nec prima nec secunda nec prime nec secunde primi paralella; sed secunda tercii, debuerat autem esse prima tercii, sed conversa esse paralella prime primi. Sed uno quadrante, quo secunda spera mota est, retracta est, et poli tercii inclinantur a primo.

Secunda alium quadrantem currente a prima secundi, sed conversa, paralella est prime primi et prime tercii, et est centrum eius in centro circinantis linee, et poli magis inclinantur quam primo, et est tunc magna inclinatio. Sic ergo ad eundem locum simul et eandem positionem revertitur. Ex quo manifestum est centro ferentis semper in eadem parte <a> centro mundi vel prime spere posito, quocumque eius inclinentur poli, nunquam easdem superficies fieri.

Oportet ergo ita ratione constituere, ut centrum ferentis circa centrum mundi volvatur, et poli eius circa polos signorum, ut hic quidem modo superius in aliis, modo inferius in aliis sit signis. Hoc autem fieri nequit, nisi quartam quoque imaginemur magnam speram, que ferens dicatur, et eam, quam ferentem dicebamus, ita dividat in duas, sicut illa referentem et referens primam. Tangit ergo alta huius quarte superficies prime in uno puncto altam superficiem, infima infimam, secunde radius polorum paralellus prime radio, tercie radius hunc oblique secans et tercium quartus.

Ponantur in una linea primum centrum mundi et prime, secundum secunde

---

7 Tercii] Tercia **8** debuerat] debue (L) **22** altam] alteram **23** hunc oblique] hac obliqua

When the first and second planes of the second sphere revolve by a semi-circle, both planes of the third axis revolve. But we will not discuss the latters' own motion; because, when the second axis reaches its position, the third axis has similarly returned to the same position.

Let us assume the planes arranged as we have said; then the furthest distance will touch the surface of the first sphere. Let the second sphere revolve by a quadrant, then the third one has revolved by a semi-circle; and the first plane of the second axis is no longer parallel [to the first plane of the first axis], but crosses it, whereas its second plane is now parallel. And neither the first nor the second plane of the third axis is parallel any longer to the first or the second plane of the first axis. Now, in place of the first plane of the third axis, the second plane of the third axis had to become parallel to the first plane of the first axis. However, it has become retracted by the one quadrant by which the second sphere has moved, and also the poles of the third axis have become inclined from the first axis.

But when the second sphere rotates by another quadrant, the first plane of the second axis becomes parallel to the first plane of the first axis and to the first plane of the third axis, and the centre of the latter is at the centre of the dirigent line, and the poles are more inclined than before; this is also the moment of the greatest inclination. In this way, thus, whenever it returns to the same position, it will also return to the same orientation. It is obvious from this, as the centre of the deferent lies always on the same side from the centre of the world, or of the first sphere, that wherever its poles incline there can never occur identical planes.[18]

One therefore has to arrange it such that the centre of the deferent revolves around the centre of the world, and its poles around the poles of the ecliptic, such that it is sometimes higher in some of the signs and sometimes lower in other signs. But this is impossible, unless we imagine a fourth great sphere, which we shall call 'deferent' and which divides the one that we previously called deferent into two parts in the same way as the latter divides the referent and as the referent divides the first one. Thus, the high surface of this fourth sphere touches the high surface of the first sphere, and its lowermost surface the lowermost one, in one point; and the axis through the poles of the second sphere is parallel to the axis of the first sphere, whereas the axis of the third one crosses that of the second one, and the fourth axis crosses the third axis, in an oblique angle.

Let us place in one line first the centre of the world and of the first sphere;

[18] Alternative translation: '...the planes never become equal.' The final part of this argument is somewhat obscure.

spere, cui annua nomen sit ex eodem quod in orientem semel in anno volvatur, tercia sit referens et eius centrum in eadem linea et quarte ferentis. Erunt ergo trium sperarum centra in una spera parte. Centrum annue sit centrum circinantis linee diametrum rotundum.

5 Volvitur annua quadrante uno, fertur secum centrum referentis. Volvitur et referens dimidio circulo, et fert secum centrum ferentis. Volvitur et ferens dimidio quadrante annue. Refertur referens uno quadrante, et est in opposito. Ferens, quia habet cursum equum referenti, nichil profecisset, nisi quadrantem annue progressa esset. Semper enim sub illa posita est eisdem partibus ad se invicem suppositis. 10 Progreditur annua altero quadrante et cum ea ferens et referens dimidio, sed refertur quadrante. Tunc sunt omnes simul in uno loco circuli signorum posite, et centrum referentis iunctum centro prime et ferentis in altera opposita parte positum, medium ferentis et annue centrorum centrum prime et referentis.

Movetur annua, movet referentis centrum in orientem quadrante. Movetur re- 15 ferens, movet centrum ferentis in centrum annue. Movetur annua, movet centrum referentis in primum locum. Movetur referens, movet centrum ferentis in eum, a quo eduxerat, locum. Hoc in virgulis radiorum formam habentibus aperte poterit monstrari, et quomodo radius ferentis fiat paralellus bis in anno radiis prime et annue, et inclinatio inclinati circuli variabilis ad austrum et septemtrionem.

20 ¶Verum quoniam occulta res est, circulorum quoque necessariam adhibeamus imaginationem. Imaginamur ergo, positis ita omnium centris, ut centrum tarde quidem et mundi prius et solis sit ab una parte, sed annue distans a primo, ostendendi rem causa, uno gradu; et in eadem linea, tantumdem ab hoc distans centrum referentis tantumdemque a referentis centro in eadem centrum ferentis, circulum f. 42r 25 transeuntem per polos | tarde, et punctum, quem in ea tangunt superficies alte

---

8 nisi] *add.* (L) **11** circuli] circulo (L)

second comes the centre of the second sphere, which is called the ‘annual’ sphere, since it makes one revolution eastwards per year; the third one is the referent with its centre on the same line as well as the centre of the deferent, which comes fourth. Thus, there will be the centres of three spheres on the same side of the sphere. The centre of the annual sphere shall also be the centre of the dirigent line for the diameter of the epicyclic sphere.

When the annual sphere revolves by one quadrant, the centre of the referent will be taken along with it; also the referent revolves by a semi-circle and takes along with itself the centre of the deferent; also the deferent revolves by a semi-circle during a quadrant of the annual sphere. The referent is taken by one quadrant and <itself> has <adopted> the opposite <orientation>. The deferent, having the same speed as the referent [in the opposite direction], would not have moved at all, had it not proceeded by the quadrant of the annual sphere. For it is always oriented as the latter, in the same mutually coinciding directions. When the annual sphere proceeds by another quadrant, the deferent and the referent proceed at the same time by a semi-circle, but taken along by a quadrant. Then they are all located at the same place of the ecliptic simultaneously, and the centre of the referent coincides with the centre of the first sphere, while the centre of the deferent is placed on the other, opposite, side, and the centres of the first and the referent spheres will be in the middle between the centres of the deferent and the annual spheres.

When <subsequently> the annual sphere moves, it moves the centre of the referent sphere eastwards by a quadrant; and when the referent moves, it moves the centre of the deferent onto the centre of the annual sphere. When the annual sphere moves <further>, it moves the centre of the referent to the initial place; and when the referent moves <further>, it moves the centre of the deferent to the place from where it had initially moved the latter away. This could be demonstrated easily with small sticks having the form of the axes, and also how twice in a year the axis of the deferent sphere becomes aligned parallel to the axes of the first and the annual spheres, and also the changing inclination of the inclined circle to the south and to the north.

But since the matter is complex, we must also use imagination of circles. With the centres of all spheres positioned such that there is first the centre of the slow sphere and of the world, while that of the sun lies in one direction, and the centre of the annual sphere lies spaced from the first, for the purpose of demonstration, by one degree, and on the same line and spaced by the same distance is the centre of the referent and, again on the same line and by the same distance remote from

annue, referentis et tarde, et centrum eius centrum mundi.

Imaginamur secundum transeuntem per polos annue et punctum, quo tangit superficiem altam prime, centrum huius annue, et est totus in superficie primi tangens in uno puncto, quo se exterius spere iungunt; et tercium transeuntem per polos referentis tercie, qui inclinentur a polis annue duobus gradibus, et punctum, quo ipsa, annua et tarda iunguntur, <et> centrum referentis huius. Et quartum transeuntem per polos ferentis, qui inclinantur a polis referentis duobus gradibus, totidemque a polis annue, ut sit eorum locus in superficie celi triangulus equis lateribus constans, et super punctum, quem ferens tangit in superficie tarde, et centrum ferentis. Erit ergo hic circulus in superficie circulorum annue et tarde, quoniam poli eius in eadem. Tangunt igitur se omnes hii .d. circuli in uno puncto, quo se omnes alte sperarum superficies iungunt.

Horum circuitionem ostendendi gratia ea, que dicimus, ponamus. Cancer in medio celi positus sit, stet firmamentum none spere nec volvatur uno die. In hoc volvatur annua, referens et ferens toto unius anni cursu, annua uno circuitu, referens et ferens duobus, secunda et quarta in orientem, tercia contra. Longinque longinquitatis punctus in Cancro positus, si in meridionali circulo, et in eodem tarde, annue ferentisque circuli, alter, cuius poli inclinantur, oblique hos dividat; inclinatus et circulus signorum in eadem sunt superficie, quia eorum radii paralelli. Polus referentis septemtrionalis in occidente est distans a polis septemtrionalibus annue et ferentis duobus gradibus sub triangulo equilatero.

Volvitur annua in orientem quadrante uno, et ponit medietatem circuli, que erat in meridionali, et Cancri circulo in circulo dividenti Libram a Scorpio. Tulit secum

---

3 altam] alteram **7** inclinantur] inclinatur (L) **10** annue et tarde] *transposuit* (L) **20** Polus] Poli (L) **22** annua in orientem] annuam in oriente (L)

the centre of the referent, the centre of the deferent, we thus imagine a circle that passes through the poles of the slow sphere and through the point on this sphere that is touched by the high surfaces of the annual sphere, the referent sphere and the slow sphere, and whose centre is that of the world.

We also imagine a second circle that passes through the poles of the annual sphere and through the point where the latter touches the high surface of the first sphere; its centre is that of the annual sphere and it lies completely in the plane of the first circle, touching <the latter> in one point where the spheres contact each other on their outsides. We also imagine a third circle which passes through the poles of the third sphere, being the referent, which poles are inclined from the poles of the annual sphere by two degrees, and through the point where this [referent] sphere, the annual sphere and the slow sphere contact each other, and whose centre is that of the referent sphere. We further imagine a fourth circle which passes through the poles of the deferent, which are inclined from the poles of the referent by two degrees and [shifted] by the same amount from the poles of the annual sphere, such that their place on the surface of the heaven consists of an equilateral triangle, and through the point on the surface of the slow sphere that is touched by the deferent, and whose centre is that of the deferent. This circle will thus be in the plane of the circles of the annual sphere and of the slow sphere, because its poles are in that same plane. Accordingly, all these four circles touch each other in one point, in which all outer surfaces of the spheres contact each other.

For a demonstration of their circuit let us assume the following. Let Cancer be placed in the middle of the heaven, and let the firmament of the ninth sphere stand still and not revolve for one day. During that day the annual sphere, the referent sphere and the deferent sphere shall revolve through the entire course of one year; the annual sphere one revolution, the referent and the deferent spheres two revolutions, the second and the fourth spheres eastwards, and the third one in the opposite direction. Let the point of furthest distance be placed in Cancer, on the meridian circle, and on the same circle are also the circles of the slow sphere, the annual sphere and the deferent sphere, whereas the other circle, whose poles are inclined, divides them in an oblique angle. The inclined circle [i.e., the deferent] and the ecliptic are in the same plane, because their axes are parallel. The north pole of the referent is two degrees west of the north poles of the annual and the deferent spheres, in the form of an equilateral triangle.

The annual sphere turns east by one quadrant and places that half of the circle which before was on the circle of the meridian and of Cancer onto the circle that

referentem et centrum eius, quod super erat suo uno gradu positum in superficie circuli meridiei et Cancri ac Capricorni, posuitque in superficie circuli Libre et Arietis per polos signorum transeuntis a parte Libre in orientem. Septemtrionalis polus referentis, qui erat ab occidente, est ab oriente tantumdem in orientem et
5 super distans, quantum prius in occidentem et desuper distabat. Quartam enim circuli partem peragravit. Australis quoque referentis polus, qui erat ab oriente, adhuc in oriente, sed mutato loco. Quantum enim tunc erat super, tantumdem nunc sub est polo australi annue.

De ferentis motu nichil dicimus. Ipsa enim semper annue motum comitatur.
10 Referens autem dimidio circuitu peracto ferentis centrum in centrum annue retulit ab oriente in occidentem, polum australem australi supposuit, septemtrionalem septemtrionali superposito magna spatiorum distantia, inclinaturque altior pars inclinati in austrum. Volvitur alio quadrante annua et ponit centrum referentis sub suo in centro mundi et circuli alteram medietatem eius eodem polum septemtrio-
15 nalem ab oriente, australem ab occidente. Referens quoque alio dimidio circulo circumlata, centrum ferentis sub centro mundi posuit, polosque eius in eadem linea paralella radio tarde; eadem ratione sursum revertitur.

Hoc etiam linearum et circulorum circumductione videre licet. Sit linea longinque et propinque longinquitatis diametrum, et super eam centrum mundi A,
20 centrum annue B, centrum referentis C, centrum <ferentis>(L) D, equalibus distantia spatiis. Circinamus super B circulum, super eum A C E F, et secundum, et super eum centrum ferentis D, et in opposito M. Dividimus hunc in XII equales partes, et super punctos dividentes D G H I K L M N O P Q R; et alterum magnum, et super eum S T U X, et facimus diametrum ab T ad X, et sint super ipsum T I E B F P X;
f. 42v 25 | et super illud, quod hoc dividit rectis angulis, S D C B A M U. Facimus punctum C centrum, et circinamus super eum dimidium circulum ab D usque B a parte G;

1 quod] qui **11** polum] polos **18** videre licet] videlicet (L)

separates Libra from Scorpio. It has taken along with itself the referent and also the latter's centre, which had been placed above[19] its own centre by one degree while in the plane of the meridian circle and the circle of Cancer and Capricorn, and it has placed it onto the plane of the circle of Libra and Aries, which passes through the poles of the ecliptic from the side of Libra to the east. The north pole of the referent, which before was in the western part, now is in the eastern part and it stands by the same distance to the east and elevated as it was standing to the west and elevated before. For it has traversed a quarter of a circle. And the south pole of the referent, which before was in the eastern part, is now still in the east, but at a different place. For as much as it was elevated then, it is now lower than the south pole of the annual sphere.

We do not discuss the motion of the deferent. For it always follows the motion of the annual sphere. But the referent, having traversed a semi-circle, has taken the centre of the deferent westward from the east and onto the centre of the annual sphere, and it has placed the south pole [of the deferent sphere] lower than the south pole [of the annual sphere], and the [former's] north pole higher than the [latter's] north pole, by a great distance, and the upper part of the inclined circle inclines southward. The annual sphere then revolves by another quadrant and places the centre of the referent lower than its own one, to the cente of the world, and into the other half of its circle <and it places> in the same way the northern pole to the east and the southern pole to the west. The referent sphere, having also traversed another semicircle, has placed the centre of the deferent below the centre of the world and its poles on the same line parallel to the axis of the slow sphere; in the same manner it will return upwards.

This becomes apparent again from drawing lines and circles. Let a line be the diameter through the furthest distance and the closest distance, and on it let A be the centre of the world, B the centre of the annual sphere, C the centre of the referent sphere, and D the centre of the deferent sphere, all spaced by equal distances. We draw a circle around B and on it ACEF, and another circle, and on it the centre of the deferent, D, and in opposition the point M. We divide this circle into twelve equal parts, and on it the dividing points D G H I K L M N O P Q R; and another great circle, and on it STUX, and we make the diameter from T to X, and on it let there be T I E B F P X, and on that one which divides this in right angles let

[19]The orientation, 'above' and 'below', corresponds to a top-view onto the ecliptical plane from its northern pole with the apogee pointing upwards. Stephen will use the same perspective in the diagram which he describes below; cf. p. 354.

et alium super punctum E ab I usque B in eadem parte; ubi se dividunt, ponimus Y. Et circinamus dimidium circulum super punctum G ab I usque S ab exteriori parte; et alium super punctum H ab T usque D ab eadem parte; ubi se dividunt, ponimus Z. Et dimidium super punctum A ab B usque M a parte N, et alium super F ab P usque B in eadem parte; ubi se dividunt, ponimus P (ms: P). Et alium super punctum O ab M usque X <ab exteriori parte, et alium super punctum N ab P usque U ab eadem parte;> ubi se dividunt, <ponimus P>; alium quoque super punctum K ab T per H et B usque M. Et alium super punctum L ab I per B et N usque U; et alium super punctum Q ab X per O et B usque D; et alium super punctum R ab P per B et G usque S, ut sit talis figura.

---

1 ab I] A B et **4** Et] Z

there be S D C B A M U. We take C as the centre and draw around it a semi-circle from D to B on the side of G, and another one around the point E, from I to B, on the same side, and where they intersect we place the point Y. And we draw a semi-circle around G, from I to S, on the outer side, and another one around H, from T to D, on the same side, and where they intersect we place the point Z. Also a semi-circle around A, from B to M, on the side of N, and another one around F, from P to B, on the same side, and where they intersect we place the point P̱. And another one around O, from M to X, on the outer side, and another one around N, from P to U, on the same side, and where they intersect we place the point P̱. Also another one around K, from T through H and B to M, and another one around L, from I through B and N to U, and another one around Q, from X through O and B to D, and another one around R, from P through B and G to S, such that the following diagram is produced [Fig. 4.12, missing in the manuscript].

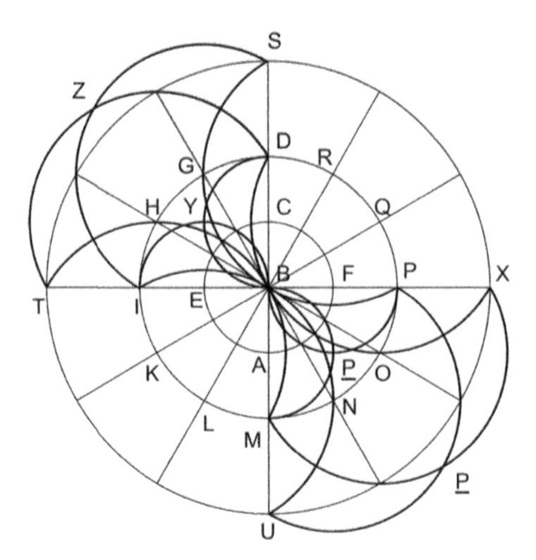

Fig. 4.12: Reconstruction of Stephen's diagram of the trajectories of Mercury's axes (missing in the manuscript). In its description in the text, the label P is used for two different points

Est igitur ea, quam prius posuimus, positione <A>[(L)] centrum tarde et in eodem poli eius; in puncto B centrum annue et poli eius, quia paralellus primo radius eius; in puncto <C>[(L)] centrum referentis, et duobus inclinati gradibus poli eius septemtrionalis in puncto Q, australis in puncto H; centrum ferentis et poli, quia radius duobus primis paralellus, in puncto D.

Volvitur annua uno quadrante, tulit ferentem et referentem in orientem, et posuit centrum ferentis in puncto E, septemtrionalem polum in puncto G, australem in puncto L transcursis arcubus QRDG et HIKL. Posuisset ergo centrum et polos ferentis, qui erant in puncto <D>, transcurso arcu DGHI in puncto I, referens dimidio peracto cursu reduxisset centrum eius in centrum B transcurso arcu IYB, et septemtrionalem polum septemtrionalis a puncto I in puncto S dimidio circulo transcurso in orientem IZS, et australis australem ab I in U transcurso arcu in occidentem IBNU. Quibus ita positis septemtrionalis ferentis polus multum inclinatur a septemtrionali annue, itidem altum a suo centro in centro posito. Dividunt ergo sese medios radii, et superior circuli inclinati pars multum in magna sua inclinatione in austrum, opposita in septemtrionem. Deinde movetur annua, et incipit

8 et] i **9** qui] que

At the position which we have previously defined as A there lies the centre of the slow sphere and, at the same place, also its poles. At point B there is the centre of the annual sphere and also its poles, because its axis is parallel to the first one. At point C there is the centre of the referent, and of its poles, which are inclined by two degrees, the northern one is at the point Q and the southern one at the point H. The centre of the deferent, and, since its axis is parallel to the first two, also its poles, lie at D.

When the annual sphere turns by one quadrant, it has taken the deferent and the referent eastwards and has placed the centre of the deferent at point E, the north pole at G and the south pole at L, the latter having traversed the arcs QRDG and HIKL. It would correspondingly also have placed the centre and the poles of the deferent, which had been at the point D, to the point I along the arc DGHI, and the referent, after having passed through half a course, would have taken the latter's centre to the centre B, along the arc IYB, and the north pole [of the referent] would have taken the north pole [of the deferent] from the point I to the point S, eastwards along the semi-circle IZS, and the south pole of the referent the south pole of the deferent from I to U, westwards along the semi-circle IBNU. When these have been thus placed, the north pole of the deferent is much inclined from

inclinatio minui. Qua currente alium quadrantem fert secum centrum referentis, et ponit in puncto A, polos septemtrionalem in puncto K, australem in puncto O, attulissetque centrum ferentis in eodem B manente polos septemtrionalem a puncto S in orientem ad punctum T, nisi reduxisset ipsum ferens in occidentem ab T ad M transcurso arcu THBM, et australem a puncto U ad X, nisi ferens retulisset in occidentem ab X usque M transcurso arcu XP̲M, centrum quoque referens ferentis tulit ab B usque M per occidentem transcurso arcu BP̲M.

Positi sunt igitur et centrum et poli ferentis in eodem puncto M, et est paralellus eius radius radiis prime tarde et annue. Est ergo superficies inclinati in superficie circuli signorum. Ab hinc incipit inclinari quoque medietas eius in oppositum. Volvatur enim annua tercio quadrante, feret secum centrum ferentis et referentis, et ponet centrum referentis in puncto F, septemtrionalem eius polum in puncto N, australem in puncto R. Posuissetque centrum ferentis in puncto P, nisi reduxisset in occidentem motus referentis ab P usque B transcurso arcu PP̲B, et polos utrosque ibidem, nisi septemtrionalem reduxisset septemtrionalis a puncto P usque punctum U transcurso arcu eius PP̲U, et australem australis ab P ad punctum S transcurso arcu PBGS. Quibus ita positis | radius ferentis radium annue medium dividit in centro a centro posito, et polus australis super est, septemtrionalis sub est. Ex quo <fit>, ut medietas inclinati, que prius inclinabatur in austrum, inclinetur in septemtrionem, altera oppositum tenente, exinde minuitur inclinatio.

Movetur annua quarto quadrante, et refert centrum referentis in orientem ad suum locum in punctum C et polos eius septemtrionalem in puncto Q, australem in puncto H. Tulissetque polos ferentis septemtrionalem in punctum X ab U, nisi septemtrionalis referentis in occidentem retulisset eum ab X usque D transcurso arcu XOBD, et australem in puncto T, nisi australis referentis retulisset ab T ad D transcurso arcu TZD. Centrum autem referentis in eodem esset puncto B, nisi movisset ipsum referens ab B ad D transcurso arcu BYD. Rediit igitur omnium positio ad eundem locum completo annue uno pleno motu ab occidente in orientem, du-

6 referens] *corr. ex* referentis (L) **16** U] N **21** quarto] tercio **23** H] B (L) Tulissetque] *corr. ex* Tulissque (L) X] X P **25** XOBD] XOB **356.28–358.1** duplici] dupli (L)

the north pole of the annual sphere and equally far up from its own centre, which is placed at the centre. Hence the axes bisect each other, and the upper part of the inclined circle is much inclined to the south, at its greatest inclination, and the opposite part, to the north. Then the annual sphere moves and the inclination begins to decrease. When the annual sphere moves by another quadrant, it carries along with itself the centre of the referent and places it at point A, and the north pole at the point K, and the south pole at the point O. And with the centre of the deferent remaining at B the referent would also have taken the north pole from the point S eastwards to the point T, if the deferent had not taken it back westwards, from T to M, along the arc THBM, and it would have taken the south pole from the point U to X, if the deferent had not taken it back westwards, from X to M, along the arc XP̱M. And the referent has taken the centre of the deferent westwards, from B to M, along the arc BP̱M.

Thus, the centre and the poles of the deferent are placed at point M, and its axis is parallel to those of the first, slow, sphere and of the annual sphere. Hence the plane of the inclined circle is in the plane of the ecliptic. From this moment its half begins to incline also to the other side. For when the annual sphere turns by a third quadrant, it will take along with itself the centre of the deferent and of the referent and it will place the centre of the referent at point F, its north pole at point N and its south pole at point R. And it would also have placed the centre of the deferent at point P, if the motion of the referent had not moved it back westwards, from P to B, along the arc PP̱B, and at the same place it would have placed both poles, if [its] north pole had not moved the north pole [of the deferent] back, from P to U, along its arc PP̱U, and its southern one the south pole, from P to point S, along the arc PBGS. When these have been thus placed, the axis of the deferent divides the axis of the annual sphere in the middle, which is placed at the centre, and the south pole is up and the north pole is down. In this way, that half of the inclined circle which before was inclined southwards now inclines northwards, whereas the other half takes the opposite position. From there the inclination decreases again.

The annual sphere moves by a fourth quadrant and takes the centre of the referent sphere eastwards, back to its position at point C, and its north pole at point Q and its south pole at point H. And it would have carried the north pole of the deferent from U to X, if the north pole of the referent had not taken it back westwards, from X to D, along the arc XOBD, and the south pole at point T, if the south pole of the referent had not taken it back, from T to D, along the arc TZD. But the centre of the referent would be at the same point B, if the referent had not moved it from B to D, along the arc BYD. Thus, the position of everything has

plici autem motu referentis in occidentem ab oriente. Nam motus quidem ferentis semper idem est cum annua, sed duplex. Annua enim ad orientem semel in anno solis, referens in occidentem bis, ferens in orientem totiens. Manifestum igitur est ex his, que dicta sunt, .d. oportere, sicut positum est, esse speram magnam Mer-
5 curii, quibus ostensis reliqua videamus. Hac eadem ratione manfestius a quovis diligente, que dicta, lectore poterit probari .c. tantum non esse.

returned to the same place, when the annual sphere has completed one full motion from west to east and with the double motion of the referent from east to west. For, the motion of the deferent is always the same as that of the annual sphere, but double; because the annual sphere turns eastwards once in a solar year, the referent westwards twice and the deferent eastwards twice. Hence it is obvious from what we have said that there must be a fourth great sphere of Mercury, as we have stated. As this has been demonstrated, we can now discuss the rest. In this way it can be clearly proved by anyone who has read the above carefully that there cannot be only three spheres.

[309] ¶*Diametrum quoque rotunditatis, cuius fines sunt longinqua et propinqua longinquitas rotunditatis, non est semper in superficie extra centrum circuli, sed nec fixum. Movetur enim super parvum circulum, qui est in fine propinque longinquitatis et fert secum superficiem rotundi circuli. Is circulus stat recte super circuli extra centrum superficiem equali angulo, et centrum eius in eadem, et dicitur motus diametri circa illum rotundum circulum inclinatum rotunditatis.*

[310] *Diametrum autem, quod hoc secat equali angulo, non servat eundem statum, et movetur in parvo circulo, qui est prope finem eius, et est <super> superficiem* extra centrum circuli *equali angulo. Atque hic motus dicitur obliquatio rotunditatis, et motus eius equalis alteri motui. Per hos itaque duos motus et motum inclinati fit latum Mercurii.*

[311] *Qui quando movetur in circulo extra centrum et incipit moveri a puncto <longinque> longinquitatis versus punctum divisionis,* est in austro, *et incipit inclinatus moveri versus circulum signorum. Statimque ubi centrum rotunditatis <ad> punctum ligantem pervenerit, iungitur inclinatus signifero.* [312] *Descendente autem centro rotunditatis in propinquam longinquitatem inclinatus incipit separari, et extenditur ad aliam partem, que est septemtrionis, et fert secum longinquam longinquitatem, inclinaturque propinqua longinquitas, in quo est rotundum centrum, in austrum. Quando est rotundum centrum in longinquitate propinqua, fit status inclinationis.*

[313] *Eo ad alterum sectionis punctum renitente, et inclinatus ad zodiacum revertitur, quo cum pervenerit, iunguntur circuli.* [314] *Transgresso autem eo et ad longinquam longinquitatem ascendente, medietas, que in austrum prius, in septemtrionem inclinatur, et opposita oppositum facit; sicque fit, ut centrum rotunditatis Mercurii semper, cum inclinatur, ad austrum habeat inclinationem.* Sed de his nunc brevius dicimus, quia in sequenti plenius exequi curamus, cum de omnium planetarum lato plenius et rotunditatibus disseremus simul. Nunc iterum istec dixisse sufficiat. De motibus autem Mercurii, quot sint, in sequentibus dicendum videtur et aliorum planetarum, cum lati eorum rationem plene exequuti erimus. Figura autem horum circulorum, quos diximus, subscribitur hic, in quibus plura diligens inspector poterit animadvertere.

---

7 hoc] hec **9** obliquatio] obligatio (L) **15** in] in in **22** cum] eum

[309] *Also the diameter of the epicyclic sphere, whose ends are the furthest distance and the closest distance, is not always in the plane of the eccentric circle, neither is it fixed. For it moves on a small circle that is at the end of the closest distance and which takes along with itself the plane of the epicycle. This circle stands upright on the plane of the eccentric circle, at a right angle, and its centre is in that plane, and the motion of the diameter around that round circle is called the 'inclination' of the epicyclic sphere.*

[310] *And the diameter that crosses the former one at a right angle does not preserve its position, but moves on a small circle which is near that diameter's end and which stands at a right angle on the plane of the* eccentric[20] circle. *This motion is called the 'slant' of the epicyclic sphere, and its motion is equal to the other. Thus, the latitude of Mercury is produced by these two motions and by the motion of the inclined circle.*

[311] *When it moves on the eccentric circle and starts moving from the point of furthest distance to the intersection point,* it is inclined south, *and the inclined circle will also start moving towards the ecliptic. And as soon as the centre of the epicyclic sphere has reached the node, the inclined circle will become tangential to the ecliptic.* [312] *But when the centre of the epicyclic sphere descends to the closest distance, the inclined circle begins to depart [from the ecliptic] and it tends to the other, that is, the northern, side, and it takes the furthest distance along with itself, whereas the closest distance, at which point there is the epicyclic centre, inclines southwards. When the epicyclic centre is at the closest distance, the inclination becomes stationary.*

[313] *When it moves back to the other intersection point, the inclined circle too will return towards the zodiac, and when it has arrived there, the circles become tangential.* [314] *But when it has passed through and ascends to the furthest distance, that half which formerly was inclined to the south now inclines to the north, and the opposite side makes the opposite; in this way, whenever the centre of the epicyclic sphere of Mercury is inclined, it has an inclination southwards.* But for the moment we speak about this only briefly, because we will provide a fuller discussion of it later, when we discuss comprehensively the latitudes and epicyclic spheres of all planets. For the moment the above shall be sufficient. But how many motions of Mercury and of the other planets exist seems to be a subject that is to be discussed later, after we will have achieved a complete understanding of their latitude.[21] But here is given a diagram of the circles that we have mentioned, in which the attentive beholder can find many things [Fig. 4.13].

[20] 'eccentric': Arab., 'ecliptic', corresponding to *Alm.* XIII,2 (H530). Stephen possibly adapted the text to his own theory of the epicyclic spheres, which he describes later.

[21] That later discussion of the latitudes includes parts of Ibn al-Haytham's following passages 315 to 321, which Stephen thus postponed.

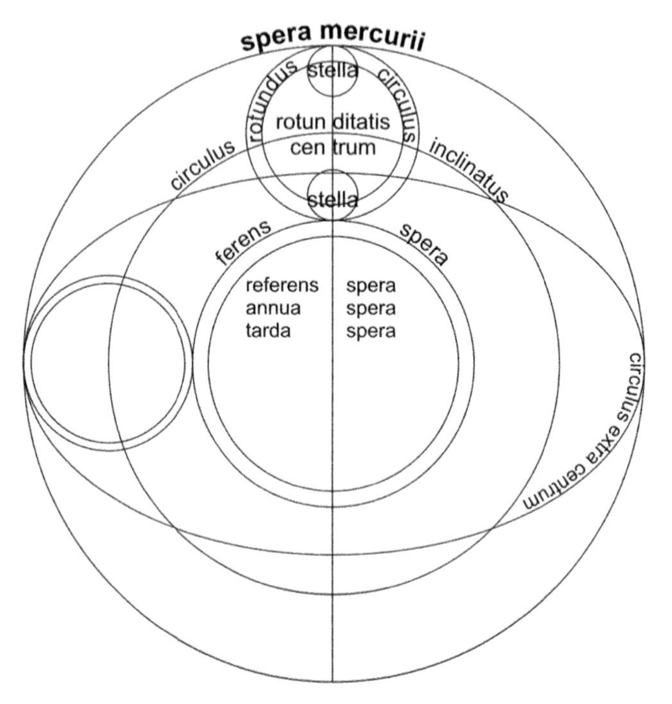

Fig. 4.13: From MS Cambrai 930, fol. 43v. Labels: "the sphere of Mercury" (spera Mercurii), "star" (stella), "epicycle" (rotundus circulus), "centre of the epicyclic sphere" (rotunditatis centrum), "inclined circle" (circulus inclinatus), "deferent/referent/annual/slow sphere" (ferens/referens/annua/tarda spera), and "eccentric circle" (circulus extra centrum)

f. 44r [322] | *Veneris quoque spera sperale corpus est, quam circueunt due sperales paralelleque superficies. Centrum earum et centrum mundi unum est. Earum altera exterior tangit interiorem spere solis superficiem, interior autem exteriorem Mercurii. Alta superficies habet circulum similem signorum circulo. Huius spere motus est super duos fixos polos suppositos polis signorum tardus ab occidente in orientem, sicut est motus spere Mercurii.*

[323] *Separatur autem de hac quedam rotunda spera duabus paralellis superficiebus circumdata. Harum et ipsius spere centrum unum est extra centrum mundi positum. Alta quidem tangit altam superficiem prime spere,* interior interiorem, *in uno puncto. Movetur autem super duos fixos polos* solis cursui equali motu *in orientem, nomen habens ferentis.*

[324] *De hac autem quedam parva spera separatur, que dicitur rotunditas Veneris. Hec continetur tota inter duas paralellas superficies ferentis, cuius motus circa suum centrum est ad orientem in duobus fixis polis. Corpus quoque stelle Veneris sperale est et solidum, positum in rotundo corpore* sic, ut huius superficies illius tangat superficiem in uno puncto.

[325] *Cum ergo movetur ferens spera, fert secum rotunditatem, facitque per motum centri rotundum imaginarium quemdam circulum. Imaginamur hunc crescere, sicut et de circulo Mercurii superius dictum est. Fiunt igitur tres paralelli circuli ferentis, quorum qui exterior et superior est, inclinatus circulus vocatur.*

[326] *Rectam quoque imaginamur lineam, que incipit a centro mundi et extenditur usque centrum ferentis crescens utrimque, donec in modum diametri superficiem ferentis utrobique tangat. Huius igitur duo capita alterum quidem longinquam, alterum autem propinquam vocamus longinquitatem.* [327] *Imaginamur etiam diametrum rotundum hinc recte linee iunctum, quando centrum eius in puncto longinque longinquitatis volvitur. Ferenti autem suo motu rotunditatem movente movetur et diametrum rotunditatis, eritque contra alium punctum diametri ferentis, quod non est centrum mundi neque ferentis. Huius longitudo a centro ferentis equalis est longitudini sui ipsius a centro mundi.*

---

12 quedam] *corr. ex* que (L)

4.3 <The Sphere of Venus>

Config. 12: The Orb of Venus

[322] *The sphere of Venus, too, is a spherical body which is surrounded by two parallel spherical surfaces. Their centre and the centre of the world are the same. One of them, that is, the outer one, touches the inner surface of the sphere of the sun, whereas the inner one, the outer surface of Mercury. The high surface bears the similar circle of the ecliptic. The motion of this sphere is slow, from west to east, around two fixed poles that are placed below the poles of the ecliptic, as is the motion of the sphere of Mercury.*

[323] *But from this sphere a certain round sphere is separated, which is enclosed by two parallel surfaces. Their centre and that of the sphere itself is the same, and it is placed outside the centre of the world. The high surface touches the high surface of the first sphere,* and the inner surface the inner one, *in one point [each]. It moves eastwards around two fixed poles,* in a motion equal to the course of the sun, *and it is named the deferent.*

[324] *From this sphere, however, a certain small sphere is separated, which is called the epicyclic sphere of Venus. It is completely contained between the two parallel surfaces of the deferent, and its motion is eastwards around its centre about two fixed poles. Also the body of the star of Venus is spherical and solid, and placed inside the epicyclic body,* such that its surface touches that of the latter in a single point.

[325] *Thus, when the deferent sphere moves, it takes the epicyclic sphere along with itself and, by the round motion of the centre [of the epicyclic sphere], it produces a certain imaginary circle. We imagine this circle to expand, just as has been said above about the circle of Mercury. Then three parallel circles are produced on the deferent, the highest and outermost of which is called the inclined circle.*[22]

[326] *We also imagine a straight line which begins from the centre of the world and extends until the centre of the deferent, and which is prolonged in both directions, until it touches at both sides the surface of the deferent like a diameter. Its both ends we thus call, in the case of one of them, the furthest distance and, the other one, the closest distance.* [327] *We imagine also the epicyclic diameter aligned with this straight line from here, when the centre of the epicyclic sphere revolves at the point of furthest distance. But when the deferent, by its own motion, moves the epicyclic sphere, also the diameter of the epicyclic sphere is moved, and it will point towards a different point on the deferent's diameter, which is neither the centre of the world nor that of the deferent. The distance of this point from the centre of the deferent is equal to its distance from the centre of the world.*

[22] Stephen has reformulated the sentence to define the inclined sphere. In the Arabic, the deferent orb is assumed to be already defined; cf. Langermann's note 5 to this passage.

[328] *Quando igitur imaginamur <lineam>(L) incipientem ab hoc puncto extendi usque ad centrum rotunditatis, iungetur recte diametro rotunditatis. Hec ergo linea dicitur circinans diametrum rotunditatis, et punctus circinans.*

[329] *Prima igitur spera currente super duos, qui subsunt polis signorum, moventur ab ea omnes Veneris spere. Movetur igitur superficies inclinate spere et longinqua longinquitas et quicquid in ea est ab occidente in orientem tardo motu, centum solis annis gradus transitu.*

[330] *Ferens quoque spera movetur super duos fixos polos circa suum centrum rotundo motu in orientem ferens secum rotundum et corpus stelle Veneris per motum rotunditatis.*

[331] *Et centrum ferentis est fixum et non movetur nisi primo tardo motu. Movetur etiam rotunditas super suos duos fixos polos rotundo motu circa suum centrum movens secum corpus Veneris eodem motu.*

[332] *Fitque per motum centri stelle imaginarius circulus, centrum cuius rotundum, nomen habens rotundus circulus. Is autem inclinatur extra superficiem circuli extra centrum. Motus vero centri stelle semper est in rotundo circulo. Unde ut eo in alta parte huius circuli posito stella in orientem feratur, in ima vero in occidentem.*

[333] *At vero diametrum rotunditatis, cuius duo sunt capita longinqua et propinqua longinquitas, movetur super parvum circulum ferens secum superficiem rotundi circuli; alterum vero diametrum, quod secat hoc recto angulo, movetur ea ratione, que posita est in diametris rotunde spere Mercurii, preter divisionem a superficie extra centrum. Quod enim est hic propinqua longinquitas, in spera Mercurii longinqua longinquitas. Divisio quoque per secundum diametrum illic ab uno puncto sectionis, hinc ab opposito.* [334] | (f. 44v) Hoc autem diametrum aliud secat angulo recto, *motus quorum super duos est parvos circulos, non circa sed extra centrum, sicut est in motu longinquitatis diametri, dum motus equalis est motui longitudinis.*

---

15 habens] habemus (L) **17** ima] una

[328] *When we thus imagine a line starting from this point and extending until the centre of the epicyclic sphere, it will be aligned straight with the diameter of the epicyclic sphere. Hence this line is called the dirigent line of the diameter of the epicyclic sphere, and the point is called the dirigent point.*[23]

[329] *Thus, when the first sphere moves around the two poles which are below the poles of the ecliptic, all the spheres of Venus will be moved by it. Therefore, the plane of the inclined sphere and the furthest distance, and whatever is inside it, is moved from west to east with a slow motion, by one degree in a hundred solar years.*

[330] *The deferent sphere also moves about two fixed poles around its centre with a round eastward motion, taking along with itself the epicyclic sphere and, by the motion of the epicyclic sphere, the body of the star Venus.*

[331] *And the centre of the deferent is fixed and does not move, except by the first slow motion.*[24] *Also the epicyclic sphere moves about its two fixed poles around its centre in a round motion, thereby moving along with itself the body of Venus by the same motion.*

[332] *By this motion of the centre of the star an imaginary circle is produced, whose centre is the epicyclic centre, and which is named 'epicycle.'*[25] *This circle is inclined from the plane of the eccentric circle.*[26] *But the motion of the centre of the star is always on the epicycle. As a result, when the star is located in the high part of that circle, it is taken eastwards, and when in the low part, westwards.*

[333] *But the diameter of the epicyclic sphere, whose two ends are the furthest distance and the closest distance, moves on a small circle and takes along with itself the plane of the epicycle; whereas, the other diameter, which crosses the former at a right angle, moves in the same way as has been described for the diameters of the epicyclic sphere of Mercury, except for the separation from the eccentric plane. For what is the closest distance here, is the furthest distance in the sphere of Mercury. And the separation according to the second diameter occurred from one intersection point there, but from the opposite point*[27] *here.* [334] But this diameter crosses the other one at a right angle, *and their motion is on two small circles, not around but outside the centre, as in the motion of the diameter through the furthest distance, while the motion equals the motion in longitude.*

---

[23] In the Arabic text this is the first occurrence of an expression equivalent to 'equant', which Ibn al-Haytham avoided in the previous discussion of Mercury; cf. Langermann's note to this passage.

[24] Arab. add.: 'contrary to the situation of the star Mercury.'

[25] Arab. emphasises that in the Arabic terminology the epicycle and the epicyclic sphere are termed identically, *falak al-tadwīr*. Stephen omitted this passage, as his Latin terminology distinguishes clearly between 'spheres' and 'circles'.

[26] Arab. add.: 'similar to the situation of the orb of Mercury.'

[27] Arab. add.: 'which is the tail.'

[335] *Inclinate vero spere motus sicut et inclinate Mercurii, cuius inclinatio eius partis, que centrum rotunditatis habet, sive in longinqua seu in propinqua sit longinquitate, in septemtrione est. Hinc igitur fit, ut centrum rotunditatis Veneris inclinatum semper in septemtrionem inclinetur, sicut Mercurii in austrum.* Inde est, quod in eorum inveniendo lato in canonis regulis quoddam latum sumi precepimus, quod quidem si Mercurii, esset in austrum semper, si Veneris, in septemtrionem inclinatum esset.

[336] *Sunt ergo motus Veneris omnes* VIII. *Primus quidem totius spere motus in orientem tardus. Secundus ferentis motus in orientem. Tercius, qui per eundem fit in circulo signorum, dictus motus longitudinis. Motus vero rotunditatis circa suum <centrum>*[(L)] *quartus est, quintus autem motus linee circinantis rotunditatis diametrum, sextus inclinati diametri rotunditatis,* VII^{us} *diametri rotunditatis, quod recto angulo primum secat. Octavus inclinati circuli superficiei motus est.*

[337] Quibus autem que recte se habent excepto sperarum numero, qua de re similis est probatio atque in tractatu Mercurii inducta est, totidem et hic ponende sunt, quot Mercurium habere dictum est. *Ideo autem in huius tractatu spere breves fuimus, quoniam motus Veneris et spere et circuli similes per omnia fere Mercurio. Quoniam ergo <in>*[(L)] *Mercurium omnia large dixeramus, breves sumus hic propter multimodam similitudinem.* Imago igitur Veneris quoniam imagini Mercurii simillima est in speris, circulis lineisque, sufficit illius descripsisse figuram.

[335] *The motion of the inclined sphere is like the motion of the inclined sphere of Mercury, [except that] its inclination on that side which bears the centre of the epicyclic sphere, be it at furthest or closest distance, is northwards. As a consequence, thus, the inclined centre of the epicyclic sphere of Venus is always inclined northwards, just as that of Mercury is always inclined southwards.* For this reason, when determining their latitude we have prescribed in the *Rules of the Canon* that a certain latitude is to be taken, which, if it is that of Mercury, should always be inclined southwards, and if it is that of Venus, northwards.

[336] *In total, there are thus eight motions of Venus. The first one, the slow motion of the entire sphere eastwards; the second one, the eastward motion of the deferent; the third one, the motion that is produced by the former on the ecliptic, which is called the motion in longitude. The motion of the epicyclic sphere around its centre is the fourth one, whereas the fifth one is the motion of the dirigent line of the diameter of the epicyclic sphere; the sixth one is that of the inclined diameter of the epicyclic sphere; the seventh one, that of the diameter of the epicyclic sphere, which crosses the first at a right angle. The eighth one is the motion of the plane of the inclined circle.*

[337] The above is correct, except for the number of spheres, as can be proved by an argumentation analogous to the one introduced in the discussion of Mercury[28]; and we must assume the same number here as Mercury has been said to have. *But we have kept ourselves brief in the discussion of this sphere, because the motions of Venus and the spheres and the circles are in every aspect similar to Mercury. Thus, as we explained everything in length for Mercury, we keep ourselves brief here because of the similarity in many respects.*[29] Therefore, as the picture of Venus is most similar to that of Mercury regarding the spheres, circles and lines, it is sufficient that a diagram of Mercury has been given.[30]

[28]This invalidates much of Ibn al-Haytham's preceding argumentation. As in the case of Mercury, Stephen has again first translated Ibn al-Haytham's description before refuting parts of it.

[29]Arab. add.: 'All that which we have not presented [fully] in this summary is, however, epitomized in it, given that it is set forth in the preceding [discussion].'

[30]'Therefore... given': this replaces the phrase in Arab., 'This is the figure of the orbs of Venus,' and the diagram given there.

[338] ¶*Trium autem planetarum, Saturni scilicet, Iovis et Martis, spere simillime sunt ad invicem in numero et divisionibus sperarum, motu quoque et circulis,* fere *sicut dixisse Ptolomeus in sua probatur sintaxi. Nos igitur de similibus loquentes unius sub sermonis disputatione simul omnes amplectimur.* [339] *Harum ergo queque stellarum propriam habet speram, quam dividunt due sperales paralelleque superficies, centrum omnium unum est quod et mundi. Spere autem hee ad invicem iuncte sunt.* [340] *Prime enim, que est Saturni, spere alta quidem superficies iungitur ime superficiei spere fixarum, ima altam tangit Iovis spere, Iovis interior altam Martis, illius ima solis exteriorem.*

[341] *Earum singule moventur super duos fixos polos recte suppositos polis signorum tardo motu in orientem.* [342] *Separatur autem de unaquaque spera extra centrum,* sicut in spera solis supra dictum est, *quam circundant due sperales paralelleque superficies, centrum quarum earundem sperarum est. Moventurque circa suum centrum super suos fixos <polos>*^(L)^ *nequaquam polis signorum suppositos.* Suo quoque motu Saturni quidem spera in die .b. sexagenariis secundis, Iovis .d. sexagenariis .oi. secundis, Martis .ma. sexagenariis .lf. secundis *dicte ferentes.*

[343] *De corporibus autem cuiusque harum separantur* due *parve spere inter duas cuiusque paralellas superficies incluse, que dicuntur rotunditates, sicut in Mercurii et Veneris speris dictum est. Earumque rotundum illius stelle dicitur, in cuius ferenti spera continetur. Hee quoque moventur sub centro super suos fixos polos equali motu.* [344] *Stellarum corpora ipsa quoque speralia sunt, posita queque in corpore sue rotunditatis moveturque motu eiusdem.*

---

7 ima] una **18** speris] spere (L) **20** sue] *corr. ex* suo (L)

4.4 <The Spheres of the Superior Stars>

Config. 13: The Orbs of the Superior Stars

[338] *The spheres of the three planets,*[31] *that is, of Saturn, Jupiter, and Mars, are most similar to one another in the number and the divisions of the spheres and also in the motion and the circles,* almost[32] *as Ptolemy is shown to have said in his Almagest. As we are thus speaking about similar objects, we grasp them all together in a single discussion.* [339] *Each of these stars has its own sphere, which is defined by two spherical and parallel surfaces, the centre of all of which is the same as that of the world. But these spheres are tangent to one another.* [340] *For the high surface of the first sphere, which is that of Saturn, is tangent to the low surface of the sphere of the fixed stars, and the low one is tangent to the high surface of Jupiter, the inner one of Jupiter to the high one of Mars, and the low one of the latter to the outer surface of the sun.*

[341] *Each one of them moves around two fixed poles that are placed straight below the poles of the ecliptic with a slow eastward motion.*[33] [342] *But from each one an eccentric sphere is separated,* as has been said above for the sphere of the sun, *which is enclosed between two spherical and parallel surfaces, whose centre is that of their same spheres [i.e., of the spheres they enclose, respectively]. And they move around their centre about their fixed poles, which are not placed below the poles of the ecliptic.* The sphere of Saturn moves by its own motion every day by 0;2,- degrees,[34] that of Jupiter by 0;4,59 degrees, and that of Mars by 0;31,26 degrees, *and they are called deferents.*

[343] *But from the bodies of each of them,* two[35] *small spheres are separated and confined between the two parallel surfaces of each, which are called epicyclic spheres, as has been said for the spheres of Mercury and Venus.*[36] *And each of them is called the epicyclic sphere of that star in whose deferent sphere it is contained. And they are also moved around the centre about their fixed poles by a constant motion.* [344] *The bodies of the stars themselves are also spherical, each one placed in the body of its epicyclic sphere and moved by the latter's motion.*

[31] 'three planets': Arab., 'three outer planets'.

[32] Stephen is clearly aware of developments and modifications in the astronomical theory after Ptolemy.

[33] Arab. add.: 'as in the motions of Venus and Mercury.'

[34] Here and in passage 358, 'arc-seconds' in Saturn's motion are indicated but not specified by a number. This could mean a value of 'nil' or, if one reads the abbreviation 'scd' as a singular, 'one' arc-second. In passage 358, the abbreviated 'et' between minutes and seconds may be the relic of a numeral.

[35] 'two': Arab., 'one'. Stephen may be referring to both the epicyclic sphere and the body of the planet within it, or he may account already for his own theory of the epicyclic spheres as nested couples; see p. 398:6f.

[36] 'as... Venus': Arab., 'similar to what we presented earlier.'

[345] *Quando igitur movetur ferens proprio motu* in orientem, *movet secum rotundum, facitque per centrum eius imaginarium circulum, cuius centrum ferentis est, et dicitur ferens. Imaginantibus autem nobis per superficiem eius totum secari mundum fiunt inclinati circuli in superficiebus sperarum paralelli ferentis, sicut in imagine sperarum* 5 *Mercurii et Veneris dictum est, nisi quod in his fixi,* illi mobiles, *probantur.*

f. 45r [346] *Imaginamur rectam lineam, que incipiens a centro mundi extenditur ad | ferentis centrum et crescens utrimque protelatur usque fines inclinati circuli utrobique secans ferentem in duobus diametrice oppositis punctis circulum. Eorum alter est longinqua longinquitas, alter vero longinquitas propinqua,* [347] *et longinque quidem longinquitatis* 10 *punctus harum stellarum manens in eadem positione, hic punctus quidem in septemtrione, ille vero in austro.*

[348] Centro autem rotunditatis longinque longinquitatis in puncto posito vel opposito, *rotundum diametrum diametro iungitur recte inclinati seu ferentis. Ferente autem suo motu rotundum movente rotundi variatur positio diametri, estque oppositum alii* 15 *puncto, quam sit centrum ferentis aut mundi, sicut in spera Veneris se res habet.* [349] *Imaginamur igitur diametrum usque ad illud extendi punctum, diciturque punctus et linea circinans diametrum rotunditatis. Sic Veneris et Mercurii spere continent.*

[350] *Et prima quidem spera cuiusque stellarum trium movetur, sicut dictum est, tardo motu in orientem centum solis annis uno gradu super suos polos, qui polis signorum* 20 *suppositi sunt. Hee ergo movent suo motu quicquid inter ipsas est. Moventur igitur hoc motu et superficies inclinatorum circulorum et longinque et propinque longinquitates et centra ferentium sperarum.*

[351] *Ferens autem movetur circa suum centrum equali motu in orientem, et movet suo motu rotundum et per rotunditatis motum corpus stelle.*

25 [352] *Rotunditas quoque movetur circa suum centrum* in suis polis *ferens secum corpus stelle, et centrum stelle facit imaginarium circulum per motum rotunditatis, centrum*

1 orientem] oriente (L) **14** rotundi] rotundum (L) **19** centum] centrum polis] polos (L)

[345] *Therefore, when the deferent moves* eastwards *by its own motion, it moves along with itself the epicyclic sphere and, by the latter's centre, produces an imaginary circle whose centre is that of the deferent and which is called deferent. But if we imagine the entire world dissected by the plane of that circle, inclined circles will be produced on the surfaces of the spheres that will be parallel to the deferent circle, as has been said in the picture of the spheres of Mercury and Venus, except that the circles are proved to be fixed in the present cases,* whereas those other ones are moving.

[346] *We imagine a straight line which begins from the centre of the world and extends until the centre of the deferent and is prolonged in both directions until the limits of the inclined circle, thereby crossing the deferent circle on both sides at two diametrically opposite points. One of them is the furthest distance, and the other one, the closest distance,* [347] *and the point of the furthest distance of these stars remains in the same position, that is, one point in the north and the other one in the south.*[37]

[348] When the centre of the epicyclic sphere is placed at the point of furthest distance or at the opposite point,[38] *the epicyclic diameter is aligned straight with the diameter of the inclined circle or deferent. But when the deferent moves by its own motion the epicyclic sphere, the position of the epicyclic diameter is changed and it points towards a point different from the centre of the deferent or that of the world, as is the case in the sphere of Venus.* [349] *We therefore imagine the diameter extended until that point, and the point and the line will be called the dirigent ones of the diameter of the epicyclic sphere. This corresponds to what is found in the spheres of Venus and Mercury.*

[350] *The first sphere of each of the three stars moves, as has been said, in a slow eastward motion around its poles, which are placed below the poles of the ecliptic, by one degree in a hundred solar years. By their motion, these spheres thus move whatever is inside them. Hence, by this motion the planes of the inclined circles and the furthest and closest distances, and also the centres of the deferent spheres, are moved.*

[351] *But the deferent moves around its centre with a constant eastward motion, and by its own motion it moves the epicyclic sphere and, by the motion of the epicyclic sphere, also the body of the star.*

[352] *Also the epicyclic sphere moves around its centre,* about its poles, *taking along with itself the body of the star, and by the motion of the epicyclic sphere the centre of the*

[37] Arab. repeats from passage 345: 'because the inclination of the inclined orbs of these stars neither changes nor moves. They are inclined in one manner: the furthest distance to the north and the closest distance to the south.' The passage is also omitted in L1.

[38] Arab. implies from the previous passage that the epicyclic sphere must stand at one of the apsides to align with the deferent's diameter. Stephen repeats that detail here in analogy to the corresponding description of the previous planets.

cuius rotunditatis est, dictus rotundus circulus. Inclinatur autem hic rotundus a superficie circuli extra centrum sicut et in duabus aliis stellis Mercurio Venerique. Qua de re in sequentibus plenius dicemus. [353] *Centrum autem cuiusque trium planetarum movetur super suum,* illumque quem facit secundum, *rotundum. Unde fit, ut quando est in* 5 *alta parte rotundi circuli, moveatur in orientem, quando in humili, in occidentem.*

[354] *Diametrum autem illius circuli, cuius sunt longinqua longinquitas et propinqua rotundi termini, inclinatur aliquando in septemtrionem, aliquando in austrum, moveturque super unum parvum circulum ferens secum, quocumque inclinetur, superficiem rotundi circuli.*

10 [355] *Separatur diametrum hoc de superficie extra centrum circuli, quando centrum rotunditatis per motum ferentis incipit removeri a puncto septemtrionalis sectionis. Statim autem post ipsam separationem incipit inclinari in septemtrionem.* [356] *Huius diametri motus super illum parvum circulum non est circa centrum, sed extra.* [357] *Tempus autem, quo diametrum hoc totum circuit circulum, equale est tempori circuitus longitudinis* 15 *motus* eiusdem stelle *in circulo signorum. Ex quo fit, ut motus huius in quadrante sui parvi circuli equalis sit temporis spatio motui longitudinis eiusdem in signorum circulo sub uno transmeando quadrante.*

[358] *Motus autem harum trium stellarum sunt* VI. *Primus est motus totius spere tardus in orientem; secundus ferentis eodem,* quod est Saturni .b. sexagenariis et se- 20 cundis et de aliis sicut supra diximus. *Hec defert rotunditatem in orientem. Tercius est motus, qui per secundum fit, in circulo signorum in longitudine. Motus autem rotunditatis circa suum centrum quartus est. Quintus vero circinantis diametrum rotunditatis motus est circa circinans centrum. Sextus est motus diametri rotunditatis inclinatio, nunc ad austrum, nunc ad septemtrionem.*

25 [359] *Atqui hec sunt, que de trium planetarum,* Saturni scilicet, Iovis et Martis, *speris ac circulis percipienda videbantur.* Verum de lato eorum subtili investigando, f. 45v retrogradatione quoque ac rectificatione et Mercurii simul | et Veneris diligentius iam pertractabimus. *Eorum vero, que de his dicta sunt, imaginem supponimus.*

---

5 orientem] occidentem occidentem] orientem **16** temporis] tempori (L)

star produces an imaginary circle whose centre is that of the epicyclic sphere and which is called epicycle. But this epicycle is inclined from the plane of the eccentric circle, as in the case of the two other stars, Mercury and Venus. We will speak about this in more detail in the following. [353] *The centre of each of the three planets moves on its own epicycle,* which it produces in turn by following it. *As a result, when it is in the high part of the epicycle, it moves eastwards, and when in the deep one, westwards.*

[354] *But the diameter of that circle, whose ends are the furthest distance and the closest distance of the epicycle, inclines sometimes northwards and sometimes southwards, and it moves on one small circle, taking along with itself the plane of the epicycle, wherever it inclines.*

[355] *This diameter diverges from the plane of the eccentric circle as soon as the centre of the epicyclic sphere begins to move away from the northern intersection point by the motion of the deferent. Immediately after this separation it begins to incline north.* [356] *The motion of this diameter on that small circle is not around, but outside, the centre.*[39] [357] *But the time in which this diameter traverses an entire circuit is equal to the time of one circuit of the motion in longitude* of the same star *on the ecliptic. It follows from this, that its motion in one quadrant of its small circle has an equal time-span as its longitudinal motion while traversing one quadrant.*

[358] *The motions of these three stars are six. The first one is the slow motion of the entire sphere eastwards; the second one is that of the deferent, in the same direction,* which is 0;2,- degrees for Saturn, and for the other planets as we have said above. *This sphere takes the epicyclic sphere eastwards. The third one is the motion that is produced by the second one, in longitude on the ecliptic. But the motion of the epicyclic sphere around its centre is the fourth one. The fifth one, however, is the motion of the dirigent of the diameter of the epicyclic sphere around the dirigent centre. The sixth motion is the inclination of the diameter of the epicyclic sphere, sometimes northwards and sometimes southwards.*

[359] *This is what seemed necessary to understand about the spheres and the circles of the three planets,* namely Saturn, Jupiter and Mars. Now we will thoroughly work through the difficult investigation of their latitude, the retrogradation and rectification,[40] including at the same time also those of Mercury and Venus. *But we give a diagram of what has been said about them [Fig. 4.14].*

[39] Arab. add.: 'as in the case of the motion in longitude.'

[40] Arab. add.: 'It is a general exposition of them, like the exposition of the first stars. For that reason, we have preferred to be brief rather than lengthy.'

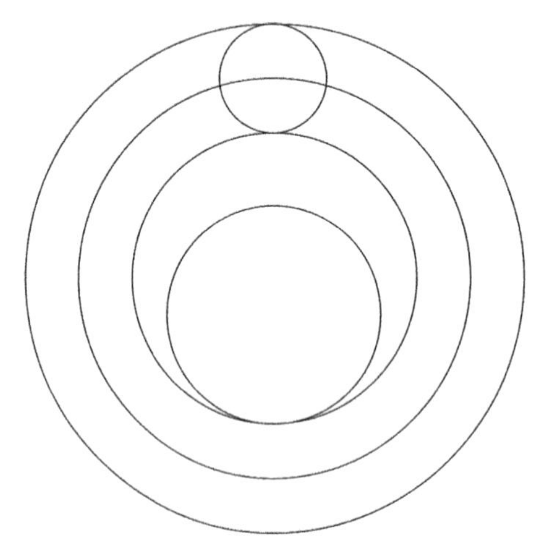

Fig. 4.14: From MS Cambrai 930, fol. 45v

<**R**>(L)ectificandi .e. planetarum locum ratio in tabulis quidem sub numerorum inconcussa veritate disposita est. Verum que causa sit, ponimus. Cum enim .e. in rectitudine cuiusque sub numeris tabule posite, quarum primam et quartam aliquando auferri, aliquando iungi, terciam auferri tantum, quintam iungi tantum secunda rectificante precipimus, non immerito quibus, quare sit, dubitatio non parva gignitur. Nos igitur omnium questiones breviter absolvemus.

Facimus igitur circulum similem signorum circulo atque super eum A B C D et super centrum E. Circinamus etiam inclinatum circulum et super eum H B G D et super centrum E, sicut est in sperali. Deinde facimus circulum extra centrum et super eum H I L M, et super centrum N. Et punctus H est longinqua longinquitas spere extra centrum, et punctus L longinquitas propinqua. Faciemus autem I circuli extra centrum punctum rotundi circuli centrum, et circinamus super eum rotundum circulum, super quem ponimus R Q P. Et exiet linea NIP et linea EIR, et ponemus stellam in rotundo circulo in puncto Q, et exiet linea EQO, super quam videtur motus stelle in circulo signorum. Facimus quoque M circuli extra centrum et circinamus super eum circulum rotundum, super quem ponimus K MA X, et exiet linea NMK et linea EM.MA, et ponimus stellam in rotunditatis puncto X, et exeuntem lineam EXP, super quam videtur motus stelle in circulo signorum. Videtur autem per huiusmodi figuram.

18 exeuntem] exeunte (L)

4.5 <The Rectification, the Retrogradation, and the Latitudes of the Planets>

The method of rectifying the position of the five planets has been arranged in tables with the incontrovertible truth of numerical precision. But we will now explain the reason of it. For, as there are five columns for the numerical rectification of each, the first and the fourth of which we prescribed to be sometimes subtracted, sometimes added, the third only to be subtracted, the fifth only to be added, while the second provides the correction, it is understandable that for some people severe uncertainty arose about why this is so. We will therefore briefly answer the questions of them all.

We thus draw the similar circle of the ecliptic, and on it A B C D, and on the centre, E. We also draw the inclined circle, and on it H B G D, and on the centre, E, as in the spherical one. Then we produce the eccentric circle, and on it H I L M, and on the centre, N. Point H is the furthest distance of the eccentric sphere, and point L is the closest distance. And we will make the point I, of the eccentric circle, the centre of the epicycle, and we draw the epicycle around it, on which we put R Q P. And the lines NIP and EIR will come out, and we will place the star on the epicycle at point Q, and the line EQO will come out, on which the motion of the star on the ecliptic is seen. We also make M on the eccentric circle and we draw around it the epicycle, on which we put K MA X, and the lines NMK and EM.MA will come out, and we place the star at the point X of the epicyclic sphere, and we make the outcoming line EXP, on which the motion of the star on the ecliptic is seen. This is seen from a diagram of the following type [Fig. 4.15].

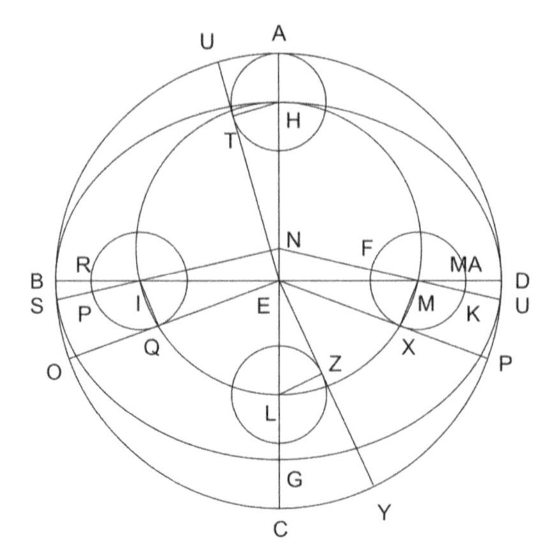

Fig. 4.15: Reconstruction of the heavily corrupted diagram in MS Cambrai 930, fol. 47v. In the manuscript, all circles and lines are drawn very carelessly and the labels have been added

Quando est punctus A contra longinquam longinquitatem in circulo signorum et stella in puncto Q rotundi circuli, cuius centrum est I, erit locus eius inter longinquam et propinquam longinquitates. Linea autem, que egreditur de puncto N, qui superius dictus est esse centrum circuli extra centrum, et transit super punctum I, designat punctum longinque longinquitatis rotundi circuli in puncto P, diciturque equalis longinqua longinquitas. Arcus autem QP est equalis numerus rotunditatis stelle; punctus vero R, quem designat linea EIR, que egreditur de centro circuli signorum E, est verus locus longinque longinquitatis rotundi circuli. Et quia punctus R est propinquior puncto A quam punctus <P>$^{(L)}$, auferimus rectitudinem centri rotunditatis, que est in prima tabula rectitudinis, de cursu centri, quia est cursus centri minor numero .tr., et iungimus numero rotunditatis, et est mensura rectitudinis | inter R et P. Tunc erit arcus RPQ rectificatus numerus rotunditatis, et arcus AB similis signorum circulo est rectificatus numerus cursus centri, arcus autem ABS arcus equalis cursus centri. Quia vero stella est extra punctum <R>$^{(L)}$ et punctum P in puncto Q rotundi circuli, in prima scilicet medietate rotunditatis, erit arcus AB similis circuli, quem designat linea EIR. Hinc igitur fit, ut rotunditatis numero infra .tr. contento iungamus rectificationem stelle, que est in quarta tabula, postquam rectificata fuerit super rectificatum numerum cursus centri. His

f. 46r

3 qui] que **4** dictus] dicta **12** rotunditatis] rotundus **16** AB] AQ

When point A is towards the furthest distance on the ecliptic and the star is at point Q of the epicycle whose centre is I, its position will be between the furthest and the closest distances. But the line that originates from point N, which above has been said to be the centre of the eccentric circle, and passes through point I defines the point of furthest distance of the epicycle at point P, and it is called the equal furthest distance. And the arc QP is the equal number [i.e., corresponding to the equal motion] of the star's epicyclic sphere. But point R, which is defined by the line EIR that originates from the centre of the ecliptic, E, is the true position of the furthest distance of the epicycle. And since point R is nearer to point A than is point P, we subtract the rectification of the centre of the epicyclic sphere, which is found in the first column of the rectification, concerning the course of the centre, because the course of the centre is less than the number 180 [degrees], but we add it to the number of the epicyclic sphere. And the amount of the rectification is what is between R and P. Then the arc RPQ will be the rectified number of the epicyclic sphere; and the arc AB of the similar circle of the ecliptic is the rectified number of the course of the centre, whereas the arc ABS is the arc of the equal course of the centre. But since the star is outside the points R and P, at point Q of the epicycle, that is, in the first half of the epicyclic sphere, the arc AB on the

simul coniunctis habetur verus stelle locus in signorum simili circulo, et est longitudo eius a puncto longinque longinquitatis extra centrum circuli. Est autem hec rectificatio in mensura arcus, qui est super lineam IQ.

¶Quando autem est circulus rotundus in puncto M, qui est inter propinquam
5 et longinquam longinquitatem extra centrum circuli, eritque stella in puncto X rotundi circuli, rectificationis ratio contra ea, que dicta sunt, invenitur. Punctus enim equalis longinque longinquitatis rotunditatis est punctus K, rectificata vero longinqua longinquitate rotunditatis punctus MA; punctus U est equalis locus centri rotundi, punctus vero D est verus locus centri in simili signorum circulo. Stel-
10 la autem est super punctum X et transit propinquam longinquitatem rotunditatis, que est punctus F, quantum est arcus ab F usque X. Est autem posita stella super lineam, que tangit rotunditatem, linea vero MX est obliqua medietas rotundi circuli. Arcus vero circuli similis ABCU, quem designat linea NMKU, minor est arcu eiusdem circuli ABCD, quem designat linea EM.MA.D. Ideo ergo, quando cursus
15 circuli extra centrum maior est .tr., iungemus rectificationem centri, que est prius prima tabula, cursui centri et auferemus de rotundo numero.

Tunc erit arcus similis circuli ABCD locus rectificati centri, et arcus MA.FX est rectificatus numerus rotunditatis, et arcus ABCU locus equalis centri rotunditatis in simili circulo. Et quia stella est inter punctum F et MA rotundi circuli, que est
20 secunda medietas rotunditatis, et arcus ABCP similis circuli, quem designat linea EXP, minor est arcu eiusdem circuli ABCPD, quem terminat linea E.MA.D, ideo ergo, quando rotundus numerus excedit .tr., aufertur rectificatio stelle, que est in quarta tabula, de rectificato centri cursu; at si minor esset, sicut supra dictum est, iungeretur, rectificata tamen prius per terciam aut quintam tabulam, sicut habetur
25 in canonis regulis, quando exinde habetur longitudo stelle a longinque longinquitatis puncto in circulo signorum. Est autem mensura quarte tabule equalis linee MX eo in loco.

¶Si vero et longinqua quidem circuli signorum longinquitas sit in puncto <A>

2 hec] *add.* (L) **20** ABCP] ABOP **25** a] aut

similar circle will be the one that is defined by the line EIR. As a result, therefore, we need to add to the course of the epicyclic sphere, as it is less than 180 degrees, the rectification of the star that is in the fourth column, after it has been rectified according to the rectified number of the course of the centre. When these have all been added up, one has the true position of the star on the similar circle of the ecliptic, which is also its distance from the furthest distance of the eccentric circle. This rectification corresponds in size to the arc above the line IQ.

But when the epicycle is at point M, which lies between the closest distance and the furthest distance of the eccentric circle, and the star will be at point X of the epicycle, the method of rectification is found inverse to what has been said above. For the point of the equal furthest distance of the epicyclic sphere is point K, whereas the rectified furthest distance of the epicyclic sphere is point MA; and point U is the equal position of the centre of the epicycle, whereas point D is the true position of the centre on the similar circle of the ecliptic. Now, the star is at point X and has passed the closest distance of the epicyclic sphere, which is point F, by as much as the arc from F until X. And the star is placed on the line which touches the epicyclic sphere, whereas the line MX is oblique and half of the epicycle. But the arc ABCU of the similar circle, which is defined by the line NMKU, is smaller than the arc ABCD of the same circle, which is defined by the line EM.MA.D. Therefore, when the course of the eccentric circle is larger than 180 [degrees], we will add the rectification of the centre, which, as said earlier, is the first column, to the course of the centre, while we will subtract it from the epicyclic number.

Then the arc ABCD of the similar circle will be the position of the rectified centre, and the arc MA.FX is the rectified number of the epicyclic sphere, and the arc ABCU the equal position of the centre of the epicyclic sphere on the similar circle. And since the star is between the points F and MA of the epicycle, which is the second half of the epicyclic sphere, also the arc ABCP of the similar circle, which is defined by the line EXP, is smaller than the arc ABCPD of the same circle, which is defined by the line E.MA.D. Therefore, when the epicyclic number exceeds 180 [degrees], the rectification of the star, which is written in the fourth column, concerning the rectified course of the centre, is subtracted; whereas, if it was smaller, as has been said above, it would be added, but nevertheless rectified before by the third or the fifth column, as one finds in the *Rules of the Canon*, while from there one has the distance of the star from the point of furthest distance on the ecliptic. But the fourth column corresponds in size to the line MX at that place.

But if the furthest distance of the ecliptic is at point A and the star at point Q

et stella in puncto Q in rotundo circulo, cuius centrum est I, sitque hoc centrum rotunditatis in puncto H, linea, que exiet de puncto E et transiet per punctum N et super H et A, designabit punctum longinque longinquitatis rotunditatis in puncto A, quod designat etiam linea NIP. Tunc vero erit linea NIP in loco linee ENA, et punctus loci P in loco puncti A. Erit igitur longinqua longinquitas equalis et vera in uno puncto circuli signorum sine discordia, ideo quia linea EIR, super quam videtur verus punctus longinque longinquitatis rotunditatis, erit tunc in loco linee EHA. Cum vero erit centrum rotunditatis in loco I, qui est inter punctos H L circuli extra centrum, quod minus est medietate rotunditatis, erit locus vere longinque longinquitatis rotunditatis in puncto R et locus equalis longinque longinquitatis in puncto P, et ab eo erit initium proprii motus stelle in rotundo circulo, qui est arcus PQ, sicut prediximus. Idcirco rotundus motus crescit super arcum PQ, qui est discordia amborum motuum.

Posito autem rotunditatis centro in puncto L circuli extra centrum nulla erit linearum terminis discordia, quemadmodum et in puncto H positum est. Eundem etenim habent terminum linea ELC, que de centro mundi | egreditur, et linea NLC, que a centro circuli extra centrum initium habet. Sed linea NLC longinquam equalem longinquitatem designat, linea vero ELC rectam longinquam longinquitatem rotunditatis. Eodem ergo in loco posita est et recta et equalis in circulo signorum. Hec est ergo ratio, <qua> et in longinqua et in propinqua longinquitate planetarum cursum rectitudine privat et in horum mediis locis cogit rectificari.

¶Quare autem quarte tabule nunc tercia auferatur, sepe V^{ta} iungitur, evidens hoc modo fiet ratio. Secunde tabule numerus ab longinqua longinquitate circuli extra centrum usque .s. fere decrescit, exinde usque .tr. recrescens. Rotunditate autem posita in alta parte circuli extra centrum minus esse conspicitur eius diametrum, maius autem, cum in humili eiusdem spere medietate volvitur. Fiat enim rotundus circulus in puncto H longinque longinquitatis circuli extra centrum et super eum T, et alter in puncto L eiusdem circuli et super eum Z. Egrediantur linee <ab>$^{(L)}$ E, per T <et>$^{(L)}$ U et per Z et Y, et linee ab H per T <et ab L per Z>, medie-

4 A] E ENA] END **16** NLC] *corr. ex* N (L) **23** longinqua] obliqua **27** H] A **27–28** et super eum T] A per **28** eiusdem] eidem (L) **29** per Z et Y] C Z Y (L) ab H] A <et ab L per Z>] B

on the epicycle whose centre is at I, and if this centre of the epicyclic sphere is [assumed to be] at point H, then the line that goes out from E and passes through point N and through H and A will define the point of the furthest distance of the epicyclic sphere at point A, which point is also defined by the line NIP. But in that case the line NIP will be in the place of the line ENA, and the point of the position P will be in the place of the point A. Hence the equal and the true furthest distances will be at the same point of the ecliptic, without any difference [or: anomaly], because the line EIR, on which the true point of furthest distance of the epicyclic sphere is seen, will then be in the place of the line EHA. But when the centre of the epicyclic sphere is at the place I, which is between the points H and L on the eccentric circle, as it is [corresponding to] less than half of the epicyclic sphere, the position of the true furthest distance of the epicyclic sphere will be at point R, and the position of the equal furthest distance at P, and from there will be the beginning of the star's own motion on the epicycle, which is the arc PQ, as we have said before. Therefore, the epicyclic motion exceeds the arc PQ, which is the difference [or: anomaly] of both motions.

But when the centre of the epicyclic sphere is placed at point L of the eccentric circle, the ends of the lines will not have any difference [or: anomaly], as has been said also with regard to point H. For the line ELC, which originates from the centre of the world, and the line NLC, which has its beginning from the centre of the eccentric circle, terminate at the same point. But the line NLC defines the equal furthest distance, whereas the line ELC, the correct furthest distance of the epicyclic sphere. Hence, the correct and the equal ones are placed at the same position on the ecliptic. This is the reason why the course of the planets is deprived of rectification at the furthest distance and at the closest distance, and also why it demands rectification between these positions.

But why the third column is sometimes added to the fourth one, whereas the fifth one is sometimes subtracted from it, will become clear from the following consideration. The number in the second column decreases from the furthest distance of the eccentric circle until about 90 [degrees] and from there increases again until 180 degrees.[41] But when the epicyclic sphere is placed in the high part of the eccentric circle, its diameter appears shorter, whereas when it revolves in the low half of that circle's sphere, longer. For let the epicycle be at point H, of the furthest distance of the eccentric circle, and on it T, and another one at point

[41] Columns 1 to 5 of Stephen's tables apparently correspond in their respective function, partly simplified, to columns 3, 8, 5, 6 and 7 of Ptolemy's rectification tables in *Alm.* XI,11 (tr. Toomer).

tates scilicet diametrorum rotunditatum. Hee quidem linee, que medium signant diametri, ad suum respectum nullam augmenti vel detrimenti patiuntur varietatem. Verum in circulo signorum simili eorum consideratio aliter se habet. Quanto enim per motum ferentis rotunditas ipsa remotior est a terra, tanto brevius habetur 5 spatium circuli signorum a puncto, in quo est eius longinqua longinquitas, ante aut retro usque punctum, quem tangit linea de centro mundi directa transiens per caput angularis diametri in circulo signorum. Quanto autem propius terre per motum ferentis accedit, tanto maius spatium circuli signorum desiderat. Quod in ipsa quoque figura clare videre licet.

10 Arcus enim circuli similis ab A usque U multo brevior est arcu eiusdem circuli, qui est ab C usque Y. Est autem ille quidem in longinqua longinquitate ferentis ab eadem habens initium, hic autem in propinqua longinquitate. Huius vero varietatis medium est in medio utriusque longinquitatis utrimque, aut .s. scilicet aut .uq. gradibus. His ergo in locis IIII^{te} tabule parum additur aut aufertur, in longin- 15 quitatibus multum. Et quoniam in longinqua longinquitate brevius est diametrum, per secundam rectificata tercia tabula IIII^{te} aufertur. Quoniam autem in ima medietate superhabundat, V^{ta} per secundam rectificata pro loci, in quo est centrum rotunditatis, proportione IIII^{te} iungitur.

---

10 U] T **11** Y] *corr. ex* ZY (L) **14** in] *add.* (L) **16** ima] una **17** quo] qua

L of the same circle, and on it Z. Let lines originate from E, [one] through T and U and [another one] through Z and Y, and lines from H through T and from L through Z, that is, halves of the diameters of the epicyclic spheres. These lines, which represent a half of the diameter, do not experience any increase or decrease with regard to themselves. But this is different on the similar circle of the ecliptic. For the farther remote from the earth the epicyclic sphere itself is by the motion of the deferent, the shorter is the space on the ecliptic from the point of the circle's furthest distance, ahead or behind, until the point that is touched on the ecliptic by the line from the centre of the world and directed to pass through the end of the angular diameter. But the nearer it approaches the earth by the motion of the deferent, the more space of the ecliptic it demands. This can also be seen clearly in the diagram itself [Fig. 4.15].

For, the arc of the similar circle from A until U is much shorter than the arc of the same circle that extends from C until Y. But the former is at the furthest distance of the deferent and begins from this, whereas the latter is at the closest distance. And in the middle between the two distances [i.e., apsides] to both sides, namely at 90 or 270 degrees, there is the middle of this variation. At these positions, therefore, little is added to, or subtracted from, the fourth column, whereas at the distances [i.e., apsides] it is much. And since the diameter is shorter at the furthest distance, the third column, corrected by the second one, is subtracted from the fourth; whereas, since it exceeds in the low half, the fifth column, corrected by the second one in proportion to the position of the centre of the epicyclic sphere, is added to the fourth.

¶Quoniam autem rectificandi planetarum .e. loca rationes probabiliter ac sufficienter ostendimus, restat nunc de eisdem aperire, que causa sit standi, que retrogradiendi necessitas. Id enim est, in quo Latinitas diu caligat errore turpis ignorantie plurimum involuta. Que quidem veniam satis mereretur, tamquam a nemine sit instructa, nisi errorem suum inprobissimis defenderet ac philosophantium stultissimis cavillationibus. Ratio igitur retrogradationis planetarum est discordia rectificationis stelle in crescendo et minuendo. Nam per motum quidem ferentis in orientem semper nituntur. Verum rotunditatum motus, quarum quedam medietas et superior in orientem movetur, quedam et inferior contra, huiusmodi faciunt cursus stellarum discordiam. Quamdiu enim quivis planeta in alta sui movetur circuli rotundi medietate, remotior a nobis rectus fertur. Cum vero in ima volvitur, aut stat aut retrograditur. Quando ergo in ima medietate volvitur, si rotunditatis tantus est cursus in occidentem, quantus est ferentis in orientem, stat in eodem circuli signorum puncto planeta. Cum vero rotunditatis cursus motum ferentis superat, quantum ferens vincitur, tantum retrogradus apparet planeta.

Rectificato igitur vero loco stelle in uno die et equali unius diei cursui centri coniuncto, si quidem hic illi equalis est, stat nec movetur | planeta. Si vero rectificatus minor erit equali, certum est per rectificationem minus habere cursus hodie in circulo signorum quam heri, unde et retrogradus est. Hoc autem fieri potest, quando stella est in ima seu humili medietate rotunditatis. Quando tamen alta medietate, minor est.

Que ut planiora fiant, exemplum huius proponimus facientes circulum extra centrum et super eum A B, et super centrum D lineantes diametrum ab A usque B. Circinamus quoque circulum signorum super eum ponentes N O R P et super centrum eius E. Super punctum quoque A rotundum circinamus circulum et super eum G H I L M Q U C. Arcus LGH maior est arcu HIL. Quando itaque est initium motus stelle in rotundo circulo de puncto G ad punctum H, exinde ad punctum I, post ad punctum L, deinde ad punctos M et Q. Et exiet linea EQHO. Quando ergo erit stella in puncto H, erit motus eius in circulo signorum NO; quando vero in puncto I, motus eius est arcus NR; quando autem in puncto L, erit arcus a puncto N usque punctum P.

2 de eisdem] *corr. ex* deisdem (L) **4** mereretur] meretur **10** movetur] movetur movetur **11** ima] una **12** ima] una **16** in uno] in uno in uno **20** ima] una **22** planiora] pleniora (L) **28** EQHO] EQHA

But since we have demonstrated with sufficient plausibility the methods of rectifying the positions of the five planets, it remains now to reveal the reason for their stations and the inevitability of their retrogradation. For this is a matter about which Latinity has long been in error, covered deep in dishonourable ignorance. Never instructed by anyone she would well deserve clemency, if she did not defend her error by false and most foolish talk of people who practice philosophy. The reason for the retrogradation of the planets is the difference, in increasing or decreasing, of the rectification of the star. For they always tend eastwards by the motion of the deferent. But the motions of the epicyclic spheres, whose one, upper, half moves eastwards and the other, lower, one in the opposite direction, cause such anomaly in the course of the stars. For as long as any planet moves on the high half of its epicycle, it is farther remote from us and is taken direct. But when it revolves in the low half, it stands still or retrogrades. Thus, when it revolves in the low half, and if the course of the epicyclic sphere is as much to the west as that of the deferent is to the east, the planet stands still at the same point of the ecliptic. But when the course of the epicyclic sphere exceeds the motion of the deferent, the planet appears retrograde by as much as the deferent falls short.

Thus, when the true position of the star has been rectified for one day and has been added to the equal course of the centre during one day, and if the former is equal to the latter, the planet stands still and does not move. But if the rectified motion will be less than the equal one, it is certain that by the rectification the position today has less course on the ecliptic than yesterday, which means that it is retrograde. But this can happen when the planet is in the low, or deep, half of the epicyclic sphere; whereas, when it is in the high half, it occurs less.

For this to become clearer, we give an example [Fig. 4.16], drawing the eccentric circle, and on it A B, and on the centre [we put] D, and the diameter from A until B. We also draw the ecliptic, putting on it N O R P and on its centre, E. Around point A we also draw the epicycle and on it G H I L M Q U C. The arc LGH is larger than the arc HIL. Hence, when the motion of the star on the epicycle begins from G to H, [it continues] from there to the point I, then to the point L, and afterwards to the points M and Q. And there will go out the line EQHO. Thus, when the star will be at point H, its motion on the ecliptic will be NO; and when at I, its motion will be the arc NR, and when at point L, it will be the arc from point N until point P.

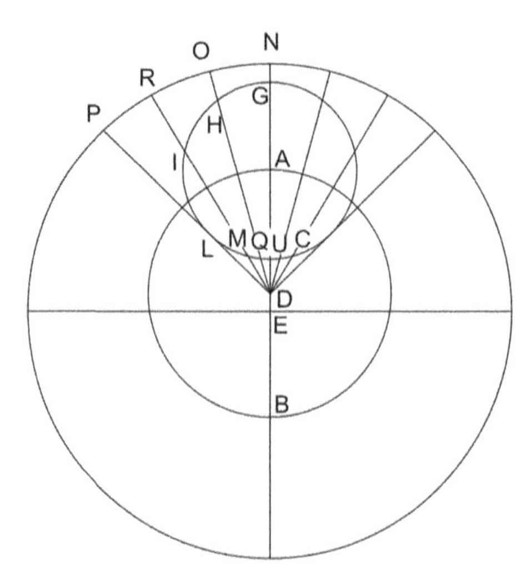

Fig. 4.16: From MS Cambrai 930, fol. 48r, partly reconstructed. In the manuscript, the diagram is drawn carelessly and without labels

Quando itaque movetur a puncto G ad punctos H I L, crescit semper motus eius in circulo signorum. Cum vero descendit a puncto L ad punctum M, est iterum locus eius in circulo signorum in puncto R; cumque a puncto M usque punctum Q recreverit, est locus eius in circulo signorum in puncto O, in quo fuerat et in puncto H. Arcus autem NR maior est arcu NO, sed et arcus NP maior est arcu NR. Sed stella descendens a puncto L, in quo fuerat in puncto P, pervenit ad punctum M, in quo est in puncto R. Minor est ergo arcus motus eius in circulo signorum, quam cum esset in puncto L.

A puncto M quando descendit ad punctum Q, est iterum in puncto O, in quo fuerat, cum esset in puncto H. Minuitur igitur arcus motus eius in circulo signorum, quia minor est arcus NO arcu NR, et arcus NR minor est arcu NP. Cum ergo fuerit <ibi>[(L)], videtur stella retro ire; ita tamen si motus eius in longum equalis, qui fit per ferentem speram et est motus centri A, minor est discordia motus, qui fit per rotunditatem in uno die. Quamdiu autem maior fuerit equalis quam discors, nulla est stelle retrogradatio, propterea quod punctus A, centrum scilicet rotundi circuli, amplius fertur versus punctum H in circulo signorum, quam sit discordia motus, que fit per motum rotunditatis ferentis centrum stelle in occidentem.

8 quam] qui **9** quando] quoque O] I (L) **15** scilicet] solis (L)

Therefore, when it moves from point G to the points H I L, its motion on the ecliptic always increases. But when it descends from point L to point M, its position on the ecliptic is again at point R; and when it has fallen back from point M until point Q, its position on the ecliptic is at point O, in which it had also been when it was at point H. But the arc NR is larger than the arc NO, and the arc NP is larger than the arc NR. But the star, descending from point L, in which it had been when it was at P, comes to point M, where it is at point R. Hence the arc of its motion on the ecliptic is smaller than when it was at L.

And when it descends from point M to point Q, it is again at point O, in which it had been when it was at point H. Hence the arc of its motion on the ecliptic decreases, because the arc NO is smaller than the arc NR, and the arc NR is smaller than the arc NP. Thus, when being there, the star appears to go backwards; this, however, occurs in this way if its equal motion in longitude, which is produced by the deferent and which is also the motion of the centre A, is less than the anomalistic motion that is produced by the epicyclic sphere in one day. But as long as the equal motion is larger than the anomalistic one, there is no retrogradation of the star, because the point A, that is, the centre of the epicycle, is taken further towards point H on the ecliptic than the amount of the anomalistic motion that is caused by the motion of the epicyclic sphere, which takes the centre of the star

Exemplum autem huius rei proponimus equalem motum Saturni, qui est motus puncti super circulum extra centrum, in uno die .b. sexagenariis in circulo signorum, incipiens a puncto <N>$^{(L)}$ et vadens versus punctum O et R et P. Quando ergo est numerus rotunditatis eius .tkd. gradus, qui est arcus GM rotundi circuli, erit rectificatio eius, que est post illum gradum in IIII^{ta} tabula, .e. gradus et .oe. sexagenarie. Motus vero Saturni in rotundo circulo est in unoquoque die .<o>g. sexagenarie. Collecta ergo rectificatione eius post illum diem, tunc scilicet quando erit cursus eius in rotundo circulo .tke. gradus, invenies in IIII^{ta} tabula .e. gradus et .oc. sexagenarias. Hoc autem minus est duabus sexagenariis rectificatione precedentis diei. Coniuncto ergo unius diei de cursu Saturni precedentis diei equali cursui, .b. scilicet sexagenariis, cum coniuncta erit his rectificatio secunda, que precedenti minor fuit .b. sexagenariis, equalis erit secunde diei rectificatus numerus numero prime.

Eo ergo die, quo hoc occurrit inquirenti recte rectificato, nullus est planete ante vel retro cursus, sed videtur stare; sed ne aliquo motu, unde et planete dicuntur. Postquam autem transierit punctum M, fiet motus eius in arcu MQ, habens scilicet numerum rotundi circuli plus quam .tl. gradus. Tunc ergo incipit decrescere rectificatio eius in unoquoque die III^{bus} sexagenariis, et hoc plus motu equali centri rotunditatis Saturni .a. sexagenaria, et tunc incipit retrocedere. Post hoc autem amplius decrescit rectificatio in unoquoque die, et amplius retrograditur | Saturnus, usque dum veniat ad punctum U rotundi circuli. Tunc est cursus rotunditatis eius .tr., minuiturque ibidem unoquoque die rectificatio eius .h. sexagenariis, et retrogradatio eius est .f. sexagenarie duabus de .h. sublatis. Cum vero pertransierit .tr., idest U, incipit rectificatio augeri. Que cum ablata fuerit de cursu centri sublatis .h. sexagenariis, quia duabus creverat, .f. restant sexagenarie retrogradationis eius in unoquoque die, usque dum perveniat ad punctum C. Quo cum pervenerit, secunda celebrata statione dirigitur in orientem. Hoc autem fit, quando numerus rotunditatis est .unf., scilicet arcus GLUC. Tunc enim est rectificatio eius precedentium quorundam et subsequentium dierum equalis eundemque locum stelle in circulo signorum designans.

Postquam transierit punctum C, IIII^{te} tabule rectificatio minus paulo .b. sexagenariis minuitur, deinde una sexagenaria, remanet quoque motus Saturni in unoquoque die prius quidem minus, post una sexagenaria in orientem. Sic ergo semper rectificatio minuitur, cursus augetur, usque dum accedat ad .xp. gradus ro-

2 .b.] .h. **10** cursu] *corr. ex* cursus (L) **14** hoc] hec **19** hoc] hec **27** celebrata statione] celebrati stationis Hoc] Hec **28** .unf.] per N F (L)

westwards.

As an example of this we consider the equal motion of Saturn, which is the motion of a point on the eccentric circle, to be on one day 0;2 degrees on the ecliptic, beginning from point N and moving towards point O and R and P. When the number of its epicyclic sphere is 114 degrees, which is the arc GM of the epicycle, its rectification in the fourth column behind that degree will be 5;55 degrees. But the daily motion of Saturn on the epicycle is 0;57 degrees. If you thus sum up its rectification after that day, that is, when its course on the epicycle will be 115 degrees, you will find in the fourth column 5;53 degrees. But this is 0;2 degrees less than the rectification of the previous day. If one thus adds Saturns course of one day to the equal course of the day before, namely 0;2 degrees, and when the second rectification is added to this, which was less than the precedent one by 0;2 degrees, the rectified number of the second day will be equal to that of the first day.

On that day, therefore, on which this occurs to an enquirer through correct rectification, the planet has no course forwards or backwards, but seems to stand still; with no motion at all, which [actually] gave them the name 'planets.' But after it has passed through point M, its motion will be the arc MQ, which has a number of the epicycle of more than 120 degrees. From then on its rectification begins to decrease by 0;3 degrees per day, and this is 0;1 degrees more than the equal motion of the centre of Saturn's epicyclic sphere, and then it starts retrograding. But after this the rectification decreases by a larger amount every day, and Saturn retrogrades faster, until it reaches point U of the epicycle. Then the course of its epicyclic sphere is 180 degrees, its rectification there corresponds to a daily reduction by 0;8 degrees, and its retrogradation, after subtraction of 0;2 degrees from the 0;8 degrees, is 0;6 degrees. But when it passes 180 degrees, that is, point U, the rectification begins to rise. And when it has been subtracted from the course of the centre, that is, [minus] 0;8 degrees increased by 0;2 degrees, its remaining retrogradation per day until it reaches point C will be 0;6 degrees. When it has reached that point, the second station will be acclaimed and it will be directed eastwards. This occurs when the number of the epicyclic sphere is 246, that is, the arc GLUC. For then its rectified positions on the preceding and the following days are equal and define the same position of the star on the ecliptic.

After it has passed point C, the rectification of the fourth column is subtracted a little less than 0;2 degrees, later 0;1 degrees, and the daily motion of Saturn will first be a bit less than, later precisely, 0;1 degrees to the east. In this way the rectification steadily decreases, while the course increases, until the epicyclic number

tundus numerus, punctum scilicet G. His igitur in locis auferuntur de rectificatione unoquoque die .f. sexagenarias, habetque motum .h. sexagenariarum in orientem, sicque fit, usque dum tangat punctum G. Hoc transgresso incipit eius rectificatio in IIII^{ta} tabula .f. sexagenariis crescere et iungitur equali motui eius .b. sexagenariarum, estque cursus eius in unoquoque die eorum .h. sexagenarie, quem directum iter dicunt. Est autem punctus G velocissimus eius cursus, citissimaque retrogradatio punctus U. Prima vero statio punctus M, secunda punctus C. Et de reliquis planetis eodem modo.

f. 48r | <**E**>[(L)]t quoniam de motu planetarum .e. in longum satis dictum est, de motu quoque in latum dicendum videtur. Latum ergo eorum, ut iam significatum est, inclinationem dicimus a circulo signorum in septemtrionem vel in austrum. Imaginamur autem circulum, qui transit per polos signorum et centrum planete et gradum aliquem in circulo signorum. Arcus huius circuli, qui est a centro stelle usque gradum circuli similis signorum circulo in eadem spera, mensura lati est. Circulum autem signorum motu solis in orientem fieri supra dictum est, propterea quod superficies spere circuli extra centrum, in qua movetur, est semper in superficie circuli signorum, nec inclinatur. Et fixe quidem stelle habent motum in orientem super polos spere signorum. Inde est, ut numquam eorum latum varietur, quia in eadem prestant superficie spere signorum eodemque loco. Que igitur sunt in circulo signorum semper eo moventur, que foris aut in septemtrionem aut in austrum inclinantur, eandem mensuram in suo motu servant longi, quod ab ipsis est usque circulum signorum in latum, numquam mutatam. Fixarum ergo queque stella aut non habet latum a circulo signorum, aut semper equale habet. At vero luna et .e. planete non sic. Motus enim eorum non semper est super polos signorum, sed super sperarum polos extra centrum in circulis, qui secant circulum signorum in diametrice oppositis punctis in austrum et septemtrionem inclinati, ideoque discordiam lati stellarum faciunt.

Et lune quidem circuli superficies extra centrum secat superficies zodiaci super duos punctos ligantis, et inclinatur ab eo in septemtrionem et austrum, inclinatioque eius unius semper mensure, nec mutatur. Superficies quoque rotundi circuli est in superficie extra centrum circuli, et non inclinatur, ideoque habet in lato unam tantum discordiam propter | inclinationem extra centrum circuli a circulo signorum. f. 48v

22 mutatam] mutato **25** secant] secatur (L)

reaches 360 degrees, that is, point G. At these places, therefore, 0;6 degrees are subtracted every day from the rectification, and <Saturn> has a motion of 0;8 degrees eastwards; and this occurs, until it reaches point G. When that point has been passed, its rectification in the fourth column begins to increase by 0;6 degrees and is added to its equal motion of 0;2 degrees, and its daily course is 0;8 degrees, which is called the 'direct' path. Hence point G represents its fastest course, and point U the fastest retrogradation. And the first station is point M, and the second one point C. The same applies analogously to the remaining planets.

Since enough has been said about the motion of the five planets in longitude, it seems necessary to speak also about the motion in latitude. 'Latitude', as it has been defined already, is how we call the inclination from the ecliptic to the north or to the south. We imagine a circle which passes through the poles of the ecliptic and the centre of the planet and through some degree on the ecliptic. The arc of this circle that is from the centre of the planet until the degree of the similar circle of the ecliptic on the same sphere is the measure of the latitude. But it has been said above that the ecliptic is produced by the eastward motion of the sun, because the plane through the eccentric sphere on which it moves is always in the plane of the ecliptic and does not incline from it. And the fixed stars have an eastward motion around the poles of the sphere of the signs. As a result, their latitude never changes, because they remain on the same surface of the sphere of the signs and at the same place. Therefore, those stars that are on the ecliptic always move on it, whereas those that are inclined to the north or to the south outside the ecliptic maintain in their motion the same amount of distance in latitude that is from them until the ecliptic, which thus never changes. Hence, every fixed star has either no latitude from the ecliptic or it has always the same one. But this is not so for the moon and the five planets. For their motion is not always around the poles of the ecliptic, but around the poles of eccentric spheres, on circles that cross the ecliptic at diametrically opposite points and are inclined southwards and northwards, and which thereby produce variation of the latitude of the stars.

And the plane of the eccentric circle of the moon crosses the plane of the zodiac at the two nodes, and it inclines from there to the north and to the south, and its inclination has always the same amount and does not change. And the plane of the epicycle is in the plane of the eccentric circle and does not incline [from it], and therefore it has only one anomaly in latitude due to the eccentric circle's inclination from the ecliptic.

Quinque vero planetarum discordia in lato non est <una> tantum, sed multiplex, ideo et extra centrum inclinatur circulus a circulo signorum et rotundus circulus eorum inclinatur a circulo extra centrum. Inclinatus enim eorum circulus signorum circulum dividit per rectum diametrum, et inclinatur ab eo medietas in austrum, altera in septemtrionem.

Huius autem divisionis locus Saturni quidem et Iovis est prope medium, quod habetur inter longinquitatem mediam et duas longinquitates discordes circuli extra centrum. Alii autem IIIes huius sectionis punctos habent super duas longinquitates medianas. Hinc ergo, quod inquirendis latis eorum Saturni quidem numero centri .o. adduntur, quoniam punctus ille, a quo incipitur medietas, que in septemtrionem inclinatur, circuli eius extra centrum, .o. gradibus ab Arietis capite remotus est positus in .l. puncto primo gradus Tauri; at vero Iovi .l. auferuntur, quia in .k. gradus Piscium in ultimo septemtrionalis medietatis incipitur. Marti nichil additur aut aufertur, quoniam in Arietis capite est. Et in Venere Marti parilis est ratio. Mercurii autem punctus septemtrionalis medietatis primus est in capite Libre.

Longinquarum autem longinquitatum inclinationis circuli extra centrum in Saturni quidem Iovis ac Martis circulis in septemtrione a circulo signorum, propinquarum longinquitatum in austrum fixe perstantes, sicut et luna. Circuli vero extra centrum Veneris et Mercurii inclinatio non est fixa, sed movetur super diametrum circuli signorum, super quem transeunt duos ligantes, eorundem leto motu in septemtrionem et austrum. Est autem reditus eorundem, a quo moventur quavis hora, <ad>$^{(L)}$ locum, post plenum annum, sicut est circuitus motus longitudinis. Est ergo longinqua seu alta medietas circuli extra centrum VI mensibus in austro, VI reliquis in septemtrione. Eodem modo mutatur propinqua medietas in austrum et septemtrionem. Manifestum est ergo superficiem extra centrum circuli bis in anno eandem esse cum superficie circuli <signorum>. Hoc autem fit in duobus punctis sectionis.

Et in Venere quidem, quando centrum rotunditatis eius est in quovis sectionum puncto, incipit inclinari medietas illa, que est <post> illum ligantem, circuli extra centrum in septemtrionem, altera in austrum. In Mercurio contra est. Centro enim rotunditatis eius in quovis puncto sectionis posito incipit inclinari ea medietas, que post illam est sectionem, in austrum, opposita in septemtrionem, ideoque est centrum rotunditatum eorum aut in superficie circuli signorum, ubi sunt et ligantes, aut inclinantur in alteram partem circuli signorum et nunquam in aliam.

3 a circulo] *add.* (L) **6** locus] locis (L) **12** .l.] .k. **30** septemtrionem] septemtrione (L)

But the anomaly in latitude of the five planets is not single, but manifold, inasmuch as both the eccentric circle is inclined from the ecliptic and their epicycle is inclined from the eccentric circle. For their inclined circle divides the ecliptic along a straight diameter, and one half inclines from it to the south, and the other one to the north.

The position of this intersection for Saturn and Jupiter is close to the middle between the mean distance [between the apsides] and the two anomalistic distances [or: apsides] of the eccentric circle. The other three have the points of this intersection at the two mean distances [between the apsides]. Therefore, when seeking their latitudes one has to add 50 to the number of Saturn's centre, because that point from which that half of the eccentric circle that is inclined northwards begins is placed 50 degrees remote from the head of Aries, at the beginning of the twentieth degree of Taurus; but from Jupiter 20 degrees are subtracted, because it begins at the 10th degree of Pisces, at the end of the northern half.[42] To Mars nothing is added or subtracted, because it is at the head of Aries; and for Venus applies the same as for Mars. The first point of the northern half of Mercury is at the head of Libra.

But the points of the furthest distances of the inclination of the eccentric circle on the circles of Saturn, Jupiter and Mars remain fixed to the north from the ecliptic, and those of the closest distances to the south, like for the moon. But the inclination of the eccentric circle of Venus and Mercury is not fixed, but it moves about a diameter of the ecliptic on which they pass through the two nodes, with their same quick motion to the north and to the south. But their return to a position from where they move at any [given] hour occurs after one complete year, as is also the course of the motion in longitude. Hence the long, or high, half of the eccentric circle is six months in the south, and the remaining six months in the north. In the same way, the shorter half changes to the south and to the north. It is therefore obvious that the plane of the eccentric circle coincides with the plane of the ecliptic twice in a year. And this happens at the two intersection points.

In the case of Venus, however, when the centre of its epicyclic sphere is at any of the intersection points, that half of the eccentric circle which follows that node begins to incline northwards, the other one southwards. For Mercury it is contrary. For when the centre of its epicyclic sphere is placed at any intersection point, that half which follows that intersection begins to incline southwards, and the other one northwards; therefore, the centre of their epicyclic spheres is either in the plane of the ecliptic, where are also the nodes, or they incline to one side

[42] Cf. *Alm.* XIII,6 (H587).

Veneris enim centrum rotunditatis nullam habet inclinationem nisi in septemtrionem, Mercurii contra centrum rotunditatis semper est in austrum inclinationem habens. Magna autem inclinatio circuli extra centrum Veneris et Mercurii fere est non in septemtrionem et austrum, quando centra rotunditatum eorum volvuntur in 5 longinquis aut propinquis longinquitatibus extra centrum circuli.

Rotundorum autem circulorum .e. planetarum inclinatio movetur, et redit in eundem locum, unde incepta est, sicut et motus longitudinis. Quod nulla constare certum est ratione, nisi duas quisque planeta rotunditates habeat. Saturnus enim et Iuppiter cum in unoquoque anno fiant retrogradi, quoque anno totum eorum 10 plenum circuitum complent, non tamen totum latum. Completur enim Saturni lati motus in .li. annis .ed. diebus .ke. horis horarum sexagenariis .ld.; Iovis autem annis .ka. .xke. diebus horis .kd. horarum sexagenariis .li.; Martis vero anno uno diebus .xlb. horis .d.; at Veneris et Mercurii anno uno horis .e. sexagenariis horarum .ni.. Temporum ergo tantorum spatium cuique planete ad complendum in longum 15 motum, qui est circuli extra centrum, et ad lati totam diversitatem perficiendam necessarium est.

f. 49r Rotunditatum autem cursus alius est. | Et Saturni quidem, Iovis ac Martis rotunditatis motus est super diametrum, quod dividit duas eius medietates, altam et humilem, estque longinqua longinquitas VI mensibus in septemtrionem, VI in 20 austrum a superficie extra centrum circuli. Diametrum autem, quod transit super duas medias longinquitates, est paralellum semper in suo motu superficiei circuli. Ideo secat superficies rotundi circuli superficiem extra centrum, et non tangit eum, tangens superficiem circuli signorum in anno bis. Hoc autem fit, quando centrum rotunditatis est in ligantibus. Initium vero inclinationis longinque et pro-25 pinque longinquitatum rotunditatis a circulo extra est in loco ligantium, quando centrum rotunditatis in ipsis volvitur, finis, quando centrum rotunditatis est aut in longinqua aut in propinqua longinquitate extra centrum circuli. Quod nulla constare ratione certum est, nisi duas quisque planeta habeat rotunditates. Qua enim positione semel statuetur longinqua longinquitas rotunditatis a circulo extra 30 centrum, ea semper erit, nisi ab alia rotunditate moveatur.

---

6 Rotundorum] Rotundo (L) **13** .d.] .ld.

of the ecliptic, but never to the other one. For the centre of the epicyclic sphere of Venus has no inclination except northwards, whereas the centre of the epicyclic sphere of Mercury is having the inclination always southwards. But the greatest inclination of the eccentric circle of Venus and Mercury is only very little to the north and to the south, while the centres of their epicyclic spheres revolve at the furthest or closest distances of the eccentric circle.

The inclination of the epicycles of the five planets, however, moves, and returns to the same position where it started from, like the motion in longitude. Certainly, this cannot occur in any way, unless each planet has two epicyclic spheres. For Saturn and Jupiter, though becoming retrograde every year and completing their full circuit [on the epicycle] every year, do not complete the entire [period in] latitude. For Saturn's latitudinal motion is completed in 29 years, 5|4 days,[43] 15 hours, and 24 minutes, and that of Jupiter in 11 years, 315 days, 14 hours, and 29 minutes, that of Mars in one year, 322 days, 4 hours, and those of Venus and Mercury in one year, 5 hours and 49 minutes. Such are the time-spans that each planet needs to complete the motion in longitude, which is that of the eccentric circle, and also to traverse the entire variety of the latitude.

But the course of the epicyclic spheres is different. The motion of the epicyclic sphere of Saturn, Jupiter and Mars is about the diameter that divides its two halves, the high one and the low one, and the furthest distance is for six months northwards and for six months southwards from the plane of the eccentric circle. But the diameter that passes through the two mean distances is always parallel in its motion to the plane of the circle. The plane of the epicycle therefore crosses the plane of the eccentric circle and is not tangent to the latter, whereas twice per year it is tangent to the plane of the ecliptic. This occurs when the centre of the epicyclic sphere is at the nodes. But at the place of the nodes there begins the inclination of the furthest distance and the closest distance of the epicyclic sphere from the eccentric circle, when the centre of the epicyclic sphere revolves in them, whereas it ends, when the centre of the epicyclic sphere is either at the furthest distance or at the closest distance of the eccentric circle. Certainly, this cannot occur in any way, unless each planet has two epicyclic spheres. For in whatever position from the eccentric circle the furthest distance of the epicyclic sphere would once be placed, it would always be in that position, unless it is moved by another epicyclic sphere.

[43]The number of days is corrupted.

Duas ergo ponere oportet, exteriorem scilicet et maiorem, que hunc faciat motum, cui eadem erit celeritas, que et spere extra centrum, secundam et interiorem, que stellam proferat. Hoc autem plurimum manifestat II[os] parvos circulos, in quibus diametrum rotunditatis superius volvi dictum est.

5 Verum quem sequimur, unam tantum posuit rotunditatem, quod non esse subtilis inspector animadvertet. He ergo utreque volvuntur, in orientem interior, et exterior contra, prima quidem, quantum motus est longitudinis in die, secunda, quantum sol sublatis prime seu circuli extra centrum cursibus in his tribus planetis, Saturno, Iove et Marte. Nam in Venere et Mercurio aliter est. Exterior enim in 10 occidentem fertur equali motu centri rotunditatis .oi. sexagenariis et .h. secundis, interior Veneris quidem .a. gradu .mf. sexagenariis .g. secundis, de quo sublato cursu prime remanent .mf. sexagenarie .oi. secunde, Mercurii .d. gradibus .e. sexagenariis .mb. secundis, de quo sublato cursu prime remanent .c. gradus .f. sexagenarie .ld. secunde. Hinc est, quod ad inveniendam trium rotunditatem eorundem 15 cursus centri de cursu solis aufertur; in reliquis non ita.

Inclinationis autem pars longinque longinquitatis rotundi circuli a circulo extra centrum in tribus est in eam partem, in quam inclinatur et circulus ipse a circulo signorum, propinqua longinquitas in oppositam. Quando ergo est centrum rotunditatis in medietate circuli extra centrum, que est in septemtrione, inclina- 20 tur eodem longinqua longinquitas rotunditatis; quando in australi, in austrum. He igitur tres stelle unum habent motum inclinatione rotunditatis, Venus vero et Mercurius duos, sed alter eorum similis est motui trium stellarum, ille scilicet, qui fit per inclinationem longinque longinquitatis et propinque rotunditatis a circulo extra centrum. Alter autem est per inclinationem duarum longinquitatum mediarum 25 rotunditatis, diciturque obliquatio.

Et initium inclinationis longinque et propinque longinquitatum rotunditatis a circulo extra centrum est in longinqua aut propinqua longinquitate circuli extra centrum, cum tangit superficies extra centrum superficiem circuli signorum; com-

---

2–3 secundam et interiorem] secunda et exterior **9** Venere] *corr. ex* nere (L) **10** secundis] s et solis (L) **13–14** secundis, de quo... inveniendam] *add. in marg.* (L) **18** in oppositam] impositum (L) **25** obliquatio] obligatio (L)

One therefore must assume two of them, namely an outer and larger one, which produces this motion and which has the same speed as the eccentric sphere, and a second and inner one, which carries the star forwards. This makes the two small circles obvious, on which the diameter of the epicyclic sphere has above been said to move.

But the author whom we are following states only one epicyclic sphere, which is impossible, as any diligent investigator will realise. These two are thus moving, the inner one to the east and the outer one in the opposite direction, and the first one by as much as the daily longitudinal motion, and the second one by as much as the sun after subtraction of the courses of the first one or of the eccentric circle in the case of these three planets, Saturn, Jupiter and Mars. For in the case of Venus and Mercury it is different; because, the outer one is taken westwards by the equal motion of the centre of the epicyclic sphere, by 0;59,8 degrees, and the inner one of Venus by 1;36,7 degrees, which if we subtract from it the course of the first one, there remain 0;36,59 degrees[44]; and that of Mercury by 4;5,32 degrees, which if we subtract from it the course of the first one, there remain 3;6,24 degrees.[45] Thus, for finding the epicyclic sphere of the three the course of their centre is subtracted from the course of the sun; whereas, this is not so for the remaining ones.

For the three planets, however, the inclination of the furthest distance of the epicycle from the eccentric circle is towards that side to which the circle itself inclines from the ecliptic, whereas the closest distance is towards the opposite side.[46] Thus, when the centre of the epicyclic sphere is in the northern half of the eccentric circle, the furthest distance of the epicyclic sphere inclines in the same direction, whereas when in the southern one, to the south. These three stars therefore have one motion by the inclination of the epicyclic sphere, whereas Venus and Mercury have two, one of which, nonetheless, is similar to the motion of the three stars, namely the one that is produced by the inclination of the furthest distance and the closest distance of the epicyclic sphere from the eccentric circle. But the other motion is produced by the inclination of the two mean distances of the epicyclic sphere, which is called 'slant'.

The beginning of the inclination of the furthest and the closest distances of the epicyclic sphere from the eccentric circle is at the furthest or the closest distance of the eccentric circle, while the plane of the eccentric circle touches[47] the plane of

[44]Cf. *Alm.* IX,4: 0;36,59,25,53,11,28°.

[45]Cf. *Alm.* IX,4: 3;6,24,6,59,35,50°.

[46]The directions are contrary to those described in *Alm.* XIII,2.

[47]That is, when the plane of the eccentric circle intersects with the plane of the ecliptic. The verb, 'tangit', could also mean that both planes are tangential to one another. Stephen uses 'tangere' for both meanings.

plementum autem eius in punctis sectionum. Initium vero <inclinationis> mediarum longinquitatum rotunditatis a superficie extra centrum circuli est in duobus punctis ligantium; finis fere in longinqua aut propinqua longinquitate extra centrum circuli. Tunc etiam finis est inclinationis circuli extra centrum a signorum circulo. Quando ergo est quevis harum duarum inclinationum in suo complemento, nichil restat alterius. Finis etenim unius initium est alterius.

Inclinationis autem pars, quando centrum rotunditatis est in ea medietate circuli extra centrum, que incipit fere a longinqua longinquitate, est alte quidem medietatis rotundi circuli a circulo ferenti, Veneris que in septemtrionem, Mercurii autem in austrum, una inclinatur in aliam. Cum vero rotunditatis centrum in alia medietate volvitur, est opposita inclinationi alterius medietatis. Maior est autem inclinatio hec in ligantibus, quo loco sunt due medie longitudines rotundi circuli in superficie extra centrum et superficie circuli signorum, heque unum sunt.

f. 49v Oblique vero inclinationis pars est, quando centrum | rotunditatis in alta medietate extra centrum circuli volvitur, eius quidem medie longitudinis rotundi circuli, que est versus orientem, Veneris quidem in septemtrionem, Mercurii in austrum, inclinatio a superficie extra centrum circuli. Ea vero media longinquitas, que est versus occidentem, est in opposito lato. Quando vero centrum rotunditatis in ima medietate extra centrum circuli, quod eius humilitatem dicimus, est obliqua inclinatio opposita inclinationi alterius medietatis. Habet autem obliqua inclinatio statum, quando centrum rotunditatis volvitur aut in longinqua aut in propinqua longinquitate extra centrum circuli; quo tempore sunt longinqua et propinqua longinquitas rotunditatis in superficie extra centrum circuli.

Mensura autem maioris lati planetarum et lune in septemtrionem et austrum a circulo signorum est lune quidem .d. graduum .ne. sexagenariarum per mensuram, qua dividitur magnus circulus in .xp. partes, Saturni trium, Iovis duorum, Martis in septemtrione .d. graduum .l. sexagenariarum, in austrum .g. graduum, Veneris, sicut dicit Ptolomeus in sua sintaxi, .f. graduum .b. sexagenariarum, sicut autem alii dixerunt astronomi, novem, Mercurii .d.. Motus ergo <Veneris>(L) et Mercurii, quoniam pares habent speras, pares sunt, ceterorumque trium similiter.

---

7 rotunditatis] *add.* (L) **9** que] quem **19** ima] una

the ecliptic, whereas its completion is at the intersection points. But the beginning of the inclination of the mean distances of the epicyclic sphere from the plane of the eccentric circle is at the two nodes, whereas the end is near the furthest or the closest distance of the eccentric circle. At that moment is also the end of the inclination of the eccentric circle from the ecliptic. Thus, whenever one of these two inclinations is at its completion, the other one has vanished. For, the end of one is the beginning of the other.

But when the centre of the epicyclic sphere is in that half of the eccentric circle which begins near the furthest distance, the side of the inclination is that of the high half of the epicycle from the deferent circle, for Venus to the north and for Mercury to the south, inclined in opposite directions. But when the centre of the epicyclic sphere revolves in the other half, it is opposite to the inclination of the other half. But this inclination is greatest at the nodes, at which place the two mean distances of the epicycle are in the eccentric plane and also in the plane of the ecliptic, as these planes are one.

But the direction of the slant, when the centre of the epicyclic sphere revolves in the high half of the eccentric circle, and for that mean distance of its epicycle which is towards the east, is an inclination from the plane of the eccentric circle to the north for Venus, whereas to the south for Mercury. The mean distance towards the west, however, is at the opposite latitude. And when the centre of the epicyclic sphere is in the low half of the eccentric circle, which we call its lowness, the slant is opposite to the inclination in the other half. But the slant is stationary, when the centre of the epicyclic sphere revolves either at the furthest or at the closest distance of the eccentric circle; at that moment the furthest and the closest distance of the epicyclic sphere are in the plane of the eccentric circle.

The greatest latitude of the planets and the moon, however, to the north and to the south, is for the moon 4;45 degrees on the scale by which the great circle is divided into 360 degrees, for Saturn it is 3,[48] for Jupiter 2,[49] for Mars 4;20 degrees to the north, but 7 degrees to the south,[50] for Venus 6;2 degrees, as Ptolemy says in his *Almagest*,[51] whereas other astronomers have said 9,[52] and for Mercury 4. The motions of Venus and Mercury, since they have equal spheres, are equal, and also those of the remaining three are analogous.

[48]*Alm.* XIII,5: N3;2° S3;5°.

[49]*Alm.* XIII,5: N2;4° S2;8°.

[50]*Alm.* XIII,5: N4;21° S7;7°.

[51]*Alm.* XIII,5 actually has 6;22°. Al-Farghānī, whom Stephen seems to follow here, gives Ptolemy's value as 6;20°. A corruption of 'l' (= 20) to 'b' (= 2) in the Latin is a possibility.

[52]Cf. Stephen's discussion of the zodiac, where he adopted the value of 9 degrees.

[319] *Sunt igitur Mercurii motus* .ka.; *primus tardus in orientem,* secundus annue eodem, *tercius referentis contra, quartus ferentis ad orientem, quintus, qui per hanc fit in circulo signorum in longum, sextus est motus rotunditatis prime ad occidentem,* VII^us^ secunde contra, VIII^us^ *est linee circinantis diametrum, nonus superficiei inclinati circuli* 5 *versus superficiem circuli signorum, dictus motus inclinationis, decimus diametri rotunditatis super parvum circulum, dictus inclinatio rotundi circuli, undecimus diametri, qui hoc secat, super parvum circulum in septemtrionem aut in austrum, dictus obliquatio rotunditatis.*

[358 *rev.*] *Trium autem* .g.; *primus tardus, ferentis secundus, tercius in longum,* 10 *quartus* et quintus *rotunditatis,* VI^us^ *circinans,* VII^us^ *inclinationis diametri rotunditatis.*

De planetis satis dictum.

---

7 obliquatio] obligatio (L)

[319][53] *Mercury thus has* eleven *motions; the first one being the slow one to the east,* the second one that of the annual sphere in the same direction, *the third one that of the referent in the opposite direction, the fourth one that of the deferent to the east, the fifth one that which is produced by the former one on the ecliptic in longitude, the sixth one is the westward motion of the first epicyclic sphere,* the seventh one that of the second [epicyclic sphere] in the opposite direction, *the eighth one is that of the dirigent line of the diameter, the ninth one is that of the plane of the inclined circle towards the plane of the ecliptic, which is called the motion of the inclination, the tenth one is that of the diameter of the epicyclic sphere on a small circle, which is called the inclination of the epicycle, and the eleventh one that of the diameter that crosses the former, on a small circle northwards or southwards, which is called the slant of the epicyclic sphere.*

[358 *rev.*][54] *The three planets, however, have* seven *motions; the first one is the slow one, that of the deferent the second one, the third one that in longitude, the fourth* and the fifth *those of the epicyclic sphere, the sixth one the dirigent one, and the seventh one that of the inclination of the diameter of the epicyclic sphere.*

About the planets there has now been said enough.

[53]The preceding discussion replaces passages 315–18 from the discussion of Mercury. Passage 317, on the eccentric motion of the diameters of the epicyclic sphere, has been omitted. Passage 319 has been adapted to Stephen's new model for the oscillating deferent planes and the motions of the epicyclic sphere, both of which are rendered by further spheres. Ibn al-Haytham's number of nine motions has thus become increased to eleven.

[54]In this paragraph, Stephen revises his previous translation of passage 358, where he follows the Arabic source by mentioning a total of only six motions.

[360] ¶*Fixarum spera est una, circundata duabus paralellis et speralibus superficiebus, centrum quarum est spere mundi. Alta eius superficies tangit altam speram, que claudit omnes et movetur cito motu, ima vero speram Saturni.* [361] *Hec movetur in orientem tardo motu, qui est in .t. annis uno gradu, super duos fixos polos. Hii sunt duo poli circuli signorum, quem solis centrum circinat, sicut dicit Ptolomeus sua et precedentium ratione inventum.*

[362] *Omnes autem stelle, que dicuntur planete, fixe sunt in corpore huius spere, ideoque non mutant positionem suam ad invicem. Moventur tamen omnes motu eiusdem spere tardo in orientem.* [363] *Queque vero earum facit suo centro per motum spere imaginarium circulum, qui omnes paralelli probantur, et sunt, quorum poli sunt signorum, moventurque super suum circulum,* [364] *et motus earum in circulo signorum et locus in eodem est, sicut de planetis diximus. Imaginamur enim circulos signorum de polis egredientes et transeuntes per centra planetarum, secantesque mundum et circulum signorum, qui est in superficie alte spere, <et> eos, qui sunt in superficiebus sperarum planetarum, super unum punctum.* [365] *Et punctus ille est locus stelle in circulo signorum, qui est in spera fixarum. Ille vero, qui superest huic in circulo signorum, qui est in alte spere superficie, est rectus locus stelle in longum. Est et motus ipsius puncti, quo movetur, motu scilicet circuli, qui transit super duos polos, est motus stelle in longum.*

[366] *Egreditur autem linea de centro mundi et vadit ad centrum stelle usque superficiem alte spere ascendens* [. . .]

---

2 Alta] Alia **3** speram] spera (L) **9** orientem] oriente (L) **14** superficiebus] *corr. ex* superfiebus (L) sperarum] *corr. ex* sperlarum (L) **17** et] *add.* (L)

4.6 <The Sphere of the Fixed Stars>

Config. 14: The Orb of the Fixed Stars

[360] *The sphere of the fixed stars is a single sphere, surrounded by two parallel and spherical surfaces whose centre is that of the sphere of the world. Its high surface is tangent to the high sphere, which encloses everything and which moves with a fast motion; whereas, its low surface is tangent to the sphere of Saturn.* [361] *This sphere moves eastwards with a slow motion, of one degree in a hundred years, about two fixed poles. These are the two poles of the ecliptic, which is described by the centre of the sun, as Ptolemy says was found through his considerations and those of his predecessors.*

[362] *But all the stars which are called planets*[55] *are fixed inside the body of this sphere; therefore, they do not change their position relative to each other. Nonetheless, they are all moved by the slow eastward motion of that sphere.* [363] *By the motion of the sphere, each of them produces with its centre an imaginary circle, which are proved to be all parallel and whose poles are those of the ecliptic, and the stars revolve each on its circle,* [364] *and their motion and their position relative to the ecliptic are [defined] as we have said for the planets. For we imagine circles that originate from the poles of the ecliptic and pass through the centres of the planets and which dissect the world and the ecliptic on the surface of the high sphere, and also those on the surfaces of the planetary spheres,*[56] *in one point.* [365] *That point is the position of the star on the ecliptic on the sphere of the fixed stars.*[57] *But that point which is above the latter on the ecliptic on the surface of the high sphere is the correct position of the star in longitude. And the motion by which this point moves, namely the motion of the circle that passes through the two poles, is the motion of the star in longitude.*

[366] *But a line originates from the centre of the world and runs to the centre of the star, and it rises until the surface of the high sphere* [...]

The rest of the text, presumably containing a translation of Ibn al-Haytham's remaining nineteen passages, 367–385, on the fixed stars and the highest sphere, is missing.

[55] 'planets': the use of the term here is contradictory and does not correspond to Stephen's previous consistent use of it to designate the wandering stars. Some unclear use of the term is also found on p. 392:15.

[56] 'planetary spheres': Arab., 'on the ecliptic orb.'

[57] 'sphere of the fixed stars': Arab., 'ecliptic orb'.

Appendix A: Glossary

The following list contains a selection of technical terms that are used by Stephen of Pisa and Antioch in the *Liber Mamonis* (MS Cambrai 930) and/or by ʿAbd al-Masīḥ Wittoniensis in the preserved parts of his translation of Ptolemy's *Almagest* (MS Dresden Db. 87, fols 1r-71r). The Latin terms are preceded by the corresponding Arabic expressions in Ibn al-Haytham's *On the Configuration of the World*, mostly taken from the more extensive lexicon in Langermann's edition. The last two columns contain terms which are used in concurrent Latin translations of related Arabic texts, i.e., the Madrid translation of *On the Configuration* (ed. Millás Vallicrosa) and *Alm.* I-IV according to Gerard of Cremona's translation (ed. Liechtenstein) from the Arabic version by al-Ḥajjāj. Only Books I-IV of Gerard's translation have been considered in order to allow for a comparison with ʿAbd al-Masīḥ's usage of terms in the same parts.

A minus (-) indicates the absence of a dedicated technical term in the respective text. Entries in italics are cases where the Antiochian terminology differs distinctively from the language used in the later western translations. Minor variations are normally not considered and no distinction has been made with regard to a prefixed or postponed position of adjectives. An exception was made in the case of the 'fixed stars', where the placing of the adjective seems to follow different preferences in the different texts.

D. Grupe, *Stephen of Pisa and Antioch: Liber Mamonis*, Sources and Studies in the History of Mathematics and Physical Sciences,
https://doi.org/10.1007/978-3-030-19234-1

Index	***On the Configuration* (Arabic)**	***Liber Mamonis***	***Almagest* I-IV (tr. ‘Abd al-Masīḥ)**	***On the Configuration* (Madrid tr.)**	***Almagest* I-IV (tr. Gerard)**
altitude	ارتفاع (irtifāʿ)	*altitudo, altum, altum altitudinis*	*altitudo*	elevacio	altitudo
a. circle	الدّائرة السّمتيّة (al-dāʾirah al-samatiyyah)	*circulus altitudinis*	-	circulus cenith, c. elevacionis, c. cenitalis (*alibi:* centralis)	-
amplitude of rising / a. of setting	سعة المشرق (saʿat al-mashriq)/ سعة المغرب (saʿat al-maghrib)	*largitas orientis, l. orientalis / l. occidentis, l. occidentalis*	*largitas orientis / -*	amplitudo orientis / a. occidentis	- / -
anomaly, motion in a.	حركة الاحتلاف (ḥarakat al-ikhtilāf)	*discordia, motus discors*	*motus discors*	motus diversitatis	diversitas
apogee / furthest distance	أوج (ʾawj) / البعد الابعد (al-buʿd al-abʿad)	*longinqua longinquitas, altitudo, altum*	*longinqua longinquitas*	aux / longitudo longior, l. longissima	longitudo longior
arc: day arc / night arc	قوس النهار (qaws al-nahār)/ قوس اللّيل (qaws al-layl)	arcus diei / a. noctis	- / -	arcus diei / a. noctis	- / -
arc-minute		*sexagenaria*	*sexagenaria*	minutum	minutum
ascendant	الطالع (al-ṭāliʿ)	ascensus	ortus	ascendens	pars oriens

Index	*On the Configuration* (Arabic)	*Liber Mamonis*	*Almagest* I-IV (tr. ʿAbd al-Masīḥ)	*On the Configuration* (Madrid tr.)	*Almagest* I-IV (tr. Gerard)
ascension	المطالع (al-maṭāliʿ)	*orientale*	*orientale*	ascensio	elevationes (temporum; pl.)
astronomy	علم الهيئة (ʿilm al-hayʾah)	*astronomia*	*astronomia*	astrologia	sciencia corporum celestium
axis	محور (maḥūr)	radius	axis	axis	axis
climate	اقاليم (āqālīm; pl.)	clima	clima	clima	clima, linea equidistans orbi equationis diei
closest distance; see ‘perigee’					
cone	مخروط (makhrūṭ)	*conus*	*conus*	piramis	pinealis (adj.)
conjunction	اجتماع (ijtimāʿ)	*sinodus, (coniunctio)*	*synodus*	coniunctio	coniunctio
corporeal	مجسّم (mujassim)	*corporalis*	*corporalis*	corporeus	corporeus
declination, inclination, obliquity *greatest decl., gr. incl., gr. obl.*	ميل (mayl) اعظم الميول (aʿẓam al-muyūl)	*inclinatio* *maior inclinatio*	*inclinacio* *maior inclinacio*	declinacio maxima declinacio	declinatio -
deferent sphere / d. circle	الفلك الحامل (al-falak al-ḥāmil)	ferens spera / f. circulus	-	orbis deferens / circulus d.	-

Index	***On the Configuration* (Arabic)**	***Liber Mamonis***	***Almagest* I-IV (tr. ʿAbd al-Masīḥ)**	***On the Configuration* (Madrid tr.)**	***Almagest* I-IV (tr. Gerard)**
degree	درجة (darajah)	*gradus*	*gradus, pars*	gradus	pars
descendant	المغارب (al-mughārib)	occasus	-	occidens	-
descension	-	occidentale	-	-	-
dirigent sphere	الفلك المدير (al-falak al-mudīr [lil-falak al-ḥāmil])	spera referens	-	orbis revolvens (orbem deferentem), o. equans (orbem d.)	-
d. centre / equant	المركز المدير (al-markaz al-mudīr [li-qaṭar falak al-tadwīr]) / مركز معدّل المسير (markaz muʿaddil al-masīr)	centrum circinantis linee	-	centrum revolvens (diametrum epicicli) / c. equans	-
d. line	الخطّ المدير (al-khaṭṭ al-mudīr [li-qaṭar falak al-tadwīr])	circinans linea (diametri/-um rotunditatis)	-	linea revolvens (epicicli diametrum)	-
dome	قبّة (qubbah)	tholus	-	torus	-

Index	***On the Configuration* (Arabic)**	***Liber Mamonis***	***Almagest* I-IV (tr. ʿAbd al-Masīḥ)**	***On the Configuration* (Madrid tr.)**	***Almagest* I-IV (tr. Gerard)**
eccenter	الفلك الخارج المركز (al-falak al-khārij al-markaz)	(spera) extra centrum	circulus forinseci centri	orbis excentricus, circulus e.	orbis ecentricus, orbis egredientis centri
eclipse - partial - total	كسوف (kusūf) - -	eclipsis, defectus - plenus	eclipsis partialiter conpleta, totaliter	eclipsis particularis universalis	eclypsis - totum eclypsatum
ecliptic / zodiac	دائرة البروج (dāʾirat al-burūj)	*circulus signorum, signifer, zodiacus, (c. eclipticus)*	*circulus signorum*	circulus signorum, orbis s. / zodiacus	orbis declivis, o. signorum (declivis)
epicycle / epicyclic sphere	فلك التّدوير (falak al-tadwīr)	*circulus rotundus (rotunditatis) / rotunditas*	*circulus rotunditatis / rotunditas*	epiciclus	orbis revolvens (stellam), o. revolutionis
equant; see 'dirigent sphere' *etc.*					
equator *- celestial* *- terrestrial*	 دائرة معدّل النهار (dāʾirah muʿaddil al-nahāri) خطّ الاستواء (khaṭṭ al-istiwāʾi)	 *circulus recti diei, c. equinoctialis* *recta linea*	 *circulus recti diei, c. rectitudinis diei, rectitudo diei, rectificacio diei* *recta linea*	 circulus equinoctialis linea equalitatis	 orbis equationis diei, circulus equalitatis recta linea
equinox, equinoctial point	نقطة الاعتدال (nuqṭat al-iʿtidāl)	equalitas, (equinoctium)	equalitas	temperancia, equalitas	equalitas, (equans)

Index	***On the Configuration* (Arabic)**	***Liber Mamonis***	***Almagest* I-IV (tr. ‹Abd al-Masīḥ)**	***On the Configuration* (Madrid tr.)**	***Almagest* I-IV (tr. Gerard)**
- vernal	الاعتدال الرّبيعيّ (al-i‹tidāl al-rabī‹īy)	*equalitas veris*	*equalitas veris*	t. vernalis, e. vernalis	e. vernalis, (equans vernale)
- autumnal	الاعتدال الخريفيّ (al-i‹tidāl al-kharīfīy)	*equalitas autumni*	*equalitas auctumpni*	t. autumpnalis, e. autumpnalis	e. autumnalis, (equans autumnale)
fixed star	الكواكب الثّابتة (al-kawākib al-thābitah; pl.)	fixa stella, fixa	fixa stella	stella fixa	stella fixa
furthest distance; see ‘apogee’					
gnomon	اشخاص المقاييس (ashkhāṣ al-maqāyīs; pl.)	- (res recte posita)	demonstratio	gnomon	gnomon
horizon	أفق (›ufuq)	orizon, circulus orizontalis, c. orizontis, c. orientis	orizon	orizon	horizon
inclination; see ‘declination’					
inclined circle	الفلك المايل (al-falak al-māyil)	*inclinatus circulus*	*inclinatus circulus*	circulus declivis	circulus declinatus
inhabited zone	معمورة (ma‹mūrah)	*habitatio*	*habitatio*	habitabilis	terra habitabilis, habitabilia terre

Index	***On the Configuration* (Arabic)**	***Liber Mamonis***	***Almagest* I-IV (tr. ʿAbd al-Masīḥ)**	***On the Configuration* (Madrid tr.)**	***Almagest* I-IV (tr. Gerard)**
latitude	عرض (ʿarḍ)	*latum*	*latum*	latitudo	latitudo
longest day		*magnus dies*	*maior dies*	longissimus dies	dies longior
longitude	طول (ṭūl)	*longum*	*longum*	longitudo	longitudo
meridian circle	دائرة نصف النّهار (dāʾirat niṣf al-nahāri)	*circulus meridiei*	*circulus meridiei*	circulus meridiei	orbis meridiei, o. meridianus
m. line	خطّ نصف النّهار (khaṭṭ niṣf al-nahāri)	*linea meridionalis*	*linea meridionalis, l. meridiei*	linea meridiei	-
motion	حركة (ḥarakah)	motus	motus	motus	motus
- equal	الحركة المستويّة (al-ḥarakah al-mustawīyah)	m. equalis	m. equalis, m. directe	m. equalis	m. equalis, m. medius
- *unequal*	الحركة المختلفة (al-ḥarakah al-mukhtalifah)	*m. inequalis*	*m. inequalis*	m. inequalis, m. inuniformis	m. diversus, m. qui apparet diversus
nadir	-	pedum punctus	-	nadir	-
nodes (dragon)	جوزهران (jawzharān)	*ligantes*	*ligantes*	puncta draconis, extremitates draconis	nodi

Index	***On the Configuration* (Arabic)**	***Liber Mamonis***	***Almagest* I-IV (tr. ʿAbd al-Masīḥ)**	***On the Configuration* (Madrid tr.)**	***Almagest* I-IV (tr. Gerard)**
- head / tail	رأس (raʾs) / ذنب (ḏanab)	caput draconis / cauda d., finis d.	ligans capitis / l. caude	capud / cauda	nodus capitis / n. caude
obliquity; see 'declination'					
opposition	استقبال (istiqbāl)	*oppositum*	*oppositum*	oppositio	oppositio
orb	فلك (falak)	- (*spera, circulus*)	- (*spera, circulus*)	orbis	orbis
parallax	-	*discordia visus*	*discordia visus*	-	-
parallel	متواز (mutawāz)	*paralellus*	*paralellicus*	equidistans	equidistans
parecliptic circle	الفلك المثّل (al-falak al-mumaththil [bi-falak al-burūj])	circulus similis (circulo signorum)	-	orbis similis orbi signorum	-
perigee / closest distance	حضيض (ḥaḍīḍ) / البعد الاقرب (al-buʿd al-aqrab)	*propinqua longinquitas, longinqua propinquitas, humilitas, imum*	*propinqua longinquitas, longinqua propinquitas, propinqua propinquitas*	hadid / longitudo propinquissima	longitudo propinquior

Index	***On the Configuration* (Arabic)**	***Liber Mamonis***	***Almagest* I-IV (tr. ‘Abd al-Masīḥ)**	***On the Configuration* (Madrid tr.)**	***Almagest* I-IV (tr. Gerard)**
prosneusis	حركة المحاذة (ḥarakat al-muḥādhah)	motus contrarius	-	motus verticationis	-
in quadrature	في التّربيعة (fīl-tarbī‘ah)	*sub quadranguli figura, sub tetragono*	*in quadrangulare*	in quadratura	in quadratura
retrograde	رجوع (rujū‘)	retrogradus	retro gradiens	retrogradus	retrogradus
slant	انحراف (inkhirāf)	obliquatio	-	reflectio	-
solstice - *summer / winter*	انقلاب (inqilāb) الانقلاب الصّيفيّ (al-inqilāb al-ṣayfīy) / الانقلاب الشّتويّ (al-inqilāb al-shitawīy)	mutatio, solsticium *m. estatis / m. hiemis*	conversio *c. estatis / c. hyemis*	tropicus tr. estivalis / tr. hyemalis	conversio, tropicus c. estivalis, tr. estivalis / tr. hyemalis
sphaera recta / s. obliqua	- / -	- / -	recta spera / inclinata sp.	- / -	orbis rectus / sphera declivis
spherical	كرّيّ (kurīy)	*speralis*	*speralis*	spericus	sphericus

Index	***On the Configuration* (Arabic)**	***Liber Mamonis***	***Almagest* I-IV (tr. ʿAbd al-Masīḥ)**	***On the Configuration* (Madrid tr.)**	***Almagest* I-IV (tr. Gerard)**
time circle	دائرة الأزمان (dāʾirat al-ʾazmān)	temporis circulus	-	circuli temporum (pl.), circulus temporalis	-
zenith z. circle; see 'altitude circle'	سمت الرّأس (samt al-raʾs)	*punctus capitis*	*punctus capitis, (punctus qui est super capita)*	cenith (capitis), punctus verticilis	- (punctum quod est supra summitatem capitum)

Appendix B: Plates

D. Grupe, *Stephen of Pisa and Antioch: Liber Mamonis*, Sources and Studies in the History of Mathematics and Physical Sciences,
https://doi.org/10.1007/978-3-030-19234-1

249

Incip liber mamonis in astronomia a stephano philosopho translatus

[illegible] in canonem astronomie quas proposueramus regulas [illegible]

tractatu promissum exsoluimus. secundum hoc opus licet arduum et subtilissi

mo ac multiplici nature celatum archano non inconsulta aut impudenti temeritate. sed fre

quenti [illegible] utilitatis ammonitione aggredior. [illegible] licet magnorum super his grauissi

morumque disputatio philosophorum [illegible] mediocres persepe maxima quemadmodum maiores [illegible]

nora. Illud quoque attendendum est plurimum quod est omnis a deo sit sapientia. ea autem uerior et sine scru

pulo fallacie concessa sit nemo nouerit. Unde et qui quos [illegible] philosophi sepe extra se maximis in rebus

[illegible] uerius et [illegible] alios qui nec philosophiam adepti essent nec ad eam aliquando posse pertingere

existimauerent. de diuini muneris larga benignitate hausisse noticiam comperimus. testes sunt plato

et aristotiles quos omnium liberalium artium fere magistros habemus. [illegible] plato in multis a ueritate dis

sonat. aristotiles mundum non esse a deo conditum [illegible] sed esse sicut nec est [illegible] est corpore umbram precessisse

[illegible] et argumentis fallacibus conatur asserere. eo minimum in loco intellectus [illegible] oculorum [illegible]

officio. quod fidelium simplicitati diuina nascitur [illegible]. [illegible] in hac de qua proposita est disputatio

questione. cum de celestibus speris dissereret octo positis de nona non ut quidam arbitrantur [illegible]

cuit. sed se ad eius noticiam nequaquam peruenisse manifestum nobis reliquit testimonium. [illegible] nulla

tenus arrogantie dictum cuipiam uideri uelim. et quod tante grauitatis et scientie et eiusdem auctori

tatis adepte philosophus ignorasse dicatur. me non latuit. [illegible] et si inter maximos locum non obtineam ad

eos tamen aspirans mediocrium inuasi disciplinam. [illegible] ille sua quibus plurima consumpta opera perpetui

tatis dum philosophantes uixerint nomen adeptus est quorum tamen plaque a maioribus omnia [illegible] a deo [illegible]

obfuscata falsitatis errore accepit. [illegible] nobis quoque [illegible] aliis derogam. sic [illegible] omnium [illegible]

[illegible] dedecet. [illegible] ab eo accepta alios docere quam ignauie silentio tegere malum. hec

[illegible] non tanta forte obtrectantium multitudo. ferociores [illegible] latinitas auctores

fertiliorque apud nos philosophie seges pullularet. [illegible] enim plurimi essent exercitus [illegible] paucos

qui benigne susciperent. pauciores [illegible] artium sectores magis exterrebantur multitudinis

immanitate quam aduna[illegible] aliquorum benigno studio. [illegible] ut quod fere plenitudinem

posset. [illegible] artium nec ceteris gentibus europa uideatur humilior. [illegible] educat et fonti

scientie sepe oblatrantes sentit et ipsis rebelles. nec hec nuncilla [illegible] gloria [illegible]. [illegible]

res tamen attulit latinis scientie odium ita quibus summe ueneraretur. debuerat rerum pro rectoribus

summe odiretur. [illegible] ex re illud quoque [illegible] non dubito quin est equitatis obseruande cum reges

ceterique bonorum ordines. qui et leges constitute sunt. [illegible] pars naturalis [illegible] sit posita [illegible]

altius omnem fere sciam [illegible] maiorem partem latinis plurimis exercitio adipiscam. ignorata semi

[illegible] quia reges esse debuerant. nomen tamen retinentes tirannosse opibus [illegible]. [illegible]

[illegible] obseruantiam delapsi eadem in lege iudicant. quod quia induxerit in usum

Fig. B.1: Médiathèque d'Agglomération de Cambrai, Ms. A 930, fol. 2r (front page)

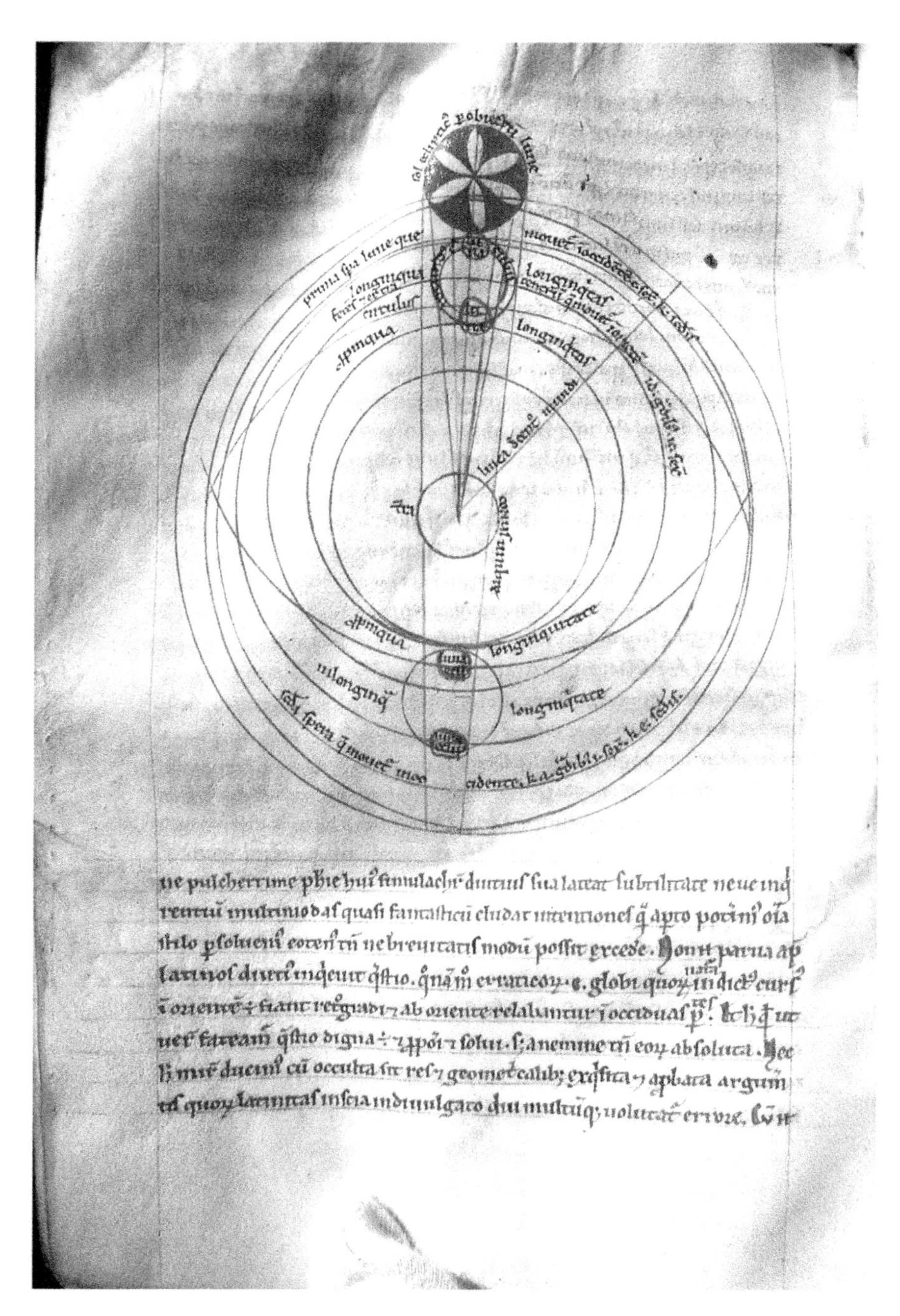

Fig. B.2: Médiathèque d'Agglomération de Cambrai, Ms. A 930, fol. 38v

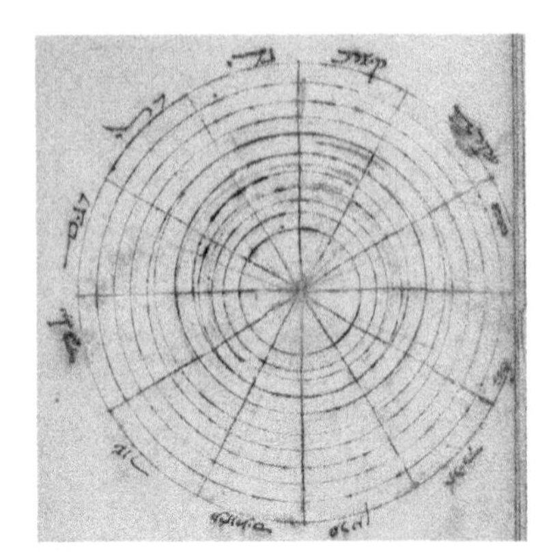

Fig. B.3: MS Parma, De Rossi 568, fol. 40v; the plane of the ecliptic

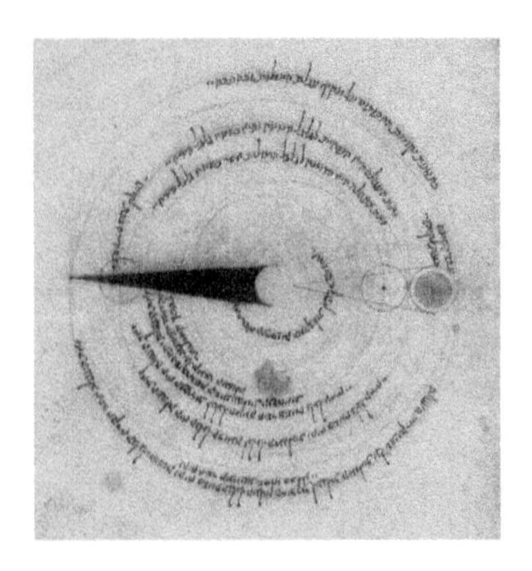

Fig. B.4: MS Parma, De Rossi 568, fol. 41r; the sphere of the moon

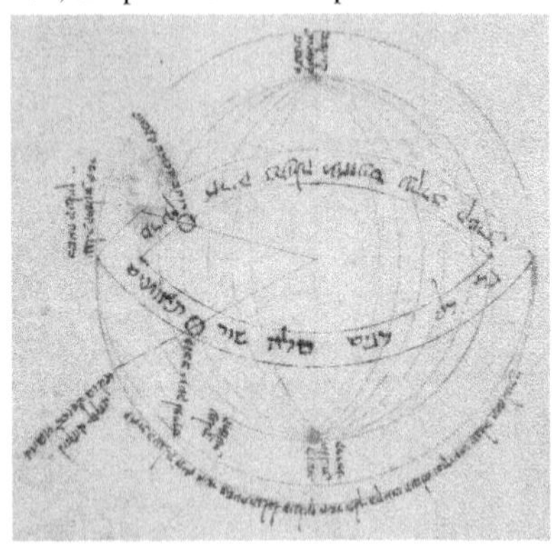

Fig. B.5: MS Parma, De Rossi 568, fol. 43r; the zodiac

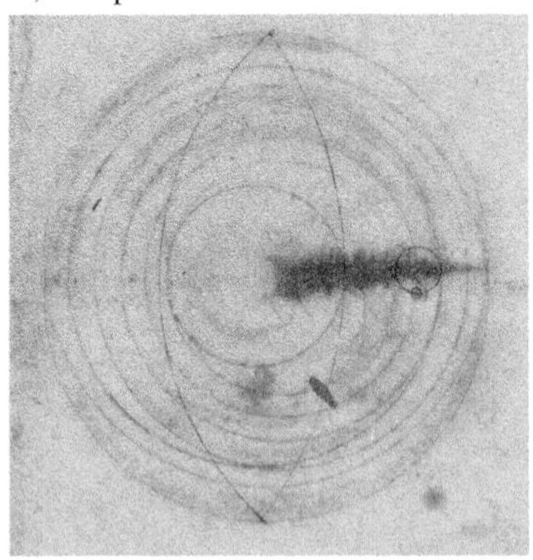

Fig. B.6: MS Parma, De Rossi 568, fol. 41v; the sphere of Mercury

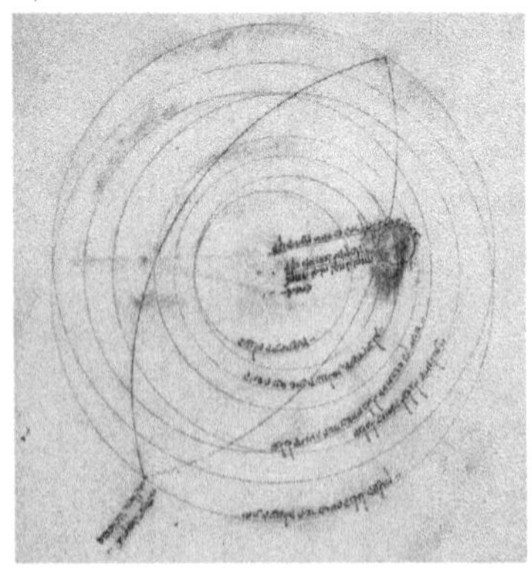

Fig. B.7: MS Parma, De Rossi 568, fol. 42v; the sphere of an outer planet

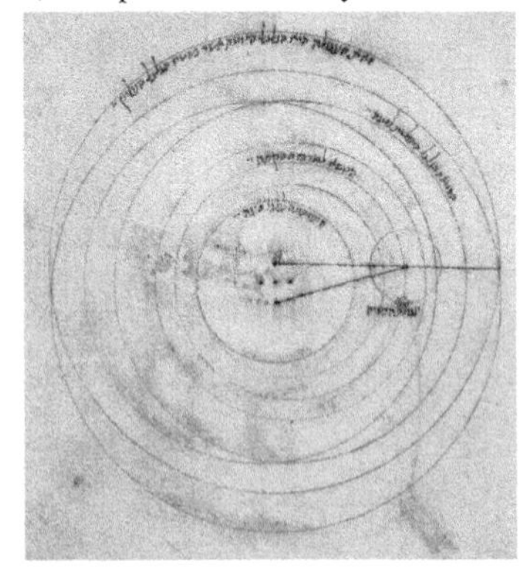

Fig. B.8: MS Parma, De Rossi 568, fol. 42r; the sphere of Venus

References

Examined Manuscripts

Cambrai, Médiathèque d'Agglomération, A 930 (Lat.; *Liber Mamonis*)

Dresden, Sächsische Landesbibliothek - Staats- und Universitätsbibliothek, Db. 87, fols 1r-71r (Lat.; *Almagest*)

Madrid, Escorial, 908 (Ar.; al-Battānī's *Ṣābiʾ Zīj*)

Oxford, Bodleian Library, Canon misc. 45 (Lat.; Alfonsine paraphrase of *On the Configuration of the World*)

Paris, Bibliothèque nationale de France, Heb 1031 (Heb.; *On the Configuration of the World*)

Paris, Bibliothèque nationale de France, Heb 1035 (Heb.; *On the Configuration of the World*)

Parma, Biblioteca Palatina, 2466 (Cod. de Rossi 568) (Heb.; *On the Configuration of the World*)

Bibliography

d'Alverny, Marie-Thérèse, 'Translations and translators', in R. L. Benson, G. Constable (eds): *Renaissance and Renewal in the Twelfth Century*, Cambridge Ma., 1982, pp. 421–462.

Aristoteles Pseudepigraphus, see Rose (ed.).

D. Grupe, *Stephen of Pisa and Antioch: Liber Mamonis*, Sources and Studies in the History of Mathematics and Physical Sciences,
https://doi.org/10.1007/978-3-030-19234-1

Ateş, Ahmed, 'Kastamonu Genel Kitaplığında bulunan bazımühim arapça ve farsça yazmalar', *Oriens* 5, 1952, pp. 28–46.

Attiya, Hussein M., 'Knowledge of Arabic in the crusader states in the twelfth and thirteenth centuries', *Journal of Medieval History* 25, 1999, pp. 203–212.

al-Battānī see Nallino (ed. and tr.).

Björnbo, Axel A., 'Die mittelalterlichen lateinischen Übersetzungen aus dem Griechischen auf dem Gebiete der mathematischen Wissenschaften', *Archiv für die Geschichte der Naturwissenschaften und der Technik* 1, 1909, pp. 385–394.

Burnett, Charles, "ʿAbd al-Masīḥ of Winchester', in L. Nauta, A. Vanderjagt (eds): *Between Demonstration and Imagination - Essays in the history of science and philosophy presented to John D. North*, Leiden, 1999, pp. 159–169.

Burnett, Charles, 'Antioch as a link between Arabic and Latin culture in the twelfth and thirteenth centuries', in I. Draelants, A. Tihon, B. v. d. Abeele (eds): *Occident et Proche-Orient: Contacts scientifiques au temps des Croisades*, Louvain-la-Neuve, 2000, pp. 1–78.

Burnett, Charles, 'Latin alphanumerical notation, and annotation in Italian, in the twelfth century: MS London, British Library, Harley 5402', in M. Folkerts, R. Lorch (eds): *Sic itur ad astra - Studien zur Geschichte der Mathematik und Naturwissenschaften, Festschrift für den Arabisten Paul Kunitzsch zum 70. Geburtstag*, Wiesbaden, 2000, pp. 76–90.

Burnett, Charles, 'The coherence of the Arabic-Latin translation program in Toledo in the twelfth century', *Science in Context* 14, 2001, pp. 249–88.

Burnett, Charles, 'The transmission of Arabic Astronomy via Antioch and Pisa in the second quarter of the twelfth century', in J. P. Hogendijk, A. I. Sabra (eds): *The Enterprise of Science in Islam: New Perspectives*, Cambridge Ma., 2003, pp. 23–51.

Burnett, Charles, 'Stephen, the Disciple of Philosophy, and the exchange of medical learning in Antioch', *Crusades* 5, 2006, pp. 113–29.

Classen, Peter, 'Kodifikation im 12. Jahrhundert - Die Constituta usus et legis von Pisa', in P. Classen (ed): *Recht und Schrift im Mittelalter*, Sigmaringen, 1977, pp. 311–317.

Classen, Peter, 'Die geistesgeschichtliche Lage: Anstösse und Möglichkeiten', in P. Weimar (ed): *Die Renaissance der Wissenschaften im 12. Jahrhundert*, Zurich, 1981, pp. 11–32.

van Dalen, Benno, 'A second manuscript of the Mumtaḥan Zīj', *Suhayl* 4, 2004, pp. 9–44.

Duhem, Pierre, *Le Système du Monde*, 2, Paris, 1914, repr. Paris 1965.

Ganszynieč, R., 'Stephanus de modo medendi', *Archiv für Geschichte der Medizin* 14, 1923, pp. 110–113.

Gautier Dalché, Patrick, 'Le souvenir de la Géographie de Ptolémée dans le monde latin médiéval (VIe-XIVe siècles)', *Euphrosyne* 27, 1999, pp. 79–106.

Gautier Dalché, Patrick, *La Géographie de Ptolémée en Occident (IVe-XVIe siècle)*, Tournhout, 2009.

Grupe, Dirk, 'The 'Thābit-Version' of Ptolemy's *Almagest* in MS Dresden Db.87', *Suhayl* 11, 2012, pp. 147–153.

Grupe, Dirk, *The Latin Reception of Arabic Astronomy and Cosmology in Mid-Twelfth-Century Antioch. The Liber Mamonis and the Dresden Almagest*, PhD dissertation, University of London, 2013.

Grupe, Dirk, 'Thābit ibn Qurra's version of the Almagest and its reception in Arabic astronomical commentaries', conference paper presented at *Ptolemy's Science of the Stars in the Middle Ages*, London, Warburg Institute, 5–7 November 2015. An article with the same title is still forthcoming in the conference proceedings. A draft version, entitled "Further witnesses of Thābit ibn Qurra's version of the Almagest (draft)", is accessible via www.academia.edu.

Grupe, Dirk, 'Stephen of Pisa's theory of the oscillating deferents of the inner planets (1h. 12th c.)', *AHES* 71, 2017, pp. 379–407.

Hartner, Willy, 'The Mercury horoscope of Marcantonio Michiel of Venice', in Arthur Beer (ed): *Vistas in Astronomy*, 1, London/New York, 1955, pp. 84–138.

Haskins, Charles H., *Studies in the History of Mediaeval Science*, Cambridge Ma., 1924.

Haskins, Charles H., *The Renaissance of the Twelfth Century*, Cambridge Ma., 1927.

Haskins, Charles H., and Dean Putnam Lockwood, 'The Sicilian translators of the twelfth century and the first Latin version of Ptolemy's Almagest', *Harvard Studies in Classical Philology* 21, 1910, pp. 75–102.

Heiberg, Johan L. (ed.), *Claudii Ptolemaei Opera quae exstant omnia*, Leipzig, 1898.

Heiberg, Johan L., 'Noch einmal die mittelalterliche Ptolemaios-Übersetzung', *Hermes* 46, 1911, pp. 207–216.

Hiestand, Rudolf, 'Un centre intellectuel en Syrie du Nord? Notes sur la personnalité d'Aimery d'Antioche, Albert de Tarse et Rorgo Fretellus', *Le Moyen-Âge* 100, 1994, pp. 7–36.

Hillenbrand, Carole, *The Crusades. Islamic Perspectives*, New York, 2000.

Hunt, Richard W., 'Stephen of Antioch', *Mediaeval and Renaissance Studies* 2, 1950, pp. 172–3.

Ibn al-Haytham, see Langermann (ed. and tr.).

Irwin, Robert, 'Islam and the Crusades 1096–1699', in Jonathan Riley-Smith (ed): *The Oxford Illustrated History of the Crusades*, Oxford, 1995, pp. 217–259.

Kedar, Benjamin Z., 'The subjected Muslims of the Frankish Levant', in James M. Powell (ed): *Muslims under Latin Rule*, Princeton, 1990, pp. 135–174.

King, David A., 'An eleventh-century summary of the *Almagest* - review of: Ibn al-Haytham's On the Configuration of the World, by Y. Tzvi Langermann', *Journal for the History of Astronomy* 26, 1995, pp. 84–85.

Kluxen, Wolfgang, 'Der Begriff der Wissenschaft', in P. Weimar (ed): *Die Renaissance der Wissenschaften im 12. Jahrhundert*, Zurich, 1981, pp. 273–293.

Kohl, Karl, 'Über den Aufbau der Welt nach Ibn al Haiṯam', *Sitzungsberichte der Physikalisch-medizinischen Sozietät in Erlangen* 54–55, 1922–23, pp. 140–179.

Köhler, Michael A., *Allianzen und Verträge zwischen fränkischen und islamischen Herrschern im Vorderen Orient. Eine Studie über das zwischenstaatliche Zusammenleben vom 12. bis ins 13. Jahrhundert*, Berlin, 1991.

Langermann, Y. Tzvi, 'A note on the use of the term orbis (falak) in Ibn al-Haytham's Maquālah fi hayʾat al-ʿālam', *Archives Internationales d'Histoire des Sciences* 32, 108, 1982, pp. 112–113.

Langermann, Y. Tzvi, 'Ibn al-Haytham', in T. Hockey et al. (eds): *Biographical Encyclopedia of Astronomers*, New York, 2007, pp. 556–557.

Langermann, Y. Tzvi (ed. and tr.), *Ibn al-Haytham's On the Configuration of the World*, New York, 1990, repr. London, 2016.

Lemay, Richard, 'De la scolastique à l'histoire par le truchement de la philologie: itinéraire d'un médiéviste entre europe et islam', in Fondazione Leone Caetani (ed): *La diffusione delle scienze islamiche nel medio evo europeo (Roma, 2–4 ottobre 1984)*, 1987, pp. 399–535.

Lemay, Richard, *Hermann de Carinthie: Astronomia (sive De Circulis) et Regule in Canonem Astronomie*, unpublished study, containing an edition of MS Cambrai 930, with introduction and commentary.

Lemay, Richard, 'Nouveautés fugaces dans des textes mathématiques du XIIe siécle. Un essai d'abjad latin avorté', in M. Folkerts, R. Lorch (eds): *Sic itur ad astra - Studien zur Geschichte der Mathematik und Naturwissenschaften, Festschrift für den Arabisten Paul Kunitzsch zum 70. Geburtstag*, Wiesbaden, 2000, pp. 376–392.

Liechtenstein, Petrus (ed.), *Almagestum Cl. Ptolemei pheludiensis alexandrini astronomorum principis opus ingens ac nobile omnes celorum motus continens*, Venice, 1515.

Macrobius, see Willis, Stahl.

Mancha, José Luis, 'La version alfonsi del fī hay'at al-'ālam (de configuratione mundi) de Ibn al-Haytam. (Oxford, Canon. misc. 45, ff. 1r-56r)', in Mercé Comes (ed): *"Ochava espera" y la "astrofísica"*, Barcelona, 1990.

Mancha, José Luis, 'Ibn al-Haytham's homocentric epicycles in Latin astronomical texts of the XIVth and XVth centuries', *Centaurus* 33, 1990, pp. 70–89.

Manitius, Karl (tr.), *Des Claudius Ptolemäus Handbuch der Astronomie aus dem Griechischen übersetzt und mit erklärenden Anmerkungen versehen*, Leipzig, 1912–13.

Mercier, Raymond, 'East and West contrasted in scientific astronomy', in I. Draelants, A. Tihon, B. v. d. Abeele (eds): *Occident et Proche-Orient: Contacts scientifiques au temps des Croisades*, Louvain-la-Neuve, 2000, pp. 325–342.

Mercier, Raymond, 'The lost zij of al-Sufi in the twelfth-century tables for London and Pisa', in: *Lectures from the conference on al-Sufi and ibn al-Nafis*, Beirut/Damascus, 1991, pp. 38–72, also published in

Mercier, Raymond, *Studies in the Transmission of Medieval Mathematical Astronomy*, VIII, Aldershot, 2004.

Millás Vallicrosa, José María, *Las traducciones orientales en los manuscritos de la Biblioteca Catedral de Toledo*, Madrid, 1942.

Morelon, Régis, 'General survey of Arabic astronomy', in Roshdi Rashed (ed): *Encyclopedia of the History of Arabic Science*, 1, London/New York, 1996, pp. 1–19.

Mozaffari, S. Mohammad, 'Planetary latitudes in medieval Islamic astronomy: an analysis of the non-Ptolemaic latitude parameter values in the Maragha and Samarqand astronomical traditions', *AHES* 70, 2016, pp. 513–541.

Mozaffari, S. Mohammad, 'An analysis of medieval solar theories', *AHES* 72, 2018, pp. 191–243.

Nallino, Carlo A. (ed. and tr.), *Al-Battānī sive Albatenii. Opus astronomicum ad fidem codicis Escurialensis arabice editum, latine versum adnotationibus instructum*, 3 pts, Milan, 1899–1907.

Neugebauer, Otto, *A History of Ancient Mathematical Astronomy*, Berlin, 1975.

Obrist, Barbara, '*'Imaginatio'* and visual representation in twelfth-century cosmology and astronomy: Ibn al-Haytham, Stephen of Pisa (and Antioch), (Ps.) Māshāʾallāh, and (Ps.) Thābit ibn Qurra', in Christoph Lüthy, Claudia Swan, Paul Bakker, Claus Zittel (eds): *Image, Imagination, and Cognition. Medieval and Early Modern Theory and Practice*, Leiden/Boston, 2018, pp. 32–60.

Otte, Gerhard, 'Die Rechtswissenschaft', in P. Weimar (ed): *Die Renaissance der Wissenschaften im 12. Jahrhundert*, Zurich, 1981, pp. 123–42.

Pedersen, Olaf, *A Survey of the Almagest*, Odense, 1974, with annotation and new commentary by A. Jones, New York, 2010.

Postl, Brigitte, *Die Bedeutung des Nil in der römischen Literatur*, Wien, 1970.

Prawer, Joshua, 'Social classes in the crusader states: The 'minorities'', in K. M. Setton, N. M. Zacour, H. W. Hazard (eds): *A History of the Crusades. Vol. V. The Impact of the Crusades on the Near East*, Madison, 1985, pp. 59–115.

Prawer, Joshua, 'Social classes in the Latin Kingdom: The Franks', in K. M. Setton, N. M. Zacour, H. W. Hazard (eds): *A History of the Crusades. Vol. V. The Impact of the Crusades on the Near East*, Madison, 1985, pp. 117–192.

Ptolemy, *Almagest*, see Heiberg (ed.), Liechtenstein, Manitius, Toomer.

Rashed, Roshdi, *Les Mathématiques infinitésimales du IXe au XIe siécle. Vol. II. Ibn al-Haytham*, London, 1993.

Rashed, Roshdi, 'The Configuration of the universe: A book by al-Ḥasan ibn al-Haytham?', *Revue d'histoire des sciences* 60, 2007, pp. 47–63.

Rose, Valentin (ed.), *Aristoteles Pseudepigraphus*, Leipzig, 1863.

Rose, Valentin, *Verzeichniss der lateinischen Handschriften der königlichen Bibliothek zu Berlin*, II, 3, Berlin, 1905.

de Roziére, Eugéne, *Cartulaire de l'Église du Saint Sépulcre de Jérusalem*, Paris, 1849.

Sabbadini, Remigio, 'Spogli Ambrosiani Latini', *Studi Italiani de Filologia Classica* 11, 1903, pp. 165–388.

Sabra, Abdelhamid I., 'Ibn al-Haytham', in C. C. Gillispie et al. (eds): *Dictionary of Scientific Biography*, 6, 1972, pp. 189–210.

Sabra, Abdelhamid I., 'One Ibn al-Haytham or two?', *Zeitschrift für Geschichte der arabischen-islamischen Wissenschaften* 12, 1998, pp. 1–51.

Samsó, Julio, 'El original arabe y la version alfonsi del Kitāb fī hay'at al-ʿālam de Ibn al-Hayṯam', in Mercè Comes (ed): *"Ochava espera" y la "astrofísica"*, Barcelona, 1990.

Samsó, Julio, '«Dixit Abraham Iudaeus»: algunas observaciones sobre los textos astronómicos latinos de Abraham ibn 'Ezra', *Iberia Judaica* 4, 2012, pp. 171–200.

Samsó, Julio, and Miquel Forcada, *Las Ciencias de los Antiguos en al-Andalus*, 2nd ed., Almeria, 2011.

Samsó, Julio, and Honorino Mielgo, 'Ibn al-Zarqālluh on Mercury', *Journal for the History of Astronomy* 25, 1994, pp. 289–296.

Sarton, George, *Introduction to the History of Science*, Baltimore, 1931.

Schnorr von Carolsfeld, Franz, *Katalog der Handschriften der Königl. Öffentlichen Bibliothek zu Dresden*, 1, Leipzig, 1882, repr. Dresden 1979.

Schramm, Matthias, *Ibn al-Haythams Weg zur Physik*, Wiesbaden, 1963.

Sezgin, Fuat, *Geschichte des arabischen Schrifttums*, 6, Leiden, 1978.

Sezgin, Fuat (ed.), *The Verified Astronomical Tables for the Caliph al-Maʾmūn. Al-Zīj al-Maʾmūnī al-mumtaḥan by Yaḥyā ibn Abī Manṣūr*, Frankfurt a. M., 1986.

Stahl, W. H. (tr.), *Commentary on the dream of Scipio by Macrobius. Transl. with an Introduction and Notes*, New York, 1952.

Stahlman, William D., *The Astronomical Tables of Codex Vaticanus Graecus 1291*, unpublished Ph.D. dissertation, Brown University, 1959.

Steinschneider, Moritz, 'Donnolo - Pharmakologische Fragmente aus dem X. Jahrhundert, nebst Beiträgen zur Literatur der Salernitaner, hauptsächlich nach handschriftlichen hebräischen Quellen (Forts.)', *Archiv für pathologische Anatomie und Physiologie und klinische Medicin* 39, 1867, pp. 296–336.

Steinschneider, Moritz, 'Notice sur un ouvrage astronomique inédit d'Ibn Haitham', *Bullettino di bibliografia e di storia della scienze matematiche e fisiche* 14, 1881, pp. 721–736.

Steinschneider, Moritz, 'Supplement a la Notice sur un ouvrage astronomique inédit d'Ibn Haitham', *Bullettino di bibliografia e di storia della scienze matematiche e fisiche* 16, 1883, pp. 505–513.

Steinschneider, Moritz, *Die hebraeischen Übersetzungen des Mittelalters und die Juden als Dolmetscher*, Berlin, 1893.

Steinschneider, Moritz, *Die europäischen Übersetzungen aus dem Arabischen bis Mitte des 17. Jahrhunderts*, Wien, 1905.

Steinschneider, Moritz, *Die arabischen Uebersetzungen aus dem Griechischen*, Leipzig, 1897, repr. Graz 1960.

Talbot, Charles H., 'Stephen of Antioch', in C. C. Gillispie et al. (eds): *Dictionary of Scientific Biography*, 13, New York, 1976, pp. 38–39.

Toomer, G. J. (tr.), *Ptolemy's Almagest*, London, 1984, 2nd ed. Princeton 1998.

Viladrich, Mercè, 'The planetary latitude tables in the Mumtaḥan Zīj', *Journal for the History of Astronomy* 19, 1988, pp. 257–268.

Williams, Steven J., 'Philip of Tripoli's translation of the pseudo-Aristotelian Secretum Secretorum viewed within the context of intellectual activity in the crusader Levant', in I. Draelants, A. Tihon, B. v. d. Abeele (eds): *Occident et Proche-Orient: Contacts scientifiques au temps des Croisades*, Louvain-la-Neuve, 2000, pp. 79–94.

Willis, James (ed.), *Ambrosii Theodosii Macrobii Commentarii in somnium Scipionis*, Stuttgart, 1994.

Wolf, Armin, 'Gesetzgebung und Kodifikation', in P. Weimar (ed): *Die Renaissance der Wissenschaften im 12. Jahrhundert*, Zurich, 1981, pp. 143–171.

Yaḥyā ibn Abī Manṣūr, see Sezgin (ed.).

Index

D. Grupe, *Stephen of Pisa and Antioch: Liber Mamonis*, Sources and Studies in the History of Mathematics and Physical Sciences,
https://doi.org/10.1007/978-3-030-19234-1

Printed by Printforce, the Netherlands